Beauty therapy - the foundations

Beauty Industry Authority and Hairdressing Training Board/ Macmillan Series

Start Hairdressing! The Official Guide to Level 1 Martin Green & Leo Palladino

Hairdressing – The Foundations: The Official Guide to Level 2 Leo Palladino

Professional Hairdressing: The Official Guide to Level 3 Martin Green, Lesley Kimber & Leo Palladino

Patrick Cameron:
Dressing Long Hair Patrick Cameron & Jacki Wadeson
Dressing Long Hair Book 2 Patrick Cameron

Mahogany: Steps to Cutting, Colouring and Finishing Hair Martin Gannon & Richard Thompson

The World of Hair: A Scientific Companion John Gray

Safety in the Salon Elaine Almond

Trevor Sorbie: Visions in Hair Trevor Sorbie, Kris Sorbie & Jacki Wadeson

The Art of Hair Colouring David Adams & Jacki Wadeson

African–Caribbean Hairdressing Sandra Gittens

Men's Hairdressing Maurice Lister

Bridal Hair Pat Dixon & Jacki Wadeson

The Total Look Ian Mistlin

Science and the Beauty Business:
The Science of Cosmetics John V. Simmons
The Beauty Salon and its Equipment John V. Simmons

Manicure, Pedicure and Advanced Nail Techniques Elaine Almond

The Complete Make-up Artist: Working in Film, Television and Theatre Penny Delamar

The World of Skin Care: A Scientific Companion John Gray

Beauty Therapy – The Foundations: The Official Guide to Level 2 Lorraine Nordmann

Professional Beauty Therapy: The Official Guide to Level 3 Lorraine Nordmann, Lorraine Appleyard & Pamela Linforth

If you would like to receive future titles in this series as they are published, you can use our standing order facility. Contact your bookseller, quoting series standing order ISBN 0–333–69338–8, or write to us with your name and address, the title of the series and the series ISBN. Customer Services Department, Macmillan Distribution Ltd, Houndmills, Basingstoke, Hampshire RG21 6XS.

Beauty therapy – the foundations

The Official Guide to Level 2

Second Edition

LORRAINE NORDMANN

with contributions from Elaine Almond, Lorraine Appleyard and Pamela Linforth

© Lorraine Nordmann, Elaine Almond, Lorraine Appleyard, Pamela Linforth 1995, 1999

© Text artwork Macmillan Press Ltd 1995, 1999

All rights reserved. No reproduction, copy or transmission of this publication may be made without written permission.

No paragraph of this publication may be reproduced, copied or transmitted save with written permission or in accordance with the provisions of the Copyright, Designs and Patents Act 1988, or under the terms of any licence permitting limited copying issued by the Copyright Licensing Agency, 90 Tottenham Court Road, London W1P 0LP

Any person who does any unauthorised act in relation to this publication may be liable to criminal prosecution and civil claims for damages.

First edition 1995
Reprinted three times
Second edition 1999

Published by
MACMILLAN PRESS LTD
Houndmills, Basingstoke, Hampshire RG21 6XS
and London
Companies and representatives
throughout the world

ISBN 0–333–77577–5

A catalogue record for this book is available
from the British Library.

This book is printed on paper suitable for recycling and made from fully managed and sustainable forest sources.

10 9 8 7 6 5 4 3 2
08 07 06 05 04 03 02 01 00

Designed by Susan Clarke

Printed in Great Britain by
Jarrold Book Printing, Thetford, Norfolk

Contents

Notes on the contributors

Elaine Almond was for many years the working owner of a successful beauty salon. A science honours graduate, she was head of the biology department in a large comprehensive school before retraining as a beauty therapist, remedial masseuse and manicurist/nail technician. Well known for her magazine articles, Elaine has also written *Manicure, Pedicure and Advanced Nail Techniques* and *Safety in the Salon.*

Lorraine Appleyard is a fully qualified beauty therapist and lecturer, who has been actively involved in the beauty industry for many years. An experienced therapist, salon owner and lecturer, Lorraine specialises in nail extensions and natural nail care. She is currently lecturer in beauty and sports therapy at Bury College of Further Education.

Pamela Linforth, Principal of the Sterex Academy of Post Graduate Studies, has made her career in the beauty therapy and electrolysis industry. She is a fully qualified beauty therapist, lecturer and assessor, and runs a wide range of courses at the Sterex Academy. She has far-reaching experience both in the industry and as a lecturer.

Acknowledgements

The author and publishers would like to thank the following people and organisations for their assistance in producing this book:

Access Brand Limited
American Express
Barclays Bank plc
Beauty Industry Authority
Bellitas
Caress Ear Piercing
Cereal Partners UK
Chubb Fire Ltd
Depilex/RVB
Derrick Mullings/Media Image
Ellisons
Fire Protection Services
Alan Hartwell
Helinova Ltd
Hive of Beauty Ltd
ICL
In Line
Jolen
Judith Ifould
Inverness
National Westminster Bank
NSI UK Ltd
OPI
Hugh Rushton
Salon System
Sharp
John Simmons
Smith & Nephew, Hull
Sorisa
Star Nails
Sukar Professional Hair Removal
Unilever

Photographers Anderson and Ian Littlewood, make-up artist Jo Crowder, and illustrator Pam Young.

Dr Michael H. Beck, Consultant Dermatologist, Salford Royal Hospitals NHS Trust.

Dr Andrew L. Wright, Consultant Dermatologist, Bradford Royal Infirmary.

The author and publishers also wish to thank the following for permission to use copyright material:

The Controller of Her Majesty's Stationery Office for Crown copyright material.

Martin Green for material on working relationships and assessment, reproduced with adaptations from *Hairdressing – The Foundations*.

Last, but by no means least, many thanks to all the colleagues and stdents who were involved in the various technique photographs taken for this book.

Every effort has been made to trace all the copyright holders but if any have been inadvertently overlooked the publishers will be pleased to make the necessary arrangements at the first opportunity.

Foreword

Have you noticed when someone browses through a book they tend to start at the back and flick through the pages until they come to the front? So if you started from the back, then this is the end.

Whether you started at the front or at the back, however, makes no difference to the sheer amount of knowledge, fact and guidance contained in this, Lorraine Nordmann's excellent second edition.

This book is the first to be endorsed by The Beauty Industry Authority and I'm especially delighted that it's so rich in content. So whether you've just started your career in Beauty Therapy or are continuing your professional development, welcome to the beginning of a voyage of discovery.

Alan Goldsbro

Chief Executive, The Beauty Industry Authority

Introduction

Beauty therapy – the foundations has been written specifically for students studying for an NVQ/SVQ or similar accredited course in beauty therapy. This second edition has been revised to incorporate coverage of the current occupational standards at Level 2, which include new skills to be learnt.

The text also covers the technical skills necessary for an assistant therapist employed in beauty services. As an assistant therapist it is important that you have a comprehensive knowledge of the services carried out in your workplace. Your duties will be:

- to help maintain a safe, hygienic and efficient workplace;
- to assist in the handling of stock;
- to assist in receiving clients and to carry out other reception duties.

You will also assist the beauty therapist with some treatments. This work will include:

- preparing the treatment cubicle, the equipment and the materials;
- preparing the client for treatment.

This book will also support you with your studies if you wish to progress to Level 2.

NVQ/SVQ units at Level 2

The Level 2 qualification guarantees the skills necessary for those carrying out reception duties and providing beauty therapy services appropriate at this level. Your duties will be:

- to maintain employment standards in the workplace, supporting a healthy, safe and secure salon environment;
- to maintain good working relationships both with clients and with colleagues;
- to provide retail and practical services to customers.

Each chapter in *Beauty therapy – the foundations* is written to support you in your training, enabling you to gain the knowledge and the skills required to achieve unit competence. Whichever

NVQ/SVQ unit you are studying at a particular time, simply turn to the relevant chapter and follow the text, which is accompanied by useful tips and activities to test your understanding.

Good luck!

I hope you enjoy this book, and I wish you well with your career in beauty therapy.

Lorraine Nordmann

CHAPTER 1

Health, safety and security

TAKING CARE OF CLIENTS AND COLLEAGUES

When working in a service industry, you are legally obliged to provide a **safe and hygienic environment**. This applies whether you are working in a hotel, a department store, a leisure centre or a private beauty salon, or operating a mobile beauty therapy service. You must pay careful attention to health and safety and to security. Exactly the same is true when working in clients' homes: it is essential to follow the normal health and safety guidelines, just as you would when working in a salon.

Legal responsibilities

ACTIVITY: HEALTH AND SAFETY INFORMATION
Write to your local health authority department to ask for a pack of relevant health and safety information.

If you cause harm to your client, or put her at risk, you will be held responsible and you will be liable to **prosecution**, with the possibility of being fined.

There is a good deal of **legislation** relating to health and safety. Details are widely available, and you must be aware of your responsibilities and your rights. It is important that you obtain and read all relevant publications.

The Health and Safety at Work Act 1974

The **Health and Safety at Work Act 1974** developed from experience gained over 150 years, and now incorporates earlier legislation including the Offices, Shops and Railway Premises Act 1963 and the Fire Precautions Act 1971. It lays down the minimum standards of health, safety and welfare required in each area of the workplace – for example, it requires that business premises and equipment be safe and in good repair. It is the employer's responsibility to implement the Act and to ensure that the workplace is safe both for employees and for clients.

ACTIVITY: HEALTH AND SAFETY RULES
Discuss the rules which you feel should appear in a salon's health and safety policy.

Each employer of more than five employees must formulate a written **health and safety policy** for that establishment. This must be made available to all employees. Regular checks should be made to ensure that safety is being satisfactorily maintained.

HEALTH & SAFETY: HEALTH AND SAFETY NOTICE
Every employer is obliged by law to display a health and safety notice in the workplace.

Employees must co-operate with their employer to provide a safe and healthy workplace. As soon as they observe any **hazard**, this must be reported to the designated authority so that the problem can be put right. Hazards include:

- obstructions to corridors, stairways and fire exits;
- spillages and breakages.

The Personal Protective Equipment (PPE) at Work Regulations 1992

The **Personal Protective Equipment (PPE) at Work Regulations 1992** require managers to identify through a **risk assessment** those activities or processes which require special protective clothing or equipment to be worn. This clothing and equipment must then be made available, in adequate supplies. Employees must wear the protective clothing and use the protective equipment provided, and make employers aware of any shortage so that supplies can be maintained.

HEALTH & SAFETY: GLOVES
If you are to come into contact with body tissue fluids or with chemicals, wear protective disposable surgical gloves.

ACTIVITY: RISK ASSESSMENT
Carry out your own risk assessment. List the potentially hazardous substances handled in beauty therapy. What protective clothing should be available?

The Workplace (Health, Safety & Welfare) Regulations 1992

The **Workplace (Health, Safety & Welfare) Regulations 1992** require all at work to maintain a safe, healthy and secure working environment. The regulations include legal requirements in relation to the following aspects of the working environment:

- maintenance of the workplace and equipment;
- ventilation;
- working temperature;
- lighting;
- cleanliness and handling of waste materials;
- safe salon layout;
- falls and falling objects;
- windows, doors, gates and walls;
- safe floor and traffic routes;
- escalators and moving walkways;
- sanitary conveniences;
- washing facilities;
- drinking water;
- facilities for changing clothing;
- facilities for staff to rest and eat meals.

HEALTH & SAFETY: EUROPEAN DIRECTIVES
As a result of directives adopted in 1992 by the European Union, health and safety legislation has been updated.

1. Obtain a copy of the new directives. *Workplace (Health & Safety & Welfare) Regulations 1992.*
2. Look through the publication, and make notes on any information relevant to you in the workplace.

TIP
The salon temperature should be a minimum of 16°C within one hour of employees arriving for work. The salon should be well ventilated, or carbon dioxide levels will increase, which can cause nausea. Many substances used in the salon can become hazardous without adequate ventilation.

Lighting should be adequate to ensure that treatments can be carried out safely and competently, with the minimum risk of accident.

Professional appearance

Assistant therapist

The assistant therapist qualified to pre-foundation level will be required to wear a clean protective overall as she will be preparing the working area for client treatments, and she may be involved in preparing clients also.

Beauty therapist

Due to the nature of many of the services offered, the beauty therapist must wear protective, hygienic clothing. The cotton overall is ideal; air can circulate, allowing perspiration to evaporate and discouraging body odour. The use of a colour such as white immediately shows the client that you are clean. A cotton overall may comprise a dress, a jumpsuit or a tunic top, with coordinating trousers.

Overalls should be laundered regularly, and a fresh, clean overall worn each day.

In Line

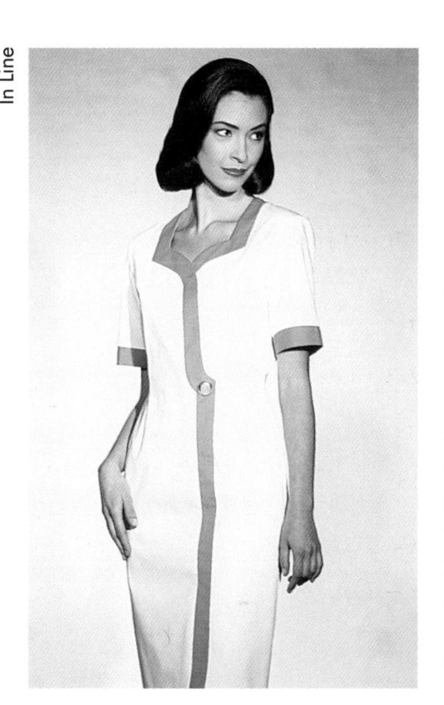

In Line

Receptionist

If a receptionist is employed solely to carry out reception duties, she may wear a different salon dress, complementary to those worn by the practising therapists. As she will not be as active, it may be appropriate for her to wear a smart jacket or cardigan. If on the other hand she is also carrying out services at some times, the standard salon overall must be worn.

General rules for employees

Make-up

Wear an attractive make-up, and use the correct skin-care cosmetics to suit your skin type. A healthy complexion will be a positive advertisement for your work.

Jewellery

Keep jewellery to a minimum, such as a wedding ring, a watch, and small earrings.

Nails

Nails should be short, neatly manicured, and free of nail polish unless the employee's main duties involve nail treatments or reception duties. Flesh-coloured tights may be worn to protect the legs.

Shoes

Wear flat, well-fitting, comfortable shoes that enclose the feet and which complement the overall. Remember that you will be on your feet for most of the day!

HEALTH & SAFETY: APRONS
For certain treatments, such as waxing, it is necessary to wear a protective apron over the overall. Assistant therapists may wear an apron whilst preparing and cleaning the working area, to protect the overall and keep it clean.

ACTIVITY: PERSONAL APPEARANCE

1. Collect pictures from various suppliers of overalls. Select those that you feel would be most practical for an assistant therapist and for a Level 2 beauty therapist. Briefly describe *why* you feel these are the most suitable.
2. Design various hairstyles, or collect pictures from magazines, to show how the hair could be smartly worn by a therapist with medium-length to long hair.

Ethics

Beauty therapy has a **code of ethics**. Although not a legal requirement, this code may be used in criminal proceedings as evidence of improper practice.

ACTIVITY: CODE OF ETHICS
As a professional beauty therapist it is important that you adhere to a code of ethical practice. You may wish to join a professional organisation, which will issue you with a copy of its agreed standards

Accidents

Accidents in the workplace usually occur through negligence by employees or unsafe working conditions.

Any accidents occurring in the workplace must be recorded on a **report form**, and entered into an **accident book**. The report form requires more details than the accident book – you must note down:

- the date;
- the name of the person or people involved;
- the accident details;

ACTIVITY: AVOIDING ACCIDENTS
Discuss potential causes of accidents in the workplace. How could these accidents be prevented?

HEALTH & SAFETY: BREAKAGES AND SPILLAGES
When dealing with hazardous breakages and spillages, the hands should always be protected with gloves. To avoid injury to others, broken glass should be put in a secure container prior to disposing of it in a waste bin.

- the injuries sustained;
- the action taken.

Accidents can damage stock, resulting in breakage of containers and spillage of contents. Breakage of glass can cause cuts; spillages may cause somebody to slip and fall. Any breakages or spillages should therefore be dealt with immediately and in the correct way.

You must determine whether the spillage is a potential hazard to health, and what action is necessary. To whom should you report it? What equipment is required to remove the spillage? How should the materials be disposed of?

Disposal of waste

Waste should be disposed of in an enclosed waste bin fitted with a polythene bin liner, durable enough to resist tearing. The bin should be regularly sanitised with disinfectant in a well-ventilated area: wear protective gloves while doing this.

Contaminated waste, such as wax strips, should be disposed of as recommended by your local authority. Items which have been used to pierce the skin, such as disposable milia extractors, should be safely discarded in a disposable **sharps container**. Again, contact your local authority to check on disposal arrangements.

First aid

The **Health and Safety (First Aid) Regulations 1982** state that workplaces must have first-aid provision. An adequately stocked first-aid box should be available. This should contain:

- basic first-aid guidance leaflet (1);
- assorted sterile adhesive dressings (20);
- individually-wrapped triangular bandages (6);
- safety pins (6);
- sterile eyepads, with attachments (2);

A first-aid kit

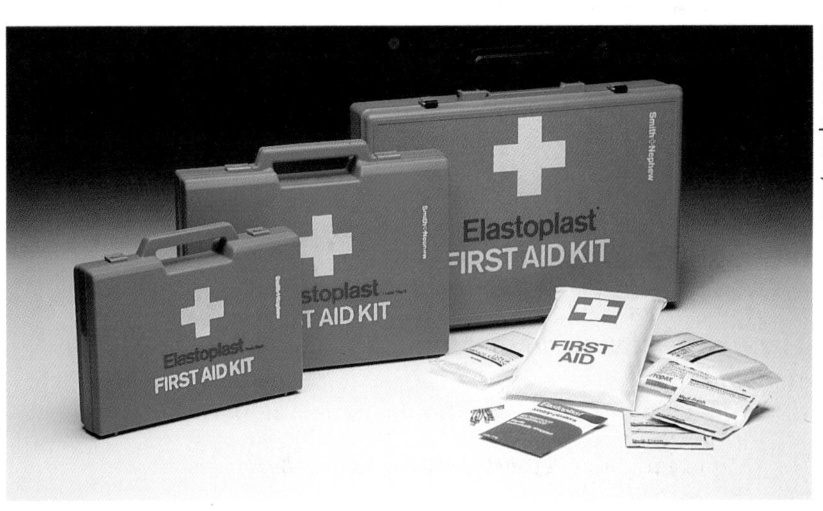

Smith & Nephew, Hull

- medium-sized individually-wrapped sterile unmedicated wound dressings, 10cm × 8cm (6);
- large individually-wrapped sterile unmedicated wound dressings, 13cm × 9cm (2);
- extra-large sterile individually-wrapped medicated wound dressings, 28cm × 17.5cm (3);
- individually-wrapped moist cleaning wipes.

If bathing eyes where tap water is not readily available, sterile water should be stored in sealed containers.

Control of Substances Hazardous to Health (COSHH) Regulations 1988

These regulations were designed to make employers consider the substances used in their workplace and assess the possible risks to health. Many substances that seem quite harmless can prove to be hazardous if used or stored incorrectly.

Hazardous substances are usually identified through the use of known symbols, examples of which are shown on the right. Any substance in the workplace that is hazardous to health must be identified on the packaging and stored and handled correctly.

Hazardous substances may enter the body via:

- the eyes;
- the skin;
- the nose (**inhalation**);
- the mouth (**ingestion**).

Each beauty product supplier is legally required to make available guidelines on how materials should be used and stored; these will be supplied on request.

HMSO

Electricity at Work Regulations 1989

These regulations state that every piece of electrical equipment in the workplace should be tested every 12 months by a qualified electrician.

You can compile your own safety checklist for your workplace. Report to your supervisor if you see any of these potential hazards:

- exposed wires in flexes;
- cracked plugs or broken sockets;
- worn cables;
- overloaded sockets.

ACTIVITY: IDENTIFYING HAZARDS
Make a list of potential electrical hazards in the workplace.

Provision and Use of Work Equipment Regulations 1992

These regulations lay down the important health and safety controls on the provision and use of work equipment. They state the duties for employers and for users, including the self-employed. They affect both old and new equipment. They identify the requirements in selecting suitable equipment and in

ACTIVITY: COSHH ASSESSMENTS
Carry out a COSHH assessment on selected treatment products used in the salon. Consider manicures, nail extensions, waxing, and facial and eye treatments.

maintaining it. They also discuss the information provided by equipment manufacturers, and instruction and training in the safe use of equipment. Specific regulations address the dangers and potential risks of injury that could occur during operation of the equipment.

Inspection and registration of premises

The Local Authority Environmental Health Department enforces the Health and Safety at Work Act, and an **environmental health officer** visits and inspects local business premises.

If the inspector identifies any area of danger, it is the responsibility of the employer to remove this danger within a designated period of time. The inspector issues an **improvement notice**. Failure to comply with the notice will lead to prosecution. The inspector also has the authority to *close* a business until he or she is satisfied that all danger to employees and public has been removed. Such closure involves the issuing of a **prohibition notice**.

Certain treatments carried out in beauty therapy, such as ear piercing, pose additional risk as they may produce blood and body tissue fluid. Inspection of the premises is necessary before such services can be offered to the public. The inspector will visit and observe that the guidelines listed in the **Local Government (Miscellaneous Provisions) Act 1982** relating to this area are being complied with. When the inspector is satisfied, a **certificate of registration** will be awarded.

Fire

ACTIVITY: FIRE DRILL
Each workplace should have a fire drill regularly. This enables staff to practise so that they know what to do in the event of a real fire. What is the fire drill procedure for *your* workplace?

The **Fire Precautions Act 1971** states that all staff must be aware of and trained in fire and emergency evacuation procedures for their workplace. The **emergency exit route** will be the easiest route by which staff and clients can leave the building safely.

A **smoke alarm** should be fitted to forewarn you of a fire, and **fire doors** should be fitted within the premises to help control the spread of flames.

HEALTH & SAFETY: FIRE!
If there is a fire, never use a lift. A fire quickly becomes out of control. You do not have very long to act!

HEALTH & SAFETY: FIRE EXITS
Fire exit doors must remain unlocked during working hours, and be free from obstruction.

Firefighting equipment must be available, located in a specified area. The equipment includes fire extinguishers, blankets, sand buckets, and water hoses. Firefighting equipment should be used only when the cause of the fire has been identified – using the *wrong* extinguisher could make the fire worse.

Chubb Fire Ltd

Firefighting equipment

ACTIVITY: FIRE EXTINGUISHERS
Since January 1997 all fire extinguishers have been red. Other colours and symbols indicate their contents and uses. Make sure you know the meaning of each of the colours and symbols.

Cause of fire and choice of fire extinguisher

Cause	*Extinguisher and colour code*	
Electrical fire	Carbon dioxide (CO_2) extinguisher	Black
Solid material fire (paper, wood, etc.)	Water extinguisher	Red
Flammable liquids	Foam extinguisher	Cream
Vaporising liquids	BCF extinguisher	Green
	Dry-powder extinguisher	Blue

Note: green and blue extinguishers can be used with all types of fire

HEALTH & SAFETY: USING FIRE EXTINGUISHERS
The vapours emitted when using vaporising liquid extinguishers 'starve' a fire of oxygen. They are therefore dangerous when used in confined spaces, as people need oxygen too!

Fire blankets are used to smother a small, localised fire or if a person's clothing is on fire. **Sand** is used to soak up liquids if these are the source of the fire, and to smother the fire. **Water hoses** are used to extinguish large fires caused by paper materials and the like – buckets of water may be used to extinguish a small fire. *Turn off the electricity first!*

Never put yourself at risk – fires can spread quickly. Leave the building at once if in danger, and raise the alarm by telephoning the emergency services on the emergency telephone number, **999**.

ACTIVITY: CAUSES OF FIRES
Can you think of several potential causes of fire in the salon? How could each of these be prevented?

Other emergencies

Other possible emergencies that could occur relate to fumes and flooding. Learn where the water and gas stopcocks are located. In the event of a gas leak or a flood, the stopcocks should be switched off and the appropriate emergency service contacted.

Insurance

Public liability insurance protects employers and employees against the consequences of death or injury to a third party while on the premises.

Secondly, every employer must have **employer's liability insurance**. This provides financial compensation to an employee should she be injured as a result of an accident in the workplace.

ACTIVITY: BUILDING STAMINA

Ask your tutor for guidelines before beginning this activity.

1. Write down all the foods and drinks that you most enjoy. Are they healthy? If you are unsure, ask your tutor.
2. How much exercise do you take weekly?
3. How much sleep do you regularly have each night?
4. Do you think you could improve your health and fitness levels?

ACTIVITY: THE IMPORTANCE OF POSTURE

1. Which treatments will be performed sitting, and which standing?
2. In what way do you feel your treatments would be affected if you were *not* sitting or standing correctly?

PERSONAL HEALTH, HYGIENE AND APPEARANCE

Your appearance enables the client to make an initial judgement about both you and the salon, so make sure that you create the correct impression! Employees in the workplace should always reflect the desired image of the profession that they work in.

Diet, exercise and sleep

A beauty therapist requires stamina and energy. To achieve this you need to eat a healthy, well-balanced diet, take regular exercise, and have adequate sleep.

Posture

Posture is the way you hold yourself when standing, sitting and walking. *Correct* posture enables you to work longer without becoming tired; it prevents muscle fatigue and stiff joints; and it improves your appearance.

Good standing posture

If you are standing with good posture, this will describe you:

- head up, centrally balanced;
- shoulders slightly back, and relaxed;
- chest up and out;
- abdomen flat;
- hips level;
- fingertips level;
- bottom in;
- knees level;
- feet slightly apart, and weight evenly distributed.

Good sitting posture

Sit on a suitable chair or stool with a good back support:

- sit with the lower back pressed against the chair back;
- keep the chest up and the shoulders back;
- distribute the body weight evenly along the thighs;
- keep the feet together, and flat on the floor;
- do not slouch, or sit on the edge of your seat.

Manual Handling Operations Regulations 1992

These regulations apply in all occupations where manual lifting occurs. The employer is required to carry out a risk assessment of all activities undertaken which involve manual lifting.

The risk assessment should provide evidence that the following have been considered:

- risk of injury;
- the manual movement involved in performing the activity;
- the physical constraint the load incurs;

- the environmental constraints imposed by the workplace;
- workers' individual capabilities;
- action taken in order to minimise potential risks.

Manual lifting and handling

Always take care of yourself when moving goods around the salon. Do not struggle or be impatient: get someone else to help. When **lifting**, lift from the knees, not the back. When **carrying**, balance weights evenly in both hands and carry the heaviest part nearest to your body.

> **TIP**
> When you unpack a delivery, make sure that the product packaging is undamaged.

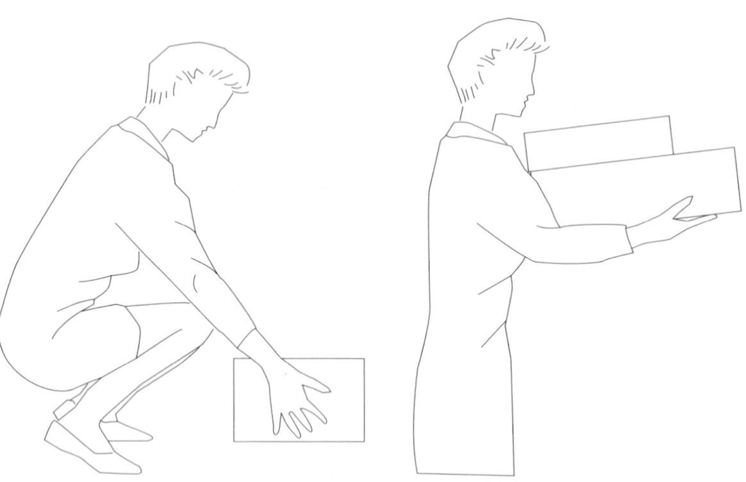

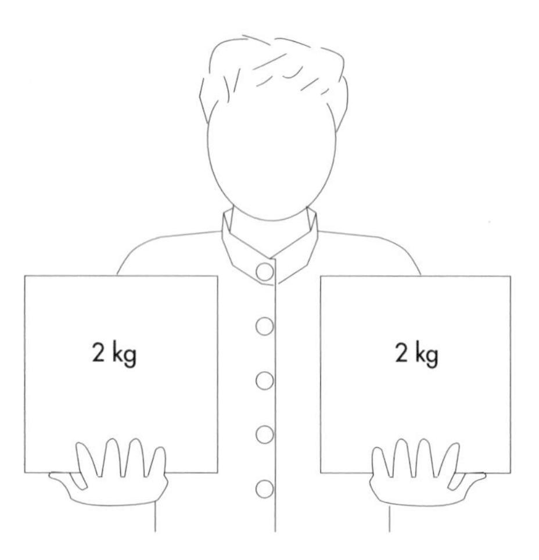

left Lifting a box
centre Carrying several boxes
right Carrying equal weights in both hands

Personal hygiene

It is vital that you have a high standard of personal **hygiene**. You are going to be working in close proximity with people.

Bodily cleanliness is achieved through daily showering or bathing. This removes stale sweat, dirt and bacteria which cause body odour. An anti-perspirant or deodorant may be applied to the underarm area to reduce perspiration and thus the smell of sweat.

Clean underwear should be worn each day.

Hands

Your hands and everything you touch are covered with germs. Although most are harmless, some can cause ill health or disease. Wash your hands regularly, especially after you have been to the toilet and before eating food. You must also wash your hands before and after treating each client, and during treatment if

> **ACTIVITY: HAND HYGIENE**
> What further occasions can you think of when it will be necessary to wash your hands when treating a client?

> **HEALTH & SAFETY: SOAP AND TOWELS**
> Wash your hands with liquid soap from a sealed dispenser. Don't refill disposable soap dispensers when empty: if you do they will become a breeding ground for bacteria.
> Disposable paper towels or warm-air hand dryers should be used to dry the hands.

HEALTH & SAFETY: PROTECTING THE CLIENT

If you have any cuts or abrasions on your hands, cover them with a clean dressing to minimise the risk of secondary infection. Disposable gloves may be worn for additional protection.

Certain skin disorders are contagious. If the therapist is suffering from any such disorder she must not work, but must seek medical advice immediately.

HEALTH & SAFETY: PROTECTING YOURSELF

You will be wise to have the relevant inoculations, including those against tetanus and hepatitis, to protect yourself against ill health and even death.

necessary. Washing the hands before treating a client minimises the risk of cross-infection, and presents to the client a hygienic, professional, caring image.

Feet

Keep your feet fresh and healthy by washing them daily and then drying them thoroughly. Deodorising foot powder may then be applied.

TIP

When working, avoid eating strong-smelling highly-spiced food.

Teeth

Avoid bad breath by brushing your teeth at least twice daily and flossing the teeth frequently. Use breath fresheners and mouthwashes as required to freshen your breath. Visit the dentist regularly, to maintain healthy teeth and gums.

HEALTH & SAFETY: LONG HAIR

If long hair is *not* taken away from the face, the tendency will be to move the hair away from the face repeatedly with the hands, and this in turn will require that the hands be washed repeatedly.

Hair

Your hair should be clean and tidy. Have your hair cut regularly, and shampoo and condition your hair as often as needed.

If your hair is long, wear it off the face, and taken to the crown of the head. Medium-length hair should be clipped back, away from the face, to prevent it falling forwards.

Hygiene in the workplace

Infections

ACTIVITY: PREVENTING THE SPREAD OF INFECTION

Think of different ways in which infection could be spread in the salon. In each case, what could be done to prevent the spread?

Effective hygiene is necessary in the salon to prevent *cross-infection* and *secondary infection*. These can occur through poor practice, such as the use of implements that are not sterile. Infection can be recognised by the skin being red and inflamed, or pus being present.

Cross-infection occurs because some micro-organisms are contagious – they may be transferred through personal contact, by touch, or by contact with an infected instrument that has not been sterilised. **Secondary infection** can occur as a result of injury to the client during the treatment, or if the client already has an open cut, if bacteria penetrate the skin and cause infection. **Sterilisation** and **sanitisation** procedures (below) are used to minimise or destroy the harmful micro-organisms which could cause infection – bacteria, viruses and fungi.

Infectious diseases that are contagious **contra-indicate** beauty treatment: they require medical attention. People with certain other skin disorders, even though these are not contagious, should likewise not be treated by the beauty therapist, as treatment might lead to secondary infection.

Sterilisation and sanitisation

Sterilisation is the total destruction of all living micro-organisms. **Sanitisation** is the destruction of some, but not all, micro-organisms. Sterilisation and sanitisation techniques practised in the beauty salon involve the use of *physical* agents, such as radiation and heat; and *chemical* agents, such as antiseptics, disinfectants and vapour fumigants.

Radiation

A quartz mercury-vapour lamp can be used as the source for **ultra-violet light**, which destroys micro-organisms. The object to be sanitised must be turned regularly so that the UV light reaches all surfaces. (UV light has limited effectiveness, and cannot be relied upon for sterilisation.)

The UV lamp must be contained within a closed cabinet. This cabinet is an ideal place for storing sterilised objects.

Sorisa

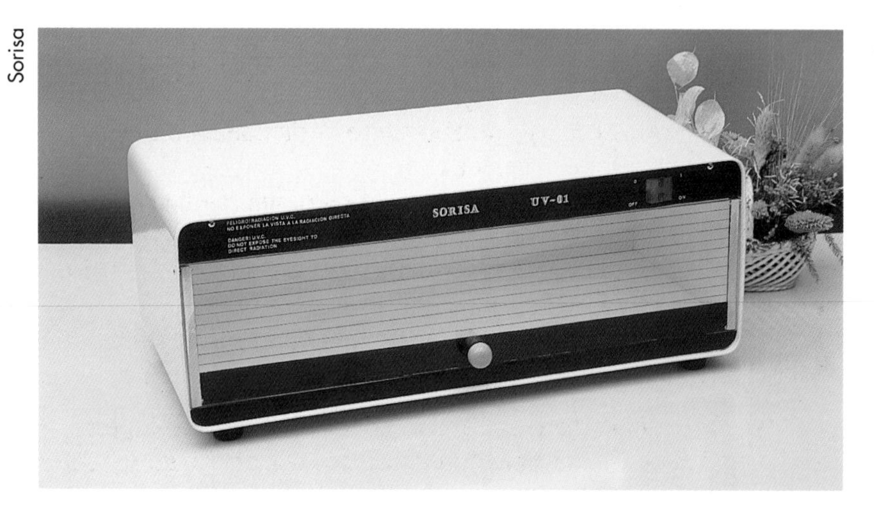

An ultra-violet light cabinet

> **HEALTH & SAFETY: STERILISATION RECORD**
> *Ultra-violet light is dangerous, especially to the eyes.* The lamp must be switched off before opening the cabinet. A record must be kept of usage, as the effectiveness of the lamp decreases with use.

Heat

Dry and moist heat may both be used in sterilisation. One method is to use a dry **hot-air oven**. This is similar to a small oven, and

Sorisa

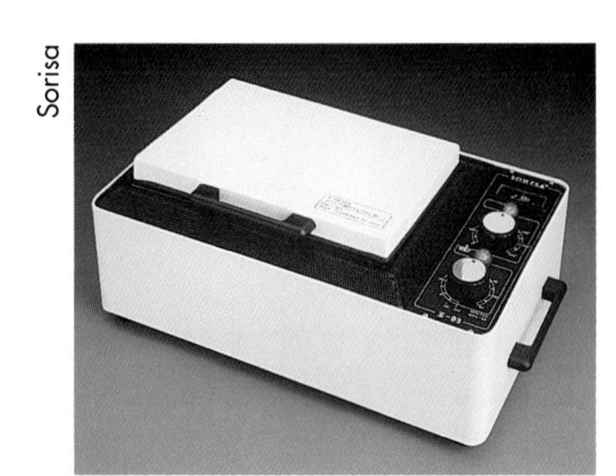
A dry-heat sterilising cabinet

Depilex

An autoclave

heats to 150–180 °C. It is seldom used in the salon.

More practical is a **glass-bead steriliser**. This is a small electrically-heated unit which contains glass beads: these transfer heat to objects placed in contact with them. This method of sterilisation is suitable for small tools such as tweezers and scissors.

Water is boiled in an **autoclave** (similar to a pressure cooker): because of the increased pressure, the water reaches a temperature of 121–134 °C. Autoclaving is the most effective method for sterilising objects in the salon.

HEALTH & SAFETY: USING AN AUTOCLAVE

- Not all objects can safely be placed in the autoclave. Before using this method, check whether the items you wish to sterilise can withstand this heating process.
- To avoid damaging the autoclave, always use distilled deionised water.
- To avoid rusting, metal objects placed in the sterilising unit must be of good-quality stainless steel.

Gases

Gases used in sterilising include **ethylene oxide** and **formaldehyde**. These chemicals are hazardous to handle, however, and are therefore unpopular.

Disinfectants and antiseptics

If an object *cannot* be sterilised, it should be placed in a chemical **disinfectant** solution such as **quaternary ammonium compounds (quats)** or **glutaraldehyde**. A disinfectant destroys most micro-organisms, but not all.

An **antiseptic** prevents the multiplication of micro-organisms. It has a limited action, and does not kill all micro-organisms. Because it is milder than a disinfectant, it can be used directly on the skin.

All sterilisation techniques must be carried out safely and effectively:

1. Select the appropriate method of sterilisation for the object. *Always* follow the manufacturer's guidelines on the use of the sterilising unit or agent.
2. Clean the object in clean water and detergent to remove dirt and grease. (Dirt left on the object may prevent effective sterilisation.)
3. Dry it thoroughly with a clean, disposable paper towel.
4. Sterilise the object, allowing sufficient time for the process to be completed.
5. Place tools that have been sterilised in a clean, covered container.

HEALTH & SAFETY: USING DISINFECTANT

Disinfectant solutions should be changed as necessary.

After removing the object from the disinfectant, rinse it in clean water to remove traces of the solution. (These might otherwise cause an allergic reaction on the client's skin.)

TIP

Before sterilisation, surgical spirit may be used to clean small objects.

Keep several sets of the tools you use regularly, so that you can carry out effective sterilisation.

General salon hygiene rules

- *Health and safety* Follow the health and safety guidelines for the workplace.
- *Personal hygiene* Maintain a high standard of personal hygiene. Wash you hands with a detergent containing **chlorhexidine**.
- *Cuts on the hands* Always cover any cuts on your hands.
- *Cross-infection* Take great care to avoid cross-infection in the salon. *Never* treat a client who has a contagious skin disease or disorder, or any other contra-indication.
- *Use hygienic tools* Never use an implement unless it has been effectively sterilised or sanitised, as appropriate.
- *Disposable products* Wherever possible, use disposable products.
- *Working surfaces* Clean all working surfaces (such as trolleys and couches) with a chlorine preparation, diluted to the manufacturer's instructions. Cover all working surfaces with clean, disposable paper tissue.
- *Gowns and towels* Clean gowns and towels must be provided for each client.
- *Laundry* Dirty laundry should be placed in a covered container.
- *Waste* Put waste in a suitable container lined with a disposable waste bag. A yellow **'sharps' container** should be available for waste contaminated with blood or tissue fluid.
- *Eating and drinking* Never eat or drink in the treatment area of the salon.

HEALTH & SAFETY: DAMAGED EQUIPMENT
Any equipment in poor repair must be repaired or disposed of. Such equipment may be dangerous and may harbour germs.

HEALTH & SAFETY: CUTS ON THE HANDS
Open, uncovered cuts provide an easy entry for harmful bacteria, and may therefore lead to infection. Always cover cuts.

HEALTH & SAFETY: USING CHEMICAL AGENTS
Always protect your hands with gloves before immersing them in chemical cleaning agents, to minimise the risk of an allergic reaction.

SKIN DISEASES AND DISORDERS

The beauty therapist must be able to distinguish a healthy skin from one suffering from any skin disease or disorder. Certain skin disorders and diseases **contra-indicate** a beauty treatment: the treatment would expose the therapist and other clients to the risk of cross-infection. It is therefore vital that you are familiar with the skin diseases and disorders with which you may come into contact in the workplace.

HEALTH & SAFETY: SKIN PROBLEMS
If you are unable to identify a skin condition with confidence, so that you are uncertain whether or not you should treat the client, *don't!* Tactfully refer her to her physician before proceeding with the planned treatment.

Bacterial infections

Bacteria are minute single-celled organisms of varied shapes. Large numbers of bacteria inhabit the surface of the skin and are harmless (**non-pathogenic**); indeed some play an important positive role in the health of the skin. Others, however, are harmful (**pathogenic**) and can cause skin diseases.

Dr M. H. Beck

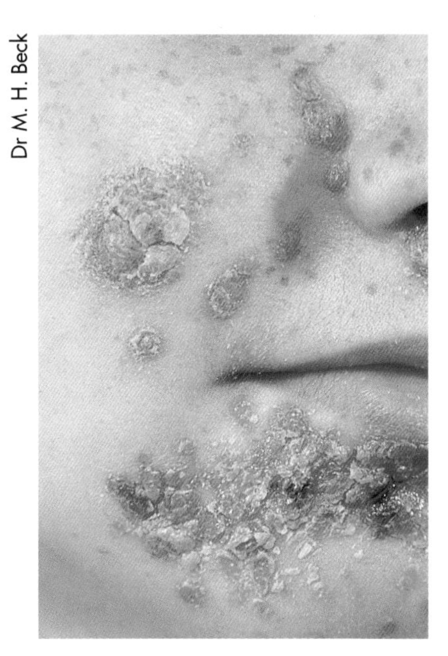

Impetigo

Impetigo

An inflammatory disease of the surface of the skin.
Infectious? Yes.
Appearance: Initially the skin appears red and is itchy. Small thin-walled blisters appear; these burst and form into crusts.
Site: The commonly affected areas are the nose, the mouth and the ears, but impetigo can occur on the scalp or the limbs.
Treatment: Medical – usually an antibiotic or an antibacterial ointment is prescribed.

Conjunctivitis or pink eye

Inflammation of the mucous membrane that covers the eye and lines the eyelids.
Infectious? Yes.
Appearance: The skin of the inner conjunctiva of the eye becomes inflamed, the eye becomes very red and sore, and pus may exude from the area.
Site: The eyes, either one or both, may be infected.
Treatment: Medical – usually an antibiotic lotion is prescribed.

Hordeola or styes

Infection of the sebaceous glands of eyelash hair follicles.
Infectious? Yes.
Appearance: Small lumps containing pus.
Site: The inner rim of the eyelid.
Treatment: Medical – usually an antibiotic is prescribed.

HEALTH & SAFETY: BOILS
Boils occurring on the upper lip or in the nose should be referred immediately to a physician. Boils can be dangerous when near to the eyes or brain.

Furuncles or boils

Red, painful lumps, extending deeply into the skin.
Infectious? Yes.
Appearance: A localised red lump occurs around a hair follicle; it then develops a core of pus. Scarring of the skin often remains after the boil has healed.
Site: The back of the neck, the ankles and the wrists.
Treatment: Medical.

Dr A. L. Wright

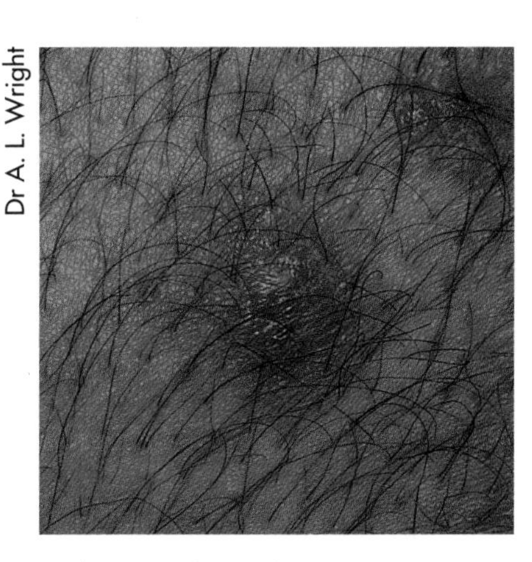

Boil

Carbuncles

Infection of numerous hair follicles.
Infectious? Yes.
Appearance: A hard, round abscess, larger than a boil, which oozes pus from several points upon its surface. Scarring often occurs after the carbuncle has healed.
Site: The back of the neck.
Treatment: Medical – usually involving incision, drainage of the pus, and a course of antibiotics.

Paronychia

Infection of the tissue surrounding the nail.
Infectious? Yes.
Appearance: Swelling, redness and pus in the cuticle and in the area of the nail wall.
Site: The cuticle and the skin surrounding the nail.
Treatment: Medical.

Viral infections

Viruses are minute entities, too small to see even under an ordinary microscope. They are considered to be **parasites**, as they require living tissue in order to survive. Viruses invade healthy body cells and multiply within the cell: in due course the cell walls break down, liberating new viral particles to attack further cells, and thus the infection spreads.

Herpes simplex

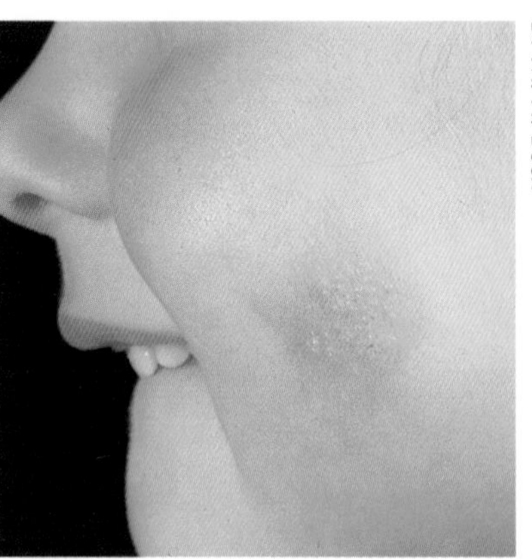
Dr M. H. Beck

Herpes simplex

This is a recurring skin condition, appearing at times when the skin's resistance is lowered through ill health or stress. It may also be caused by exposure of the skin to extremes of temperature or to ultra-violet light.
Infectious? Yes.
Appearance: Inflammation of the skin occurs in localised areas. As well as being red, the skin becomes itchy and small vesicles appear. These are followed by a crust, which may crack and weep tissue fluid.
Site: The mucous membranes of the nose or lips; herpes can also occur on the skin generally.
Treatment: There is no specific treatment. A proprietary brand of anti-inflammatory antiseptic drying cream is usually prescribed.

Herpes zoster or shingles

In this painful disease, the virus attacks the sensory nerve endings. The virus is thought to lie dormant in the body and be triggered when the body's defences are at a low ebb.
Infectious? Yes.
Appearance: Redness of the skin occurs along the line of the affected nerves. Blisters develop and form crusts, leaving purplish-pink pigmentation.
Site: Commonly the chest and the abdomen.
Treatment: Medical – usually including antibiotics. Any lasting pigmentation may be camouflaged with cosmetics.

Verrucae or warts

Small epidermal skin growths. Warts may be raised or flat, depending upon their position. There are several types of wart: plane, common, and plantar.

ACTIVITY: AVOIDING CROSS-INFECTION

1. List the different ways in which infection can be transferred in the salon.
2. How can you avoid cross-infection in the workplace?

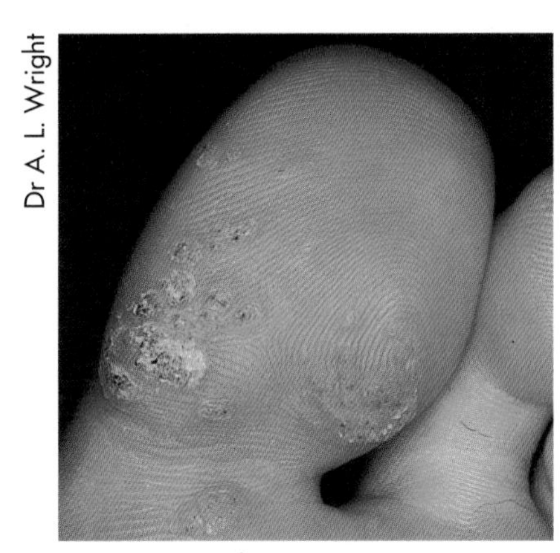

Dr A. L. Wright

Verrucae – plantar warts

Infectious? Yes.
Appearance: Warts vary in size, shape, texture and colour. Usually they have a rough surface and are raised. If the wart occurs on the sole of the foot it grows inwards, due to the pressure of body weight.
Site:
- plane wart: the fingers, either surface of the hand, or the knees;
- common wart: the face or hands;
- plantar wart: the sole of the foot.

Treatment: Medical – using acids, solid carbon dioxide, or electrocautery.

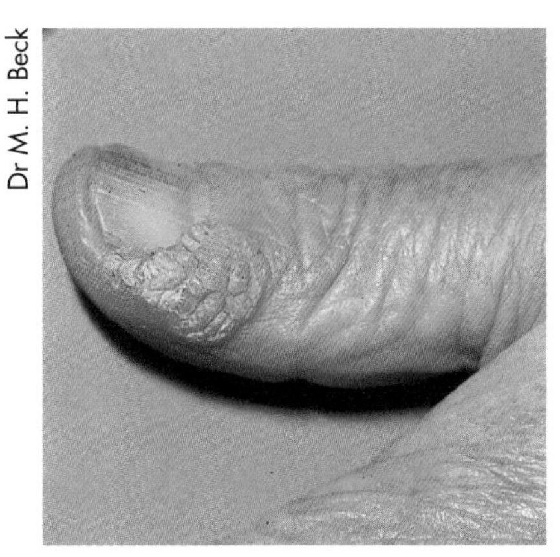

Dr M. H. Beck

A wart

Infestations

Scabies or itch mites

A condition in which an animal parasite burrows beneath the skin and invades the hair follicles.
Infectious? Yes.
Appearance: At the onset, minute papules and wavy greyish lines appear, where dirt has entered the burrows. Secondary bacterial infection may occur as a result of scratching.
Site: Usually seen in warm areas of loose skin, such as the webs of the fingers, and the creases of the elbows.
Treatment: Medical – an anti-scabetic lotion.

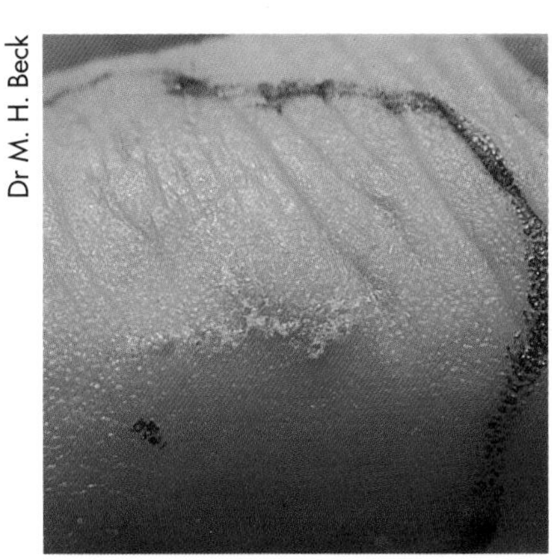

Dr M. H. Beck

Scabies burrow

Pediculosis capitis or head lice

A condition in which small parasites infest scalp hair.
Infectious? Yes.
Appearance: The lice cling to the hair of the scalp. Eggs are laid, attached to the hair close to the skin. The lice bite the skin to draw nourishment from the blood; this creates irritation and itching of the skin, which may lead to secondary bacterial infection.
Site: The hair of the scalp.
Treatment: Medical – an appropriate lotion.

Pediculosis pubis

A condition in which small parasites infest body hair.
Infectious? Yes.
Appearance: The lice cling to the hair of the body. Eggs are laid, attached to the hair close to the skin. The lice bite the skin to draw nourishment from the blood; this creates irritation and itching of the skin, which may lead to secondary bacterial infection.
Site: Pubic hair, eyebrows and eyelashes.
Treatment: Medical – an appropriate lotion.

Pediculosis corporis

A condition in which small parasites live and feed on body skin.
Infectious? Yes.
Appearance: The lice cling to the hair of the body. Eggs are laid, attached to the hair close to the skin. The lice bite the skin to draw nourishment from the blood; this creates irritation and itching of the skin, which may lead to secondary bacterial infection. Where body lice bite the skin, small red marks can be seen.
Site: Body hair.
Treatment: Medical – an appropriate lotion.

Fungal diseases

Fungi are microscopic plants. They are parasites, dependent upon a host for their existence. Fungal diseases of the skin feed off the waste products of the skin. Some fungi are found on the skin's surface; others attack the deeper tissues. Reproduction of fungi is by means of simple cell division or by the production of spores.

Tinea pedis or athlete's foot

A common fungal foot infection.
Infectious? Yes.
Appearance: Small blisters form, which later burst. The skin in the area can then become dry, giving a scaly appearance.
Site: The webs of skin between the toes.
Treatment: Thorough cleansing of the area. Medical application of fungicides.

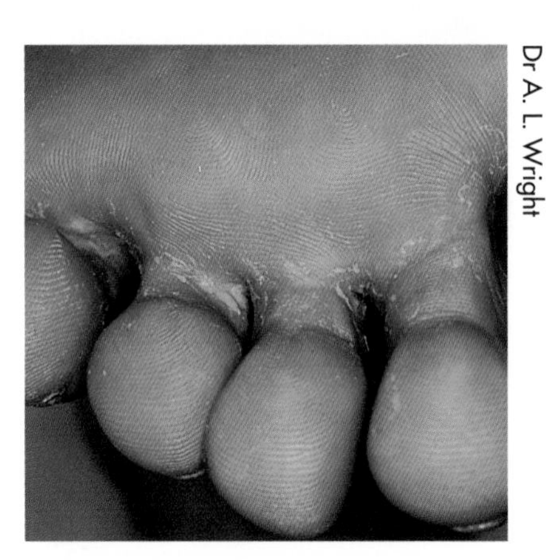
Dr A. L. Wright

Tinea pedis

Tinea corporis or body ringworm

A fungal infection of the skin.
Infectious? Yes.
Appearance: Small scaly red patches, which spread outwards and then heal from the centre, leaving a ring.
Site: The trunk of the body, the limbs and the face.
Treatment: Medical – using a fungicidal cream, griseofluvin.

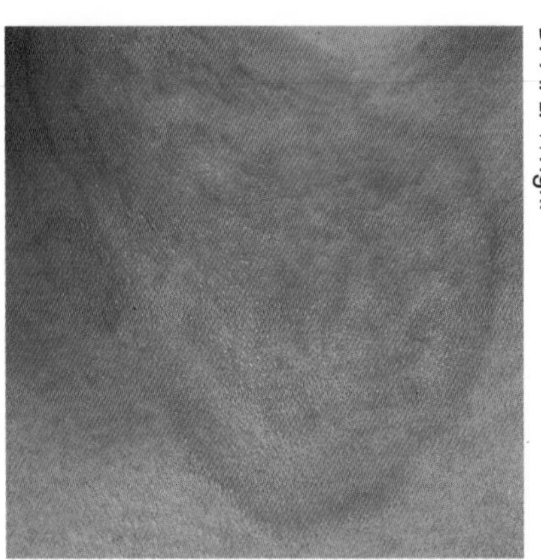
Dr A. L. Wright

Tinea corporis

Tinea unguium

Ringworm infection of the fingernails.
Infectious? Yes.
Appearance: The nail plate is yellowish-grey. Eventually the nail plate becomes brittle and separates from the nail bed.
Site: The nail plates of the fingers.
Treatment: Medical application of fungicides.

Sebaceous gland disorders

Milia

Keratinisation of the skin over the hair follicle occurs, causing sebum to accumulate in the hair follicle. This condition usually accompanies dry skin.
Infectious? No.
Appearance: Small, hard, pearly-white cysts.
Site: The upper face or close to the eyes.
Treatment: The milium may be removed by the beauty therapist or by a physician, depending on the location. A sterile needle is used to pierce the skin of the overlying cuticle and thereby free the milium.

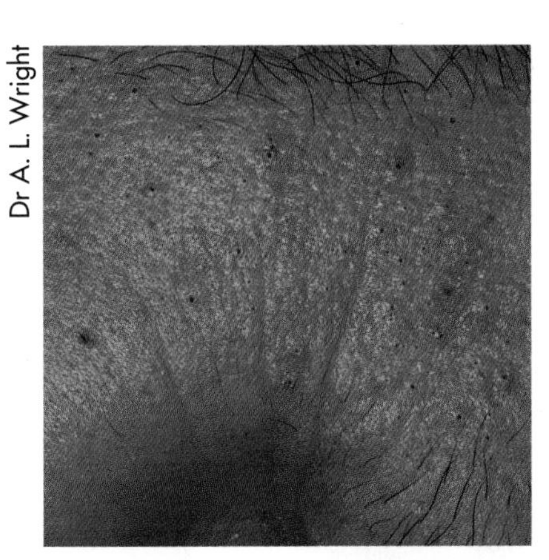

Dr A. L. Wright

Comedones or blackheads

Comedones or blackheads

Excess sebum and keratinised cells block the mouth of the hair follicle.
Infectious? No.
Site: The face (the chin, nose and forehead), the upper back and chest.
Treatment: The area should be cleansed, and an electrical vapour treatment or other pre-heating treatment should be given to relax the mouth of the hair follicle; a sterile comedo extractor should then be used to remove the blockage. A regular cleansing treatment should be recommended by the beauty therapist to limit the production of comedones.

Seborrhoea

Excessive secretion of sebum from the sebaceous gland. This usually occurs during puberty, as a result of hormonal changes in the body.
Infectious? No.
Appearance: The follicle openings enlarge and excessive sebum is secreted. The skin appears coarse and greasy; comedones, pustules and papules are present.
Site: The face and scalp. Seborrhoea may also affect the back and the chest.
Treatment: The area should be cleansed to remove excess grease. Medical treatment may be required – this would use locally applied creams.

Steatomas, sebaceous cysts or wens

Localised pockets or sacs of sebum, which form in hair follicles or under the sebaceous glands in the skin. The sebum becomes blocked, the sebaceous gland becomes distended, and a lump forms.
Infectious? No.
Appearance: Semi-globular in shape, either raised or flat, and hard or soft. The cysts are the same colour as the skin, or red if secondary bacterial infection occurs. A comedo can often

be seen at the original mouth of the hair follicle.
Site: If the cyst appears on the upper eyelid, it is known as a **chalazion** or **meibomian cyst.**
Treatment: Medical – often a physician will remove the cyst under local anaesthetic.

Acne vulgaris

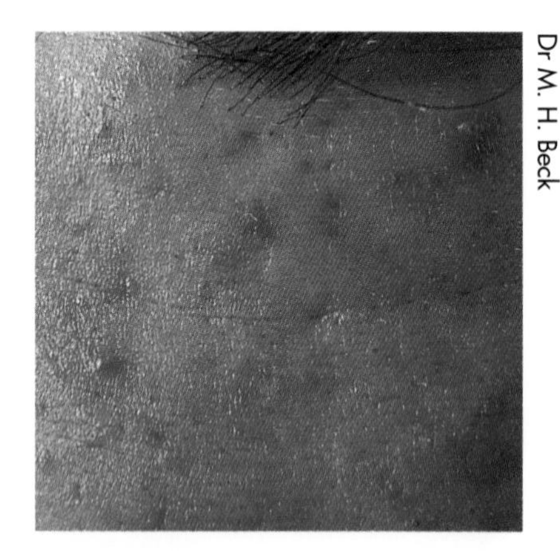

Dr M. H. Beck

Acne vulgaris

Hormone imbalance in the body at puberty influences the activity of the sebaceous gland, causing an increased production of sebum. The sebum may be retained within the sebaceous ducts, causing congestion and bacterial infection of the surrounding tissues.
Infectious? No.
Appearance: Inflammation of the skin, accompanied by comedones, pustules and papules.
Site: Commonly on the face, on the nose, the chin and the forehead. Acne may also occur on the chest and back.
Treatment: Medical – oral antibiotics may be prescribed, as well as medicated creams. With medical approval, regular salon treatments may be given to cleanse the skin deeply, and also to stimulate the blood circulation.

Acne rosacea

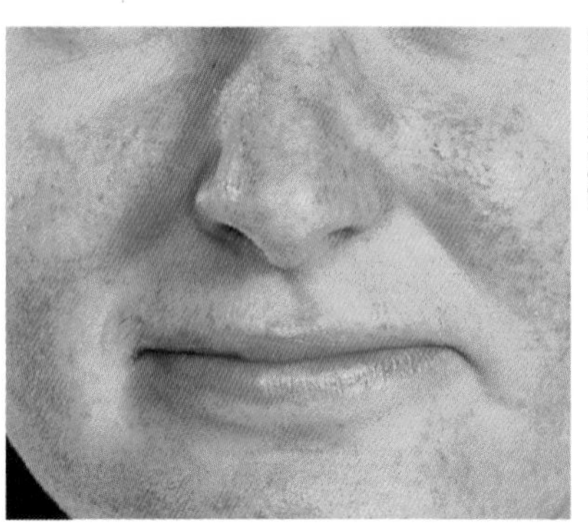

Dr M. H. Beck

Acne rosacea

Excessive sebum secretion combined with a chronic inflammatory condition, caused by dilation of the blood capillaries.
Infectious? No.
Appearance: The skin becomes coarse, the pores enlarge, and the cheek and nose area become inflamed, sometimes swelling and producing a butterfly pattern. Blood circulation slows in the dilated capillaries, creating a purplish appearance.
Treatment: Medical – usually including antibiotics.

Pigmentation disorders

Pigmentation of the skin varies, according to the person's genetic characteristics. In general the darker the skin, the more pigment is present, but some abnormal changes in skin pigmentation can occur.

- **hyperpigmentation** – increased pigment production;
- **hypopigmentation** – loss of pigmentation in the skin.

> **TIP**
> Hypopigmentation may result from certain skin injuries, disorders or diseases.

Ephelides or freckles

Multiple small pigmented areas of the skin. Exposure to ultra-violet light (as in sunlight) stimulates the production of melanin, intensifying their appearance.
Infectious? No.
Appearance: Small, flat, pigmented areas, darker than the surrounding skin.

Site: Commonly the nose and cheeks of fair-skinned people. Freckles may also occur on the shoulders, arms, hands and back.
Treatment: Freckles may be concealed with cosmetics if required. A sun block should be recommended, to prevent them intensifying in colour.

Lentigines

Pigmented areas of skin, slightly larger than freckles, which do not darken on exposure to ultra-violet.
Infectious? No.
Appearance: Brown, slightly raised, pigmented patches of skin.
Site: The face and hands.
Treatment: Application of cosmetic concealing products.

Chloasmata or liver spots

Increased skin pigmentation in specific areas, stimulated by a skin irritant such as ultra-violet. The condition often occurs during pregnancy, and usually disappears soon after the birth of the baby. It may also occur as a result of taking oral contraceptive pills. The female hormone oestrogen is thought to stimulate melanin production.
Infectious? No.
Appearance: Flat, smooth, irregularly shaped, pigmented areas of skin, varying in colour from light tan to dark brown. Chloasmata are larger than ephelides, and of variable size.
Site: The back of the hands, the forearms, the upper part of the chest, the temples and the forehead.
Treatment: A barrier cream or a total sun block will reduce the risk of the chloasmata increasing in size or number, and thereby becoming more apparent.

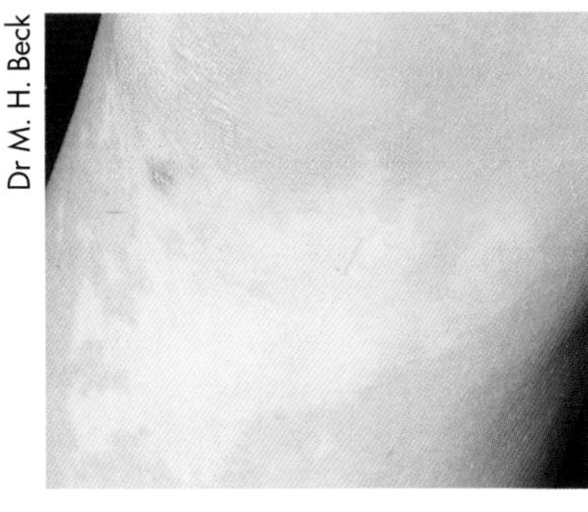
Dr M. H. Beck

Vitiligo

Vitiligo or leucoderma

Patches of completely white skin which have lost their pigment, or which were never pigmented.
Infectious? No.
Appearance: Symmetrically shaped patches of skin, lacking pigment.
Site: The face, the neck, the hands, the lower abdomen, and the thighs. If vitiligo occurs over the eyebrows, the hairs in the area will lose their pigment also.
Treatment: Camouflage cosmetic concealer can be applied to give even skin colour; or skin-staining preparations can be used in the de-pigmented areas. Care must be taken when the skin is exposed to ultra-violet light, as the skin will not have the same protection in the areas lacking pigment.

Albinism

The skin is unable to produce the melanin pigment, and the skin, hair and eyes lack colour.

Infectious? No.
Appearance: The skin is usually very pale pink and the hair is white. The eyes also are pink, and extremely sensitive to light.
Site: The entire skin.
Treatment: There is no effective treatment. Maximum skin protection is necessary when the client is exposed to ultraviolet light, and sunglasses should be worn to protect the eyes.

Vascular naevi

There are two types of naevus of concern to beauty therapists: vascular and cellular. **Vascular naevi** are skin conditions in which small or large areas of skin pigmentation are caused by the permanent dilation of blood capillaries.

TIP

If there is a vascular skin disorder, avoid overstimulating the skin or the problem will become *more* noticeable and the treatment may even cause further damage.

Erythema

An area of skin in which blood capillaries have dilated, due either to injury or inflammation.
Infectious? No.
Appearance: The skin appears red.
Site: Erythema may affect one area (locally) or all of the skin (generally).
Treatment: The cause of the inflammation should be identified. In the case of a skin allergy, the client must not be brought into contact with the irritant again. If the cause is unknown, refer the client to her physician.

Dilated capillaries

Capillaries near the surface of the skin that are permanently dilated.
Infectious? No.
Appearance: Small red visible blood capillaries.
Site: Areas where the skin is neglected, dry or fine, such as the cheek area.
Treatment: Dilated capillaries can be concealed using a green corrective camouflage cosmetic, or removed by a qualified electrologist using diathermy.

Spider naevi or stellate haemangiomas

Dilated blood vessels, with smaller dilated capillaries radiating from them.
Infectious? No.
Appearance: Small red capillaries, radiating like a spider's legs from a central point.
Site: Commonly the cheek area, but may occur on the upper body, the arms and the neck. Spider naevi are usually caused by an injury to the skin.
Treatment: Spider naevi can be concealed using a camouflage

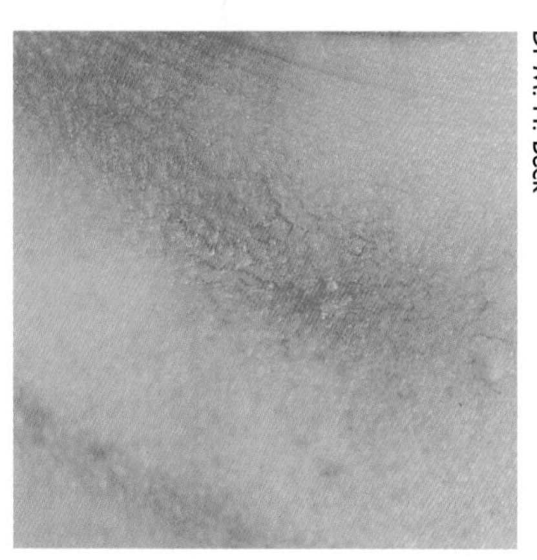
Dr M. H. Beck

Spider naevus

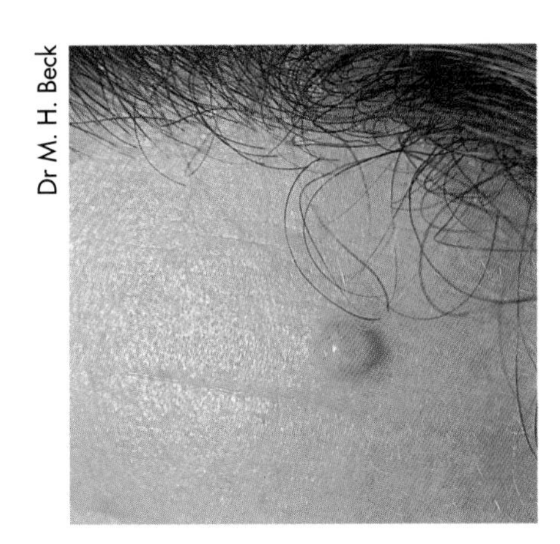
Dr M. H. Beck

Benign pigmented naevus

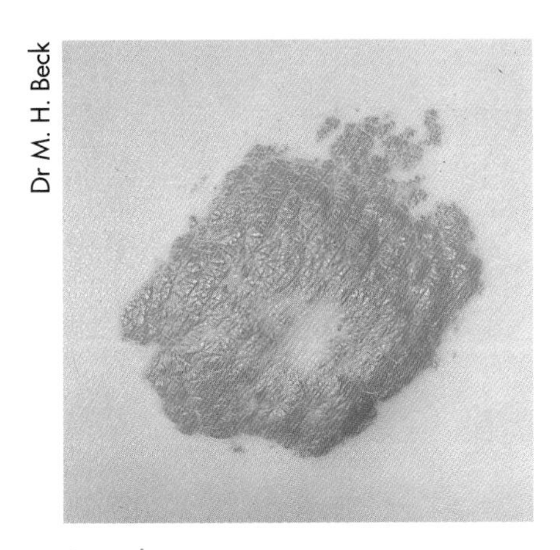
Dr M. H. Beck

Strawberry naevus

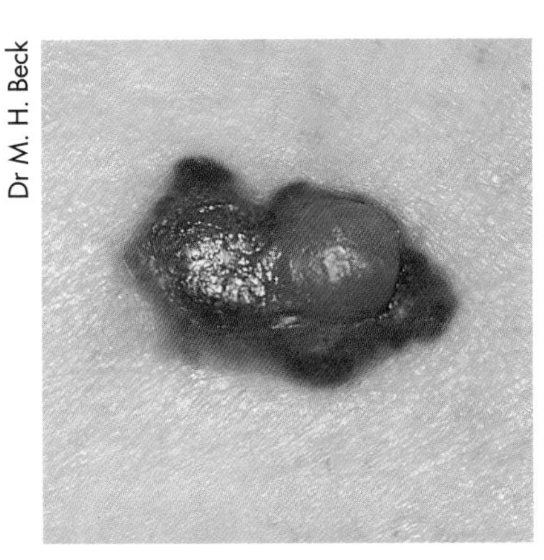
Dr M. H. Beck

Malignant melanoma

cosmetic, or treated by a qualified electrologist with diathermy.

Naevi vasculosis or strawberry marks

Red or purplish raised marks which appear on the skin at birth.
Infectious? No.
Appearance: Red or purplish lobed mark, of any size.
Site: Any area of the skin.
Treatment: About 60 per cent disappear by the age of 6 years. Treatment is not usually necessary; concealing cosmetics can be applied if desired.

Capillary naevi or port-wine stains

Large areas of dilated capillaries, which contrast noticeably with the surrounding areas.
Infectious? No.
Appearance: The naevus has a smooth, flat surface.
Site: Some 75 per cent occur on the head; they are probably formed at the foetal stage. Naevi may also be found on the neck and face.
Treatment: Camouflage cosmetic creams can be applied to disguise the area.

Cellular naevi or moles

Cellular naevi are skin conditions in which changes in the cells of the skin result in skin malformations.

Malignant melanomas or malignant moles

Rapidly-growing skin cancers, usually occurring in adults.
Infectious? No.
Appearance: Each melanoma commences as a bluish-black mole, which enlarges rapidly, darkening in colour and developing a halo of pigmentation around it. It later becomes raised, bleeds, and ulcerates. Secondary growths will develop in internal organs, if the melanoma is not treated.
Site: Usually the lower abdomen, legs or feet.
Treatment: Medical – *always* recommend that a client has any mole checked if it is changing in size, structure or colour, or if it becomes itchy or bleeds.

Junction naevi

Localised collections of naevoid cells that arise from the mass production locally of pigment-forming cells (**melanocytes**).
Infectious? No.
Appearance: In childhood junction naevi appear as smooth or slightly raised pigmented marks. They vary in skin colour from brown to black.
Site: Any area.
Treatment: None.

Dermal naevi

Localised collections of naevoid cells.
Infectious? No.
Appearance: About 1cm wide, dermal naevi appear smooth and dome-shaped. Their colour ranges from skin tone to dark brown. Frequently one or more hairs may grow from the naevus.
Site: Usually the face.
Treatment: None.

HEALTH & SAFETY: MOLES
If moles change in shape or size, if they bleed or form crusts, seek medical attention.

Hairy naevi

Moles exhibiting coarse hairs from their surface.
Infectious? No.
Appearance: Slightly raised moles, varying in size from 3cm to much larger areas. Colour ranges from fawn to dark brown.
Site: Anywhere on the skin.
Treatment: Hairy naevi may be surgically removed where possible, and this is often done for cosmetic reasons. Hair growing from a mole should be cut, not plucked: if plucked, the hairs will become coarser and the growth of further hairs may be stimulated.

Skin disorders involving abnormal growth

Psoriasis

Patches of itchy, red, flaky skin, the cause of which is unknown.
Infectious? No. Secondary infection with bacteria can occur if the skin becomes broken and dirt enters the skin.
Appearance: Red patches of skin appear, covered in waxy, silvery scales. Bleeding will occur if the area is scratched, thereby removing scales.
Site: The elbows, the knees, the lower back and the scalp.
Treatment: There is no known treatment. Medication including steroid creams can bring relief to the symptoms.

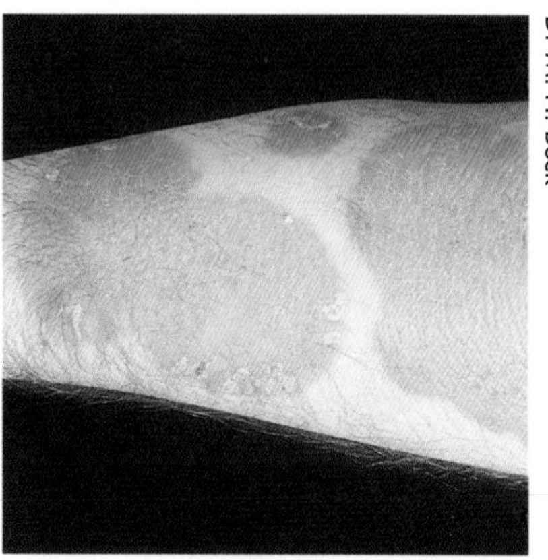
Dr M. H. Beck

Psoriasis

Seborrheic or senile warts

Raised, pigmented, benign tumours occurring in middle age.
Infectious? No.
Appearance: Slightly raised, brown or black, rough patches of skin. Such warts can be confused with pigmented moles.
Site: The trunk, the scalp, and the temples.
Treatment: Medical – the warts can be cauterised by a physician.

Verrucae filliformis or skin tags

These verrucas appear as threads projecting from the skin.
Infectious? No.
Appearance: Skin-coloured threads of skin 3–6mm long.
Site: Mainly seen on the neck and the eyelids, but may occur in other areas such as under the arms.

HEALTH & SAFETY: SKIN TAGS
Skin tags often occur under the arms. In case they are present, take care when carrying out a wax depilation treatment in this area: do not apply wax over tags.

Treatment: Medical – cauterisation with diathermy, either by a physician or by a qualified electrologist.

Xanthomas

Small yellow growths appearing upon the surface of the skin.
Infectious? No.
Appearance: A yellow, flat or raised area of skin.
Site: The eyelids.
Treatment: Medical – the growth is thought to be connected with certain medical diseases, such as diabetes or high or low blood pressure; sometimes a low-fat diet can correct the condition.

Malignant tumours

Squamous cell carcinomas or prickle-cell cancers

Malignant growths originating in the epidermis.
Infectious: No.
Appearance: When fully formed, the carcinoma appears as a raised area of skin.
Site: Anywhere on the skin.
Treatment: Medical – by radiation.

Dr M. H. Beck

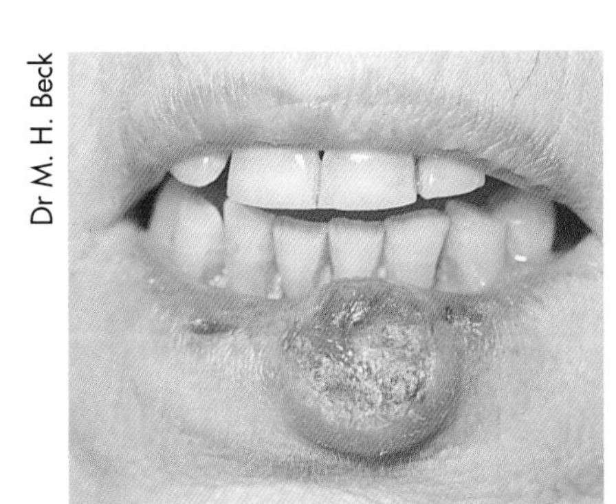

Squamous cell carcinoma

Basal-cell carcinomas or rodent ulcers

Slow-growing malignant tumours, occurring in middle age.
Infectious? No.
Appearance: A small, shiny, waxy nodule with a depressed centre. The disease extends, with more nodules appearing on the border of the original ulcer.
Site: Usually the face.
Treatment: Medical.

Dr A. L. Wright

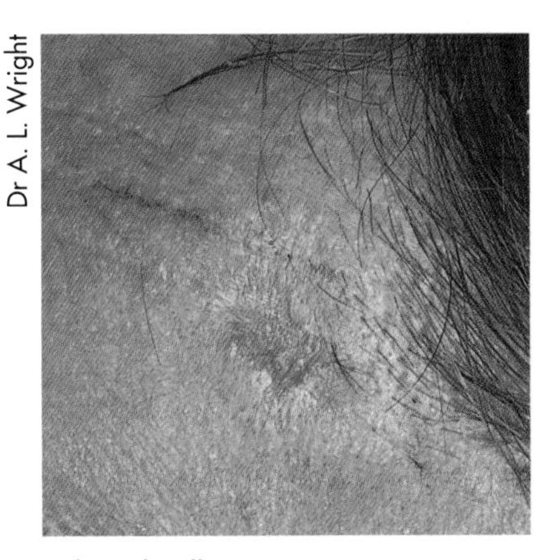

A basal-cell carcinoma

HEALTH & SAFETY: RECORD ANY KNOWN ALLERGIES
When completing the client record card, always ask whether your client has any known allergies.

Skin allergies

The skin can protect itself to some degree from damage or invasion. **Mast cells** detect damage to the skin; if damage occurs, the mast cells burst, releasing the chemical **histamine** into the tissues. Histamine causes the blood capillaries to dilate, giving the reddening we call 'erythema'. The increased blood flow transports materials in the blood which tend to limit the damage and begin repair.

If the skin is sensitive to and becomes inflamed on contact with a particular substance, this substance is called an **allergen**. Allergens may be animal, chemical or vegetable substances, and they may be inhaled, eaten, or absorbed following contact with the skin. An **allergic skin reaction** appears as irritation, itching and discomfort, with reddening and swelling (as with nettle rash). If the allergen is removed, the allergic reaction subsides.

Each individual has different tolerances to the various substances we encounter in daily life. What causes an allergic reaction in one individual may be perfectly harmless to another.

Here are just a few examples of allergens known to cause allergic skin reactions in some people:

- metal objects containing nickel;
- sticking plaster;
- rubber;
- lipstick containing eosin dye;
- nail enamel containing formaldehyde resin;
- hair and eyelash dyes;
- lanolin, the skin moisturising agent;
- detergents that dry the skin;
- foods – well-known examples are peanuts, cow's milk, lobster, shellfish, and strawberries;
- plants such as tulips and chrysanthemums.

HEALTH & SAFETY: HYPOALLERGENIC PRODUCTS
The use of hypoallergenic products minimises the risk of skin contact with likely irritants.

HEALTH & SAFETY: INFECTION FOLLOWING ALLERGY
Following an allergic skin reaction in which the skin's surface has become itchy and broken, scratching may cause the skin to become infected with bacteria.

HEALTH & SAFETY: ALLERGIES
You may suddenly become allergic to a substance that has previously been perfectly harmless. Equally, you may over time *cease* to be allergic to something.

Dermatitis

An inflammatory skin disorder in which the skin becomes red, itchy and swollen. There are two types of dermatitis. In *primary dermatitis* the skin is irritated by the action of a substance upon the skin, and this leads to skin inflammation. In *allergic contact dermatitis* the problem is caused by intolerance of the skin to a particular substance or group of substances. On exposure to the substance the skin quickly becomes irritated and an allergic reaction occurs.

Infectious? No.

Appearance: Reddening and swelling of the skin, with the possible appearance of blisters.

Site: If the skin reacts to a skin irritant outside the body, the reaction is localised. Repeated contact with the allergen will lead to a general hypersensitivity. If the irritant gains entry to the body it will be transported in the bloodstream and may cause a general allergic skin reaction.

Treatment: Barrier cream can be used to help avoid contact with the irritants. When an allergic dermatitis reaction occurs, however, the only 'cure' is the absolute avoidance of the substance. Steroid creams such as hydrocortisone are usually prescribed, to sooth the damaged skin and reduce the irritation.

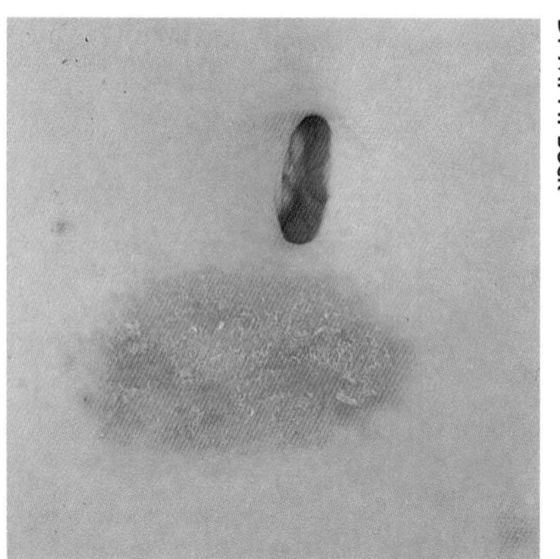
Dr M. H. Beck

Allergic contact dermatitis (to the nickel in a stud on jeans)

Eczema

Inflammation of the skin caused by contact, internally or externally, with an irritant.

Infectious? No.

Appearance: Reddening of the skin, with swelling and blisters.

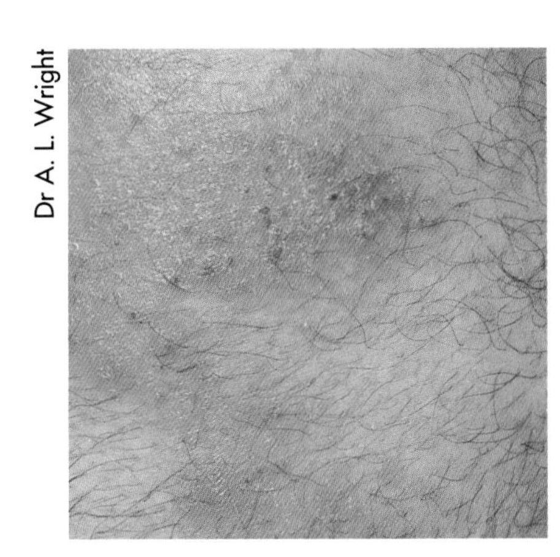

Dr A. L. Wright

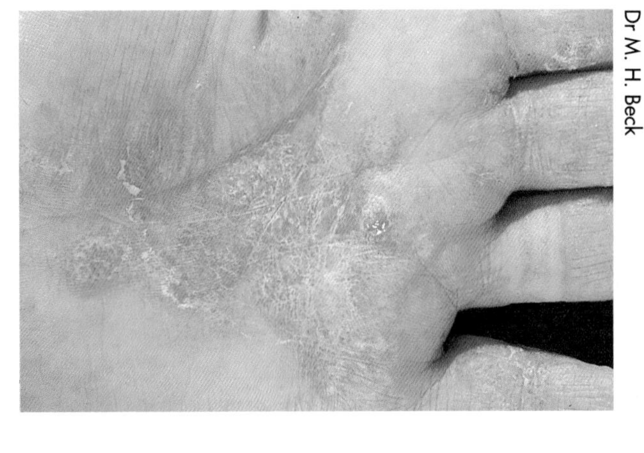

Dr M. H. Beck

Eczema

The blisters leak tissue fluid which later hardens, forming scabs.

Site: The face, the neck and the skin, particularly at the inner creases of the elbows and behind the knees.

Treatment: Refer the client to her physician. Eczema may disappear if the source of irritation is identified and removed. Steroid cream may be prescribed by the physician, and special diets may help.

Dr M. H. Beck

Urticaria

Urticaria (nettle rash) or hives

A minor skin disorder caused by contact with an allergen, either internally (food or drugs) or externally (for example, insect bites).

Infectious? No.

Appearance: Erythema with raised, round whitish skin weals. In some cases the lesions can cause intense burning or itching, a condition known as **pruritis**. Pruritis is a *symptom* of a disease (such as diabetes), not a disease itself.

Site: At the point of contact.

Treatment: Antihistamines may be prescribed to reduce the itching. The visible skin reaction usually disappears quickly, leaving no trace. Complete avoidance of the allergen 'cures' the problem.

SALON SECURITY

It is important that adequate precautions are taken to secure the premises against **theft** (during business hours) and **burglary** (outside business hours). Theft may occasionally occur in the workplace, by people posing as clients, by actual clients, and – very occasionally – by employees. Staff need to be trained in all aspects of security.

Guarding against theft

The following points will help to reduce the likelihood of theft in your salon:

- The retail area should be designed to be in full view of staff.

- Electronic cash tills make it much harder for employees to steal cash.
- From time to time (but at random), make **audits** of money and stock.
- Keep retail goods in locked **display cabinets**. Open display areas should be stocked with replica ('dummy') products.
- Only authorised staff should handle cash and have access to stock.
- A **supervisor** may be appointed to check daily cash balances.
- Cash **errors** can easily occur, if an employee charges incorrectly for a product, if a product has been incorrectly priced, if the wrong change is given, or if forged notes or invalid cheques have been accepted. However, all such errors should be traceable.
- Each employee should be provided with a **lockable cupboard** in which to keep personal belongings.
- Clients' belongings should not be left unattended. It is customary to display a notice disclaiming responsibility in the event of theft.

ACTIVITY: EVALUATING SALON SECURITY

A thief will seize any opportunity.

1. What precautions can you take to make a salon secure?
2. What security measures are used in large department stores to reduce the risk of theft?

Dealing with theft

If you suspect a client of stealing and you have reasonable evidence, you have the right to make a **citizen's arrest** under the **Police and Criminal Evidence Act 1984**. You must know your employer's policy on theft and on apprehending a thief.

TIP

The law does not allow you to prosecute a child under 10 years of age.

KNOWLEDGE REVIEW

Health, safety and security

1. Can you name *two* Acts of Parliament concerned with health, safety and hygiene in the workplace?
2. Whilst cleaning a wax heater you notice that wires in the lead are exposed. What action should you take?
3. Why must regular health and safety checks be carried out in the workplace?
4. State *five* basic requirements for maintaining personal hygiene.
5. Why is it important for the client to observe you carrying out hygienic practice when performing treatments?
6. When completing the client's record card you recognise that treatment is contra-indicated as the client has an infectious skin disorder. Name *four* such infectious skin disorders.

7 Effective sterilisation methods prevent cross-infection and secondary infection. What do you understand by these terms?
   (a) Sterilisation.
   (b) Cross-infection.
   (c) Secondary infection.
8 How can tools be kept clean after they have been sterilised?
9 What does the abbreviation COSHH stand for? Why is it important to enforce COSHH regulations in the workplace?
10 When carrying out a stock check you drop and break a bottle of hydrogen peroxide. How should you deal with this spillage? How should you dispose of the broken bottle?
11 What is the procedure for dealing with an accident in the workplace?
12 You are asked to check the contents of the workplace first-aid box. Name *six* essential items.
13 Suppose that there is a fire drill in your workplace. What is the fire evacuation procedure?
14 In the event of a real fire, after having safely evacuated the building, how would you contact the appropriate emergency service?
15 You discover a smoking bin, in which the fire has been caused by an unextinguished cigarette. How should this fire be extinguished?
16 How should a large box be lifted from floor level to be placed on a work surface?
17 When tidying the working area following a treatment, you discover that your client has forgotten her necklace. What procedure would you take to ensure that the property was returned to the client?
18 Give examples of possible breaches of salon security.
19 How can you ensure the security of the workplace premises?

Reception

RECEIVING CLIENTS

The client's **reception** gives her her first and also her final impression of the salon, whether this is on the telephone or in person when she visits the salon. First impressions count, so ensure that the client gets the *right* impression!

The design of the reception area

Location

Reception is usually situated at the front of a beauty salon; in a large department store, reception may be a cosmetic counter. It should be clean, uncluttered and inviting.

With a salon, the advantage of having reception at the front is that the window can be used to attract and capture the attention and interest of potential clients. Clients who are waiting in reception, however, may seek privacy, so the window should be attractively curtained and the seating should be situated away from the view of the main window if possible.

TIP

Have a heavy-duty footmat at the entrance to reception to protect the main floor covering from becoming marked.

Size

The entrance to reception should be large enough for wheelchair access. There should be adequate seating, and an area in which to hang clients' coats.

Depilex/RVB

A reception area

ACTIVITY: MAGAZINES
What magazines do you think would be appropriate for the beauty salon?

TIP
Video facilities in reception may be used to promote salon services. Many suitable videos are available from beauty manufacturers and suppliers.

HEALTH & SAFETY: VENTILATION
To avoid losing the custom of non-smoking clients, ensure that the air is fresh and the room adequately ventilated to remove the smell of stale tobacco.

It may be that small treatments, such as manicures, are carried out at reception. These treatments can then be seen by others, and may attract further clients.

In the event of a client arriving early for an appointment, an appropriate range of magazines should be available. These should be renewed regularly.

Smoking

The smoking policy is determined by each salon. If smoking is allowed in reception, adequate ashtrays should be provided. These should be emptied and cleaned after each use.

Decoration

The reception area should be decorated tastefully, in keeping with the décor in the rest of the salon. Attractive posters promoting proprietary cosmetic ranges may be displayed on the walls. Framed certificates of the staff's professional qualifications can be displayed, as well as health-legislation registration certificates.

Equipment

The reception area should be uncluttered. The main equipment and furnishings required for an efficient reception include the following:

- *A reception desk* The size of the desk will depend on the size of the salon; some salons may have several receptionists. The desk should include shelves and drawers; some have an in-built lockable cash or security drawer. It should be at a convenient height for the client to write a cheque. It should also be large enough to house the appointment book or computer (or both).
- *A comfortable chair* The receptionist's chair should provide adequate back support.
- *A computer* Computers are becoming more and more popular in the salon as they can perform many functions. They can be used to store data about clients, to keep appointment schedules, to carry out automatic stock control, and to record business details such as accounts and marketing information. Some computers can even be programmed to recommend specific treatments on the basis of personal data about the client!
- *A calculator* This is used for simple financial calculations, especially if the salon does not have a computer.
- *Stationery* This should include price lists, gift vouchers, appointment cards, and a receipt pad.
- *A notepad* This is for taking notes and recording messages.
- *An address and telephone book* This should hold all the frequently-used telephone numbers.

HEALTH & SAFETY: EATING AND DRINKING
The receptionist and other employees should not eat or drink at reception, nor should they smoke.

TIP
Some salons' price lists are in booklet form, detailing the treatments offered and explaining their benefits.

TIP

Visitors to the salon may leave a business card, stating the name of the company, and the representative's name, address and telephone number. These cards should be filed by the receptionist for future reference.

- *A telephone and an answering machine* The answering machine allows clients to notify you, even when the salon is closed, of an appointment request, or an unavoidable change or cancellation: you can then re-schedule appointments as quickly as possible. If you are working on your own, the answering machine avoids interruptions during a treatment, yet without losing custom.
- *Record cards* These confidential cards record the personal details of each client registered at the salon. They should be kept in alphabetical order in a filing cabinet or a card-index box, and should be ready for collection by the therapist when treating new or existing clients. Each card records:
  - the client's name, address and telephone number;
  - any medical details;
  - any contra-indications (such as allergies and contra-actions);
  - treatment aims and outcomes;
  - a base on which to plan future treatments;
  - services received, products used, and merchandise purchased.

HEALTH & SAFETY: CONTACT NUMBERS

Just in case a client should become ill whilst in the salon, a contact number should be recorded on the client's card.

A record card

Name | STRICTLY PROFESSIONAL | BEAUTY TREATMENT

Address. | Date of birth. | Postcode.

Phone Day. | Evening. | Occupation.

Doctors name and address. | Doctors Phone No.

MEDICAL HISTORY

Notes. / Allergies.	Diabetes Epilepsy Heart condition Abnormal Blood pressure Kidney Liver(esp.Hepatitis) Major operations Hormone irregularities	Hypertension Headaches/Migraine Asthema Varicose veins Verucca/Athletes foot Fainting/Giddyness Hormone replacement Very sensitive skin	Cold sores Moles Irregular skin pigmentation Prickly heat Number of children Last pregnancy Others

Are you under medical supervision
Are you taking any medication (esp. antibiotics, steroids, the pill)

FACIAL DIAGNOSIS

SKIN TYPE	Open pores	Notes.
Acne	Blocked pores	
Oily	Milia	
Normal	Dilated Capillaries	
Dry	Skin tags	
Sensitive	Moles	
Mature	Dry patches	

Eye area.

Neck area.

Recommended skin care routine and products | I have read and understood the details above.

Signed.

Dated.

Bellitas

- *Pens, pencils and an eraser* Make sure these stay at the desk!
- *A display cabinet* This may be used to store proprietary skin-care and cosmetic products, and any other merchandise sold by the salon.

ACTIVITY: DESIGNING RECORD CARDS
Design a card to be used to record the clients' requirements and the treatment details related to NVQ Level 2.

ACTIVITY: PLANNING A RECEPTION AREA
Design a reception area, to scale, appropriate to a small or large beauty salon. Discuss the choice of wall and floor coverings, furnishings and equipment, and give the reasons for their selection. Consider clients' comfort, and health and safety

HEALTH & SAFETY: CLEANING
In a large salon a cleaner may be employed to maintain the hygiene of the reception area. In a smaller salon this may be the responsibility of an apprentice, the receptionist, or a therapist. Reception must be maintained to a high standard at all times.

THE RECEPTIONIST

Receptionists should have a smart appearance and be able to communicate effectively and professionally, thereby creating the right impression.

The receptionists' duties include:

- maintaining the reception area:
- looking after clients on arrival and departure;
- scheduling appointments;
- dealing with enquiries;
- telling the appropriate therapist that a client or visitor has arrived;
- assisting with retail sales;
- operating the payment point and handling payments.

The receptionist should know:

- the name of each member of staff, her role and her area of responsibility;
- the salon's hours of opening, and the days and times when each therapist is available;
- the range of services or products offered by the salon, and their cost;
- the benefits of each treatment and each retail product;
- the approximate time taken to complete each treatment;
- how to schedule follow-up treatments.

TIP
It is a good idea for the receptionist to wear a badge indicating her name and position.

TIP
If you are engaged on the telephone when a client arrives, look up and acknowledge her presence.

HEALTH & SAFETY: FIRE DRILL
As the receptionist you should be familiar with the emergency procedure in case of fire.

Qualities of a receptionist

All clients need to feel valued. The following **interpersonal skills** are desirable in a receptionist:

- act positively and confidently;
- speak clearly;
- be friendly, and smile;
- look at the customer and maintain eye contact;
- be interested in everything that is going on around the reception area;
- give each client individual attention and respect.

ACTIVITY: RECEPTION
Pair with a colleague and share your experiences of a well managed and a badly managed reception.

ACTIVITY: NON-VERBAL COMMUNICATION
In conversation you give signals that tell others whether you are listening or not.

1 What do you think the following signals indicate?
   - A smile.
   - Head tilted, and resting on one hand.
   - Eyes looking around you.
   - Eyes semi-closed.
   - Head nodding.
   - Fidgeting.

2 Can you think of further body signals that indicate whether you are interested or not?

ACTIVITY: DO'S AND DON'TS
List *five* important do's and *five* don'ts for the receptionist.

Dealing with a dissatisfied client

The receptionist is usually the first contact with the client, and may have to deal with dissatisfied, angry or awkward customers. Considerable skill is needed if you are to deal constructively with a potentially damaging situation.

Never become angry or awkward yourself. Always remain courteous and diplomatic, and communicate confidently and politely.

1 Listen to the client as she describes her problems, without making judgement. Do not make excuses, for yourself or for colleagues.

2 Ask questions to check that you have the full background details.

3 If possible, agree on a course of action, offering a solution if you can. Check that the client has agreed to the proposed course of action. It may be necessary to consult the salon

manager before proposing a solution to the client: if you're not sure, always check first.

4. **Log** the complaint: the date; the time; the client's name; the nature of the complaint; and the course of action agreed.

ACTIVITY: RECEPTION ROLE-PLAY

With colleagues, act out the following situations, which may occur when working as a receptionist. You may wish to video the role-plays for review and discussion later.

1. A client arrives very late for an appointment but insists that she be treated.
2. A client questions the bill.
3. A client comes in to complain about a treatment given previously. (Choose a particular service.)

APPOINTMENTS

Making correct entries in the **appointment book** or salon **computer** is one of the most important duties of the receptionist. As receptionist you must familiarise yourself with the salon's appointment system, column headings, treatment times, and any abbreviations used.

Each therapist will have her name at the head of a column. Entries in columns must not be reallocated without the consent of the therapist, unless she is absent.

HEALTH & SAFETY: HEALTH AND SAFETY (DISPLAY SCREEN EQUIPMENT) REGULATIONS 1992

These regulations cover the use of visual display units and computer screens. They specify acceptable levels of radiation emissions from the screen, and identify correct posture, seating position, permitted working heights, and rest periods.

Bookings

When a client calls to make an appointment, her name and the treatment she wants should be recorded. Allow adequate time to carry out the required service (as indicated in the following chapters). Take the client's telephone number in case the therapist falls ill or is unable to keep the appointment for some other reason. If the client requests a particular therapist, be sure to enter the client's name in the correct column.

The hours of the day are recorded along the left-hand side of the appointment page, divided into fifteen-minute intervals. You must know how long each treatment takes so that you can allow sufficient time for the therapist to carry out the treatment in a safe, competent, professional manner. If you don't allow sufficient time, the therapist will run late, and this will affect all later appointments. On the other hand, if you allow too *much* time, the therapist's time will be wasted and the salon's earnings will be less than they could be. Suggested times to be allowed for each service are given in the treatment chapters.

Treatments are usually recorded in an abbreviated form. All those who use the appointment page must be familiar with these abbreviations.

Treatment	*Abbreviation*
Cleanse and make-up	C/M/up
Eyebrow shaping	E/B reshape or trim
Eyebrow tint	EBT
Eyelash tint	ELT
Eyelash perm	ELP
Manicure	Man
Pedicure	Ped
Waxing: half,	½ leg wax
three-quarter	¾ leg wax
full	Full leg wax
Bikini wax	B/wax
Underarm wax	U/arm wax
Forearm wax	F/arm wax
Eyebrow wax	E/B wax

If an appointment book is used, write each entry neatly and accurately. It is preferable to write in pencil: appointments can be amended by erasing and rewriting, keeping the book clean and clear.

Appointment cards may be offered to the client, to confirm the client's appointment. The card should record the treatment, the date, the day and the time. The therapist's name may also be recorded.

Appointments may be made up to six weeks in advance. Often clients will book their next appointment whilst still at the salon. How far ahead the receptionist is able to book appointments will vary from salon to salon.

When the client arrives for her treatment, draw a line through her name to indicate that she has arrived.

SKIN TESTS

Before clients receive certain treatments it may be necessary to carry out **skin tests**. The skin test is often carried out at reception, and the receptionist or assistant therapist will be able to perform the test once she has been trained. Every client should undergo a skin test before a permanent tinting treatment to the eyelashes or eyebrows. Further tests may be necessary, depending on the sensitivity of the client, before treatments such as artificial eyelash treatment, eyelash perming, bleaching or wax depilation. Refer to the relevant chapters to familiarise yourself with the test required.

An appointment page

DAY SATURDAY DATE 15th JANUARY

THERAPIST	JAYNE	SUE	LIZ
9.00	Mrs Young		
9.15	1/2 leg wax	Jenny Heron	
9.30	Carol Green	ELT EBT	
9.45	Full leg wax	EB trim	
10.00	B / wax		
10.15		Sandra Smith	Fiona Smith
10.30	Mrs Lord E/B wax	C / M / UP	C / M / UP
10.45			Strip lash
11.00		Mrs Jones	
11.15		U/arm wax	
11.30		F/arm wax	Carol Brown
11.45	////		ELP
12.00			\
12.15			
12.30	Nina Farrel	////	
12.45	Man.		
1.00	Ped.		
1.15		Sue Yip E/P	////
1.30	1/2 leg wax	T. Scott	
1.45	\	3/4 leg wax	
2.00	Karen Davies	U/arm wax	
2.15	facial		
2.30	\		
2.45			Pat King
3.00			C / M / UP
3.15	Anna Wood		Man
3.30	Man.		
3.45	E/B Reshape		
4.00			

TIP

Don't regard the phone ringing as an interruption. Always remember that clients matter – it is they who ensure the success of your business!

TIP

A client who has received a poor response to her telephone call may tell others about it.

TELEPHONE CALLS

In building new relationships, first impressions count. Good telephone technique can win clients; poor technique can lose them. Here are some guidelines for good technique:

- *Answer quickly* On average, a person may be willing to wait up to 9 seconds: try to respond to the call within 6 seconds.
- *Be prepared* Have information and writing materials ready to hand. It should not normally be necessary to leave the caller waiting while you find something.

- *Be welcoming and attentive* Speak clearly, without mumbling, at the right speed. Pronounce your words clearly, and vary your tone. Sound interested, and never abrupt.

Remember: the caller may be a new client ringing several salons, and her decision whether to visit *your* salon may depend on your attitude and the way you respond to her call.

Here are some more ideas about good telephone technique:

- Smile – this will help you put across a warm, friendly response to the caller.
- Alter the pitch of your voice as you speak, to create interest.
- As you answer, give the standard greeting for the salon – for example. 'Good morning, Visage Beauty Salon, Susan speaking. How may I help you?'
- Listen attentively to the caller's questions or requests. You will be speaking to a variety of clients: you must respond appropriately and helpfully to each.
- Evaluate the information given by the caller, and be sure to respond to what she has said or asked.
- Use the client's name, if you know it; this personalises the call.
- In your mind, summarise the main requests from the call. Ask for further information if you need it.
- At the end, repeat the main points of the conversation clearly to check that you and the client have understood each other.
- Close the call pleasantly – for example, 'Thank you for calling, Mrs Smith. Goodbye.'

If you receive a business call, or a call from a person seeking employment, always take the caller's name and telephone number. Your supervisor can then deal with the call as soon as she is free to do so.

TIP

If the salon uses a cordless phone or a mobile phone, keep it in a central location where it can easily be found when it rings. After each call, be sure to return the phone to its normal location.

ACTIVITY: COMMUNICATING WITH CLIENTS

Listen to experienced receptionists and notice how they communicate with clients

ACTIVITY: LISTENING TO YOURSELF

A pleasant speaking voice is an asset. Do you think you could improve your speech or manner?

Record your voice as you answer a telephone enquiry, then play it back. How did you sound? This is how others hear you!

ACTIVITY: FINDING THE RIGHT APPROACH

What telephone manner should you adopt when dealing with people who are:

- angry?
- talkative?
- nervous?

TIP

Check the salon's policy on personal calls. Usually they are permitted only in emergencies. This is so that staff are not distracted from clients, and to keep the telephone free for clients to make appointments.

Transferring calls

If you transfer a telephone call to another extension, explain to the caller what you are doing and thank her for waiting. If the

ACTIVITY: WHAT DO YOU NEED TO KNOW?

Think of different questions that you might be asked as a receptionist. Then ask an experienced receptionist what the most common requests are.

TIP

Telephone directories, codebooks and guides to charges provide a great deal of useful information. Read them carefully to make yourself familiar with the telephone services that are available.

extension to which you have transferred the call is not answered within 9 rings, explain to the caller that you will ask the person concerned to ring back as soon as possible. Take the caller's name and telephone number.

Taking messages

Messages should be recorded on a memorandum ('memo') pad. Each message should record:

- who the message is for:
- who the message was from;
- the date and the time the message was received;
- accurate details of the message;
- the telephone number or address of the caller;
- the signature of the person who took the message.

When taking a message, repeat the details you have recorded so that the caller can check that you've got it right. Pass the message to the correct person as soon as possible.

A memo

TELEPHONE MESSAGE RECEIVED

To	Angela	Date	10.6.99
From	Jenny Heron	Time	9:30 am
Number	273451	Taken by	Sandra

Please could you ring Jenny Heron regarding her appointment on Saturday.

THE PAYMENT POINT

Every beauty therapy or cosmetic business will have a policy for handling cash and for operating the payment point.

Kinds of payment point

Manual tills

With **manual tills** a lockable drawer or box is used to store cash: this may form part of the reception desk. Each transaction must be recorded by hand.

At the end of the working day, record the total cash register in a book, to ensure that accurate accounts are kept. Records of petty cash must also be kept so that the final totals will balance.

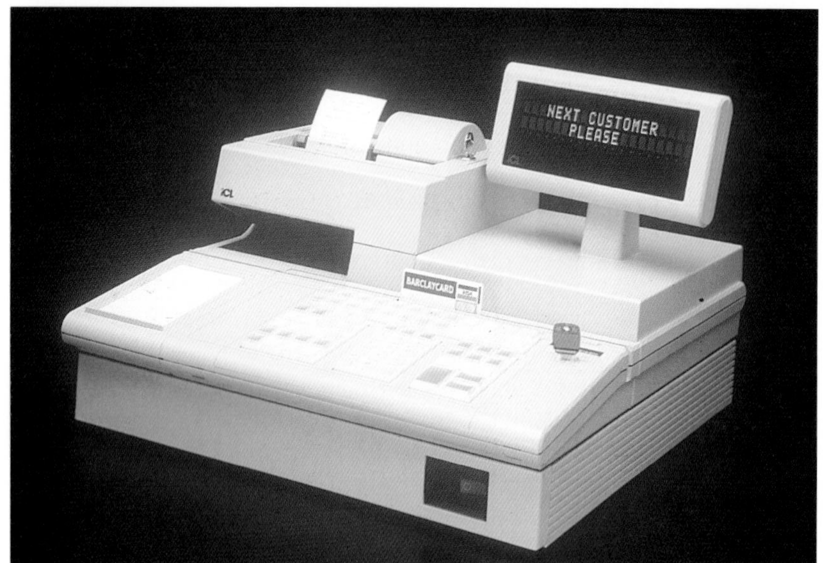

A computerised till

ICL

Automatic tills

Modern **automatic tills** use codes, one for each kind of treatment or retail sale. These are identified by keys on the till. Using these with each transaction makes it possible to analyse the salon's business each day or each week.

With each sale during the day a receipt is given to the client; the total is also recorded on the till's **audit roll**.

Automatic tills also provide **subtotals** of the amounts taken: these can be cross-checked against the amount in the till, to determine the daily **takings**:

- the **X reading** provides subtotals throughout the day, as required;
- the **Z reading** provides the overall figures at the end of the day.

Computerised cashdesks

Computerised cashdesks provide the same facilities as automatic tills, with additional features to help with the business's record-keeping, including client treatment cards and stock records.

Equipment and materials required

- *Calculator* This is useful in totalling large amounts of money or when using a manual till.
- *Credit-card equipment* If your business is authorised to accept credit cards you will need **vouchers** and a **transaction printer**.
- *A cash float* At the start of each day you need a small sum of money, comprising coins and perhaps a few notes, to provide change: this is called the **float**. (At the end of the day there will be money surplus to the float: if no mistakes have been made, this should match the takings.)

TIP

If there is too little change in the till, inform the relevant person. Running out of change would disrupt service and spoil the impression clients receive.

- *A till roll* This records the sales and provides a receipt. Keep a spare to hand. If you're using a manual cash drawer, you'll need a **receipt book**.
- *An audit roll* The retailer's copy of the till roll.
- *A cash book* This is a record of income and expenditure, for a manual till.
- *Other stationery* A date stamp and a salon name stamp (for cheques), pens, pencils and an eraser, and a container to hold these.

Security at the payment point

Having placed money in the cash drawer and collected change as required, always close the cash drawer firmly – never leave it open. Do not leave the key in the drawer, or lying about reception unattended.

Some members of staff will be appointed to **authorise** cheques and credit-card payments; one of these should initial each cheque.

Errors may occur when handling cheques or when operating an electronic or computerised payment point. Don't panic! If you can't correct the error yourself, seek assistance – but don't leave the cash drawer unattended and open.

ACTIVITY: FRAUD
Find out about your salon's policy in the case of fraudulent monetary transactions, using either cash or cards. What actions should you take?

Methods of payment

Cash

When receiving payment by **cash**, follow this sequence:

1. Check that the money offered is **legal tender** – that is, money you will be able to pay into your bank. (Your salon will probably not accept foreign currency, for example.)
2. Tell the client the amount to be paid.
3. Place the customer's money on the till ledge until you have given change, or at least state to the customer verbally the sum of money that she has given you.
4. Aloud, count the change as you give it to the client.
5. Thank the client, and give her a receipt.

TIP
Hold notes to the light to check for forgeries. You should be able to see the watermark and the metal strip that runs through the note.

Cheques

Cheques are an alternative form of payment, and must be accompanied by a **cheque-guarantee card**. This has a spending limit, usually £50 or £100. Your salon may be willing to accept cheques for larger amounts if the client can show some other identification, such as a driving licence, but you must always check first with your supervisor.

When receiving payment by cheque, follow this sequence of checks:

TIP
A cheque is only valid for 6 months from the date on the cheque.

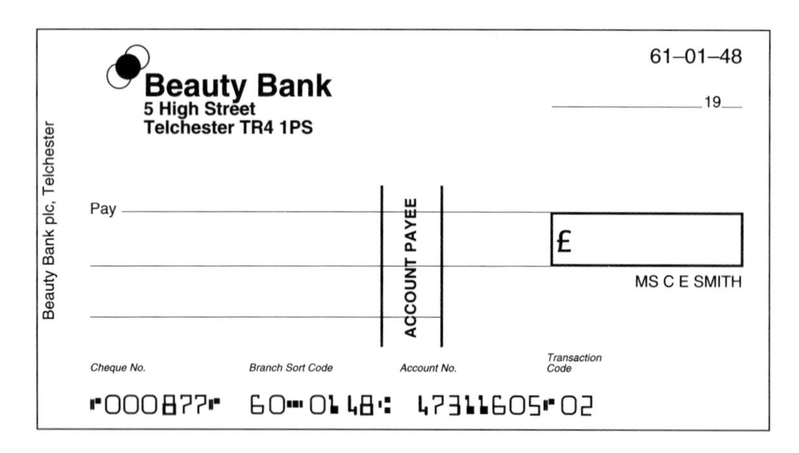

National Westminster Bank

A cheque and a cheque-guarantee card

1. the cheque must be correctly dated;
2. the cheque must be made payable to the salon (you may have a salon stamp for this);
3. the words and figures written on the cheque must match those in the box;
4. any errors or alterations must have been initialled by the client;
5. the signature on the back of the cheque card must match that on the cheque – compare these as the customer writes her signature on the cheque;
6. the bank's 'sort code' numbers on the cheque must match those on the card;
7. the date on the cheque card must be valid;
8. the value on the cheque must not exceed the cheque-card limit;
9. the cheque card number must be recorded on the back of the cheque;
10. the cheque must be signed by the client.

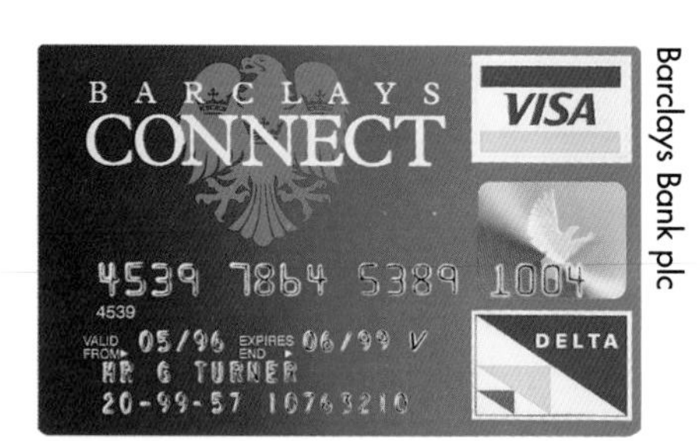

Barclays Bank plc

A debit card

Debit cards

Some businesses accept **debit cards** such as Delta, Switch or Connect. The card authorises immediate debit of the cash amount from the client's account. (This card may also be a cheque-

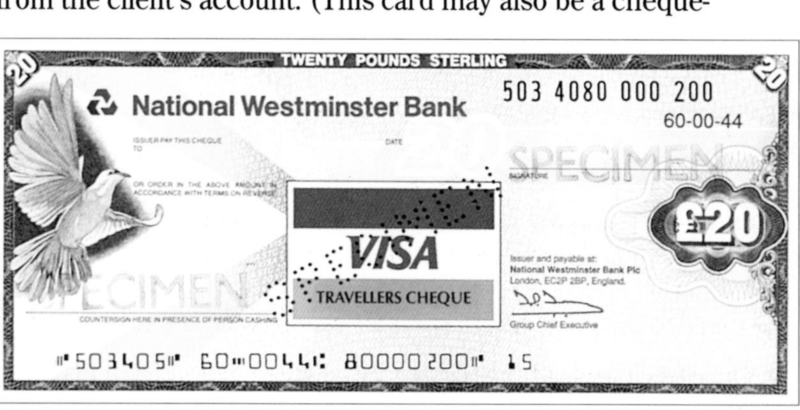

National Westminster Bank

A traveller's cheque, in sterling

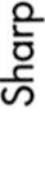

'Swiping' a card through an electronic terminal

guarantee card.) You cannot perform this kind of transaction unless your salon has an electronic terminal.

Electronic payment systems

1. Check that the terminal display is in 'SALE' mode.
2. Confirm that a sale is to be made by pressing the YES button.
3. The terminal will request that you **swipe** the card. Do this, ensuring that the magnetic strip passes over the reader head and that you retain the card in your hand. In some cases the magnetic strip cannot be read by the swipe card reader: in this situation you will have to key the complete card number into the terminal manually. This does not necessarily mean that there is any reason for suspicion, but do look carefully at it for any signs that the card has been tampered with.
4. When prompted, enter the 'AMOUNT' using the keypad to input the purchases and total them. (If you make a mistake, you can clear the figures using the CLEAR button.)
5. Press ENTER, which will automatically connect the terminal to the credit card company. A message will indicate first 'DIALLING', and then 'CONNECTION MADE'.
6. Customer details are accessed automatically. After a few moments you should receive one of two messages. If the payment is authorised you will see 'AUTH CODE', and a code number will be printed on the receipt with the other purchase details. If the transaction is declined, you will see 'CARD NOT ACCEPTED'.
7. While holding the card, tear off the two-part receipt and ask the cardholder to sign in ballpoint pen, in the space provided.
8. Check the signature matches the signature on the card, and give the customer the top, signed copy, with the card.

Travellers' cheques

Travellers' cheques also may be acceptable, provided they are in particular currencies (usually sterling). Such cheques must be compared with the client's **passport**.

Credit cards

Credit cards can be used only if your business has an arrangement with the relevant credit-card company. In this case the company will give the salon a credit limit (a **ceiling**), the maximum amount that may be accepted with the card. Any amount greater than this must be individually authorised by the credit-card company. (This is done by telephone at the time of transaction.)

When receiving payment by credit card, check these points:

1. the card logo is at the upper right corner on the front of the card;

Barclays Bank plc

A credit card

2 the hologram should have a clear, sharp image and be in the centre right of the card;

3 the date on the credit card must be valid;

4 the sex (Ms, Miss, Mrs, Mr, etc.) and the name of the customer must fit your client;

5 the cardholder's signature on the card must match the name on the front of the card;

6 the cardholder's account number should be embossed and across the width of the card;

7 the cardholder's account number must not be one of those on the credit-card company's warning list.

If you receive a card that *is* on a credit-card warning list, politely detain the customer, hold onto the card, and contact your supervisor, who will implement the salon's procedure. If the *card* is unsigned, do not allow the cardholder to sign the card unless you first get authorisation from the credit-card company's service provider. (The service provider's telephone number should be kept near the telephone.)

Having checked the details it is necessary to prepare the **sales voucher**, which comprises three copies.

1 Complete the sales details on the appropriate credit-card voucher, using ballpoint pen so that the bottom copy is legible.

2 Place the card in the transaction printer, with the front (embossed side) facing upwards.

3 Place the voucher over the card, and under the voucher guide.

4 Slide the handle from left to right, and back to its original position.

5 Remove the voucher and check that the recorded details are clear and on all copies of the voucher.

6 Ask the customer to sign the voucher. Check that the signature on the voucher matches that on the credit card.

7 Give the customer the top copy as a receipt, with her credit card.

A copy of the sales voucher is sent to the credit-card company; the third copy is kept by the salon.

Charge cards

Some businesses accept **charge cards** such as American Express. These differ from credit cards in that the account holder must repay to the card company the complete amount spent each month.

American Express

A charge card

A gift voucher

⌘ ⌘

BEAUTY STUDIO

3 Market Street — Tel: 39473
Whitely

This voucher entitles

....

to the value of £..............

Signed:

Date:

Valid for six months from date of purchase

⌘ ⌘

TIP

Sometimes the salon may publish other offers, such as a discount on producing a newspaper advertisement for the salon. The advertisement voucher is a form of payment, and must be collected.

Gift vouchers

Gift vouchers are purchased from the salon as pre-payments for beauty therapy services or retail sales.

THE IMPORTANCE OF RETAIL SALES

Retail sales are of considerable importance to the beauty salon: they are a simple way of greatly increasing the income without too much extra time and effort. Beauty therapy treatments are time-consuming and labour-intensive; selling a product in addition to providing the treatment will greatly increase the profitability.

For example: a facial treatment might take 1 hour and cost the client £15. You might then sell the client moisturiser costing £10, of which £5 might be clear profit. Supposing that you sold eight products each day, each yielding £5 profit, that would be a profit of £40 per day, or £200 per week, or £800 each month – and thus, £9600 profit over the whole year: a significant sum.

SELLING COSMETICS

When selling a client cosmetics, first find out about her needs. Consider the following:

- Is she allergic to any particular substance, contact with which should be avoided?

HEALTH & SAFETY: CONTRA-INDICATIONS

If the client has any contra-indication to the product recommend she seeks her doctor's approval before using it.

Ellisons

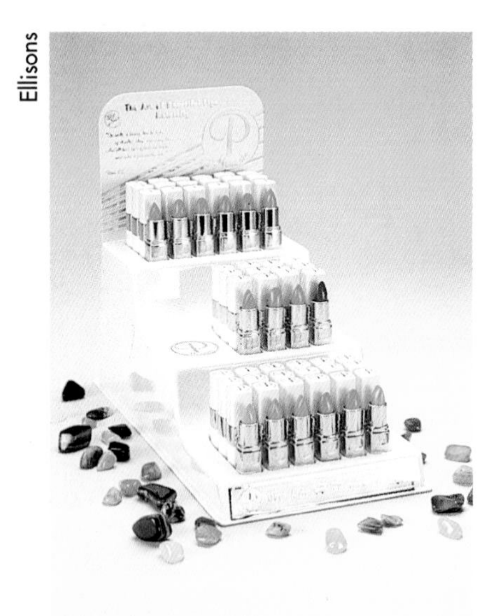

A lipstick display

Ellisons

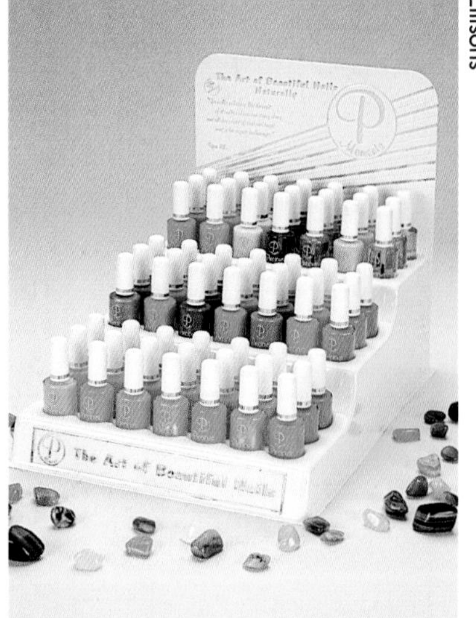

A nail varnish display

- Has she a skin disorder or nail disease which might contra-indicate use of a particular product? Contra-indication to products must always be noted and explained to the client.
- Is she planning to use the product over cuts and abrasions, or over areas with warts and moles? If so, is this safe?
- How much is she used to spending on products? Ask about what she is presently using: this will give you an idea of the types of product she has experience of using, and the sort of prices she is used to paying.

> **TIP**
>
> Encourage your client to try the testers out. Make-up can look very different on the face compared with its appearance on the palette.

Bearing in mind her needs, you can now guide her to the most suitable product. This is where your expertise and your product knowledge are so important: you can describe fully and accurately the features, functions and benefits of the products you stock.

The cosmetics themselves must be presented to the client in such a way that they seem both attractive and desirable: the presentation should encourage her to purchase them. The packaging and the product should be clean and in good condition; and **testers** should be available wherever possible so that the client can try the product on herself before purchasing it.

The final choice of product is with the client, of course, but often she will ask for a recommendation, as for example if she cannot decide between two possibilities. It is in these circumstances that your ability to answer technical questions fully, from a complete knowledge of the product, will help in closing the sale. Speaking with confidence and authority on the one product that will particularly suit her requirements may well persuade her to buy it.

If the client does not know what an **allergic reaction** is – or how to recognise it – it is important that you describe it to her (red, itchy, flaking and even swollen skin). If she experiences this sort of contra-action she should immediately stop applying the product she suspects is producing it.

Ellisons

A salon floor display of products

Ellisons

Not tested on animals

> **HEALTH & SAFETY: PATCH TESTS**
>
> If the client has not tried a product before, or if there is doubt as to how her skin will react, a skin patch test must be carried out.
>
> 1. Select either the inner elbow or the area behind the ear.
> 2. Make sure the skin is clean.
> 3. Apply a little of the product, using a spatula.
> 4. Leave the area alone for 24 hours.
> 5. If there is no reaction after 24 hours, the client is not allergic to the product: she can go ahead and use it.
>
>   If there has been any itching, soreness, erythema, or swelling in the area where the product has been applied, the client is allergic to it and should *not* use it.

TECHNIQUES IN SELLING

Preparation

The first rule of selling is: *know your products*. This applies to all retail products and to all salon services. **Information** must be supplied for clients to read, and **displays** must be set up. Be aware of your **competitors** and their current advertising displays and campaigns.

Know your products

> **TIP**
>
> Use the products on yourself. It is always good to be able to speak from experience.

Product usage must be discussed with clients, as necessary, and advice given on which product will best suit each of them. The only way to be able to do this is to memorise the complete range: all your products, including which skin types or treatment conditions each is for, what the active ingredients are, when and how each should be used, and its cost. Any questions asked must be answered with authority and confidence. Clients expect the staff in the beauty salon to be professionals, able to provide expert advice.

> **ACTIVITY: INCREASING PRODUCT KNOWLEDGE**
>
> With colleagues, discuss and note down the features, functions and benefits of a range of cosmetic products sold in your training establishment.

> **ACTIVITY: LEARNING THE PRODUCT RANGE**
>
> Learn about and memorise the product range sold in the training establishment you attend.

Ellisons

Salon reception

Information to read

The **information** available to clients can start from the window display. Use the **window stickers** if supplied with product ranges, and include stickers that advertise the salon's treatments.

Few salons use windows for product displays, but you could consider doing so.

Posters are supplied with good-quality product ranges, and most suppliers provide **information leaflets** for clients. Use the posters and **display cards** in the reception area; clients can then help themselves, and read about the products. This will generate questions – and sales.

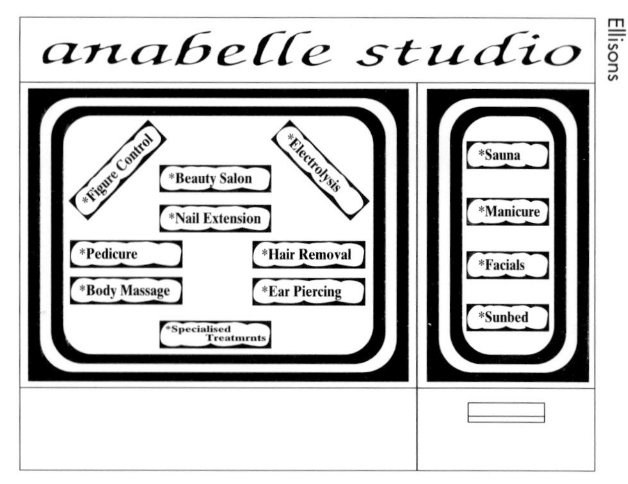

Ellisons

Window stickers

Product and retail displays

Two types of display can be used in the beauty salon. In the first, the display is there simply to be looked at, and seen as part of the decor. It should be attractive and artistically arranged, and can use dummy containers. It is not meant to be touched or sold from, so it can be behind glass or in a window display.

In the second, on the other hand, products are there to be sold. In this case products must be attractive but also accessible. The display should include testers so that clients can freely smell and touch. Each product must be clearly priced, and small signs placed beside the products or on the edge of the shelves to describe the selling points of each product.

Ellisons

'Please take one' gift token and display

HEALTH & SAFETY: MAINTAINING HYGIENE

Spatulas must be used so customers do not put their fingers into the pots either when testing or during home use. (You may also like to sell spatulas for clients to use at home.)

ACTIVITY: COLLECTING INFORMATION

Collect information leaflets from local salons and beauty product or perfume counters in department stores. Is this literature attractive? Will the presentation encourage sales?

Write to wholesalers and product companies for information about the display packs they supply with their products.

This sort of active display must always be in the part of the salon where most people will see and walk past it – the area of 'highest traffic'. A large proportion of cosmetic and perfume sales are **impulse buys**. It is no accident that perfumery departments are beside the main entrances to department stores, or right beside access points such as escalators.

Product displays must always feature in the beauty treatment area. As the beauty therapist uses the products, she can discuss and recommend them for the client. If displays are there to see and to take from, the sale can be closed even before the client returns to reception. Although in theory clients can of course change their minds between the treatment area and actually paying, in practice once they have the product in their hands they will go on to buy it.

Ellisons

Nail products

A cosmetic range

Ellisons

Ellisons

Retail packs of nail varnish

Ellisons

Showcards and client leaflets

Displays

Most small salons will design and create their own displays using the counter **display packs** provided by the product companies. Some will have a professional **window dresser** regularly to change the window displays for the best effect.

Displays must be checked and cleaned regularly – in busy salons this will usually mean daily. A window display will need to be dusted, straightened, and looked at from outside to make sure that it looks its best. The display from which products are being sold will also need to be dusted, perhaps wiped over (if testers have dripped), and straightened up. Testers need to be checked to make sure they are not sticky and spilt, and that no one has left dirty fingerprints on them.

ACTIVITY: SITING OF DISPLAYS

In your nearest large town, go into the big department stores and note where the cosmetic and perfumery departments are situated.

ACTIVITY: EVALUATING DISPLAYS

Whenever you can, look at displays and make a note of neatness, cleanliness, availability, pricing and information. Compare the best with the worst.

Ellisons

A retail make-up brush display

The range of products

It is not enough to stock just a few items and expect clients to fit in with the range you carry: different ranges must be available for each skin type, and a number of products – such as eye gel or throat cream – that will suit all skin types. Make-up and nail enamels should be attractive to all ages and types of customer. Sales must not be lost because of a lack of product range.

CUSTOMERS' RIGHTS

When selling products, you need to be aware of the following Acts, which affect the customer's legal rights.

Ellisons

A display of men's products

Trades Description Acts 1968 and 1972

The retailer must not:

- supply misleading information about products and services;
- describe products falsely;
- make false statements.

In addition, the retailer must not:

- make false comparisons between past and present prices;
- offer products at what is said to be a 'reduced' price, unless they have previously been on sale at the full price quoted for 28 days minimum;
- make misleading price comparisons.

When selling products the information supplied, both in written and verbal form, must always be accurate.

Sale of Goods Act 1979

The goods supplied – whether as part of a service, on hire, or in part-exchange – must be:

- of merchantable quality;
- fit for the purpose;
- as described.

The products sold must be of good quality and do what is claimed for them. They must also fit their description.

DEMONSTRATIONS

Planning the demonstration

Demonstrating to an audience needs particularly careful planning if the demonstration is to achieve the maximum benefit. Everything required must be in place: the products, the means to apply them, and all the relevant literature to be given out to the audience or clients.

What type of demonstration is required? Will you be working on one client, to demonstrate and sell a product, or demonstrating to a group? Is a range of products to be demonstrated, or just one item?

Single client

When demonstrating on a client have a mirror in front of her so that you can explain as you go along and the client can watch. She can then see the benefit of the product and learn how to use it at the same time. This is a simple but effective way to sell products.

TIP

An effective demonstration will always create sales. Have the product ready to sell, or give out vouchers to encourage clients to come to the counter.

A group

The presentation should include an introduction to the demonstrator and the product, the demonstration itself, and a conclusion with thanks to the audience and model. Written **promotional material** can be placed on the seats before the audience arrives, or handed out at an appropriate point during the demonstration; **samples** can be handed around the audience for them to try.

The demonstration itself must be clear, simple and not too long. The audience *must* be able to see what is being done and hear the commentary. Maximise the impact of the demonstration by giving the audience the opportunity to buy the product immediately.

If this is not possible – because the demonstration is in another room, away from the products, or at another venue, such as at a woman's club meeting – ensure that members of the audience leave with a **voucher** to exchange for the product. This should offer some incentive, such as a **discount**, to encourage potential buyers to make the effort to come to the salon and buy. Never sell the features and benefits to potential customers, creating the desire for the product, without also giving them the chance to buy it.

TIP

Always pay careful attention to your client and her responses. Remember: *question, listen, answer:*

- *Question* your client as to her needs.
- *Listen* to the answer.
- *Answer* with the relevant information.

Questions must be accurate and detailed. You are the expert: show your knowledge. Do not ask the client what sort of skin she has – she is not the expert, and will probably give the wrong answer. Instead, ask more detailed questions, such as 'Does your skin feel tight?' (which may indicate dryness), or 'Do you have spots in a particular area?' (which may indicate an oily patch).

Listening is a skill. Always listen carefully to the answers your customer gives: do not talk over her answer or interrupt. Only when she has finished should you give a considered, informed reply. You may need to ask another question, or you may be able straightaway to direct her to the best product for her needs.

KNOWLEDGE REVIEW

Reception

1. How should the client be greeted on arrival at the salon?
2. What information should be sought by the receptionist from a visitor to the reception desk?
3. What systems used in the salon record information about the client?
4. What are the main duties of the salon's receptionist?
5. What are the important details to record when taking a message?

6 If as receptionist you were unable to give appropriate information to a client, what action would you take?
7 What are the names of your immediate colleagues in the workplace? What are their responsibilities within the company structure?
8 A client wishes to make an appointment for an eyelash tint, as recommended by the beauty therapist. What information is required from the client before making the appointment?
9 How long should you allow when making an appointment for the following treatments?
   (a) Full leg wax.
   (b) Eyelash tint and eyebrow shaping.
   (c) Ear piercing.
   (d) Manicure.
10 When a client telephones the salon, her immediate assessment of the salon will be based on the receptionist's communication skills. How should you:
   (a) answer the telephone and introduce yourself?
   (b) speak on the telephone?
   (c) seek information from the caller?
   (d) finish the telephone conversation?
11 An efficient receptionist must be able to deal competently with telephone enquiries. For what different reasons may people telephone the salon?
12 A client rings up to check the time of an appointment. What information do you need from her?
13 Why is it necessary to confirm an appointment with a client verbally before she leaves the salon?
14 A client complains at reception of a skin-care product she has purchased which irritated her skin. What questions should you ask? What action should you take?
15 What is a float? When opening the payment point at the start of the day, why must the float be checked?
16 What equipment and materials do you need at the reception desk?
17 If you make an error when operating the till, why must you report this?
18 When taking cash from a client, what is the correct procedure to follow?
19 If you made a mistake in giving the client her change, how would you deal with this?
20 What should you check when receiving payment by cheque?
21 What is the procedure for receiving payment by credit card?
22 If it were very busy at reception and you realised a client had left the salon without paying, what would you do?
23 If a client presented an invalid cheque card, what should you say to her?

24 If your salon accepts vouchers in exchange for salon services, how are these handled?

Selling cosmetics

1 Why must the client's needs be ascertained before selling her cosmetics?

2 How would you describe an allergic reaction to a client if she did not know how to recognise one?

3 How should you prepare your working area:
(a) for a demonstration to one client;
(b) for a demonstration to a large audience?

4 How should cosmetic and perfume displays be cared for?

5 Why is it important to stock a range of products?

6 Why are retail sales important to the beauty salon?

7 Explain the Trades Description Acts 1968 and 1972.

8 Explain the Sale of Goods Act 1979.

9 Why is listening so important?

10 Why do you need to know if the client is allergic to a given substance?

11 Why should a patch test be carried out if the client has not used a product before, or if there is any doubt as to how her skin will react?

12 What factors should be considered when selling each of the following?
(a) Nail enamels.
(b) Facial cleansers.
(c) Foundations.

CHAPTER 3

Working relationships

WORKING RELATIONSHIPS

For any beauty therapy business to be a success requires the commitment of each employed individual to ensure quality at all levels and in all services.

For this to occur it is important both that you are effective in your job role and that you have positive **working relationships** with your colleagues and with clients.

LIAISING WITH COLLEAGUES

Good working relationships between colleagues in the workplace are essential. Each staff member, whatever their role, is valuable as part of the team in ensuring the success of the business. All staff at **induction** should be told of the function of the business, and of their role in this. They should also be told about relevant **codes of conduct**.

Job roles and responsibilities

Each team member should have a **job description**. This details:

- the job's title;
- the specific job role;
- the duties and responsibilities;
- the work location;
- any extra special circumstances affecting duties, such as attending salon promotional events.

The job description enables each member of staff to know what is expected of them and to whom they are responsible.

For reasons of safety and effectiveness it is important that you know what jobs and roles you are qualified to undertake.

ACTIVITY: ROLES AND RESPONSIBILITIES
List the different roles and responsibilities of personnel in your work place. Clients and business callers may require this information.

Effective communication between staff

Personnel problems may occur if there are ineffective communication systems. Time should be made to hold regular staff meetings where any concerns can be shared. A staff meeting can also be an exciting opportunity to share ideas to improve the efficiency of the job role and to look together at ways of improving the business.

Job description – Beauty Therapist

Location:	Based at salon as advised
Main purpose of job:	To ensure customer care is provided at all times To maintain a good standard of technical and client care, ensuring that up-to-date methods and techniques are used following the salon training practices and procedures
Responsible to:	Salon manager
Requirements:	To maintain the company's standards in respect of hairdressing/ beauty services To ensure that all clients receive service of the best possible quality To advise clients on services and treatments To advise clients on products and after-care To achieve designated performance targets To participate in self-development or to assist with the development of others To maintain company policy in respect of: • personal standards of health/hygiene • personal standards of appearance/conduct • operating safety whilst at work • public promotion • corporate image as laid out in employee handbook To carry out client consultation in accordance with company policy To maintain company security practices and procedures To assist your manager in the provision of salon resources To undertake additional tasks and duties required by your manager from time to time.

Positive relationships

There can be no place in a customer care industry for poor working relationships. A great portion of your time is spent in the workplace alongside your colleagues, and if the environment becomes stressful this will affect your effectiveness. It will also be apparent to clients, and relationships between staff members should not trouble them. Disputes must be resolved immediately.

It is important that you understand the salon's **staffing structure**. You need to know who is responsible for what, and who you should approach in various circumstances, for example if you felt that you were being treated unfairly. Any grievances should be reported to a supervisor, and you should familiarise yourself with the **grievance and appeals procedure**.

General codes of conduct

Be polite and courteous with colleagues at all times.

- Never talk down to colleagues.
- Never lose your temper with a colleague in front of a client.
- Never ridicule a colleague in front of a client or another colleague.
- If there are personality differences, do not show these in front of the client. Settle grievances as soon as possible, or your work will be affected.

ACTIVITY

What personal characteristics do you need to work effectively in a team? From your own experience, discuss times when you have worked in a team. What was your role?

Ask a colleague: are you seen by others as a team player? If not, how can you develop the necessary skills?

LIAISING WITH CLIENTS

Codes of conduct

Many organisations have a **customer care statement** which outlines the standards of service customers can expect.

Clients want to enjoy their visits to the beauty salon and they are paying for a service. It is important that during each visit they are made to feel relaxed and comfortable.

Client care

Remember that each client has a different personality and different treatment needs, requiring an individual treatment approach.

A client can be made to feel intimidated, uncomfortable or ignored – and this can happen without your saying anything! Even without speaking you communicate with your eyes, your face and our body, transmitting some of your feelings. This is called **non-verbal communication**. How you look and how you behave in front of your clients is important.

ACTIVITY: CLIENT CARE

Find out whether your organisation has a customer care statement. If so, how well do you do in providing that level of customer care? Monitor yourself against the statement, and ask colleagues for feedback.

ACTIVITY: TELEPHONE CALLS

A telephone call is often the first contact the client has with the salon and is an important method of communication.

- How should the phone be answered?
- What should you confirm if the client requires an eyelash tint?
- What action would you take if there was not sufficient time for the appointment at the time requested?
- How would you handle a complaint about a service?

Effective communication

On meeting a client, always smile, make eye contact, and greet her cheerfully – however bad your own day is! As you communicate you can:

- promote yourself, and gain the client's confidence in your professionalism and technical expertise;
- develop a professional relationship with the client;
- establish the client's needs;
- promote services and treatments.

Verbal communication occurs when you talk directly to another person, either face to face or over the telephone. Always speak clearly and precisely, and avoid slang. It is important to be a good listener: this will help you identify the client's treatment requirements and understand her personality. You can then guide the conversation appropriately.

Body language

Interpreting **body language** is an important skill: learn to notice how the client is behaving, including her voice, her eyes, her body, and her arm and hand movements. An instinctive 'feel' for customers' behaviour can be developed with experience.

Selling

When approaching potential customers in a situation such as a department store, be aware that conflicting signals may be given. For example, a person may smile and nod as if interested but, in fact, walk straight past. On the other hand, if the customer makes the first approach then she obviously has an active interest already.

Initially the customer may be formal and may even have a stiff body posture and a reserved manner. As she becomes more interested, however, her posture will relax: she may begin to lean forward. It will become obvious at this stage that she is interested, and she then will go on to nod and agree, and to listen actively.

You must use your own body language to good effect. You must be relaxed but attentive, and listen actively – nodding and shaking your head, and smiling in agreement. Use relaxed, gentle hand movements: do not twitch or turn away from the customer.

If the customer is not agreeing with you, or is not interested, she may look bored, tap her fingers, fiddle with her shopping, look away, or even look at her watch. If these signs are evident, go back to the beginning and try to find out why she is not interested. This is important when selling. Perhaps she does not want the product you have recommended? Or perhaps it is too expensive? Suggest alternatives and see whether you can get her interest again.

When the customer has decided to buy, smile – help her to feel that she has made an excellent decision. She should leave feeling proud to have purchased the product.

Conversing with a client

Having developed a professional relationship with your client, centre the conversation on her, so that she feels special. Avoid interrupting the client whilst she is speaking – be patient. A nervous client may need to be reassured. Gain her confidence by being pleasant and cheerful without chattering constantly. When asking questions, don't interrogate your client. Never talk down to her, and avoid technical jargon – instead, use commonly understood words.

Avoid all controversial topics, such as sex, religion and politics! When a relationship has been established, value it but be discreet, – clients will often share confidences with you. Never pass judgement, and ensure that you deserve clients' trust by maintaining confidentiality.

TIP

Make notes on the client's record card of topics that interest her. You can introduce these topics in conversation next time the client receives a treatment, and she will be pleased that you have taken the trouble to remember.

Working under pressure

Sometimes you will be extremely busy and you may be feeling tired and weary. This is not the client's problem, however! Remain cheerful, courteous and helpful. You should also use your initiative in helping others, for example by preparing a colleague's work area when you are free and she is busy.

You must be able to cope with the unexpected:

- clients arriving late for appointments;
- clients' treatments overrunning the allocated treatment times;
- double bookings, with two clients requiring treatment at the same time;
- the arrival of unscheduled clients;
- changes to the bookings.

With effective teamwork such situations can usually be overcome.

ACTIVITY: HOW GOOD ARE WE?

A questionnaire may be useful in monitoring client satisfaction. Questionnaires can be anonymous, and collected at a central point. Ask questions that are important to the team and the business. Collate the findings and use them to evaluate effectiveness and identify areas requiring development. Simple changes can make all the difference!

ACTIVITY: DEALING WITH THE UNEXPECTED

How would you deal with the unexpected situations listed above? With colleagues, discuss your experiences and record your ideas.

Avoiding client dissatisfaction

Some dissatisfied clients will voice their dissatisfaction; others will remain silent and simply not return to the salon. This situation can often be prevented through good customer care and effective communication.

ACTIVITY: HANDLING CUSTOMER COMPLAINTS

List five complaints that might be made by a client. How would you handle each situation to ensure a positive outcome?

- Always ensure that the client has a thorough consultation before any new service. This should be carried out by a colleague with the appropriate technical expertise.
- Regularly check the client's satisfaction. If there is any concern, make the supervisor aware of this immediately.
- Inform the client of any disruption to service – do not leave her wondering what the problem may be. Politely inform her of the situation, for example 'I'm sorry but we are running 10 minutes late – are you able to wait?' If your salon has the facilities, you may offer her a drink.
- Inconvenience caused by disruption to service can usually be compensated in some way. It is important to resolve problems and keep clients satisfied.

Customer care is vital: clients provide the salon's income and your wages. The success of the business depends upon satisfied clients.

Complaints procedure

Unfortunately problems do sometimes arise in which the client cannot be appeased. A **complaints procedure** is a formal, standardised approach adopted by the organisation to handle any complaints.

IMPROVING PERSONAL EFFECTIVENESS

In order to develop personally and to improve your skills professionally, it is important to set yourself targets against which you can measure your achievement.

To an employer it is important that you are *consistent*. You must always perform your skills to the highest standard, and present and promote a positive image of the industry and the organisation in which you are employed and which you represent.

Appraisal

Appraisal is a process whereby a supervisor identifies and discusses with individuals their strengths and weaknesses, and areas within their professional role that require further training and development. The job description is used as a key document during appraisal.

This may seem daunting, but it is an important and useful process. You can also use it to your advantage:

- identify opportunities for further or specialist training;
- identify obstructions which are affecting progression;
- identify and amend any changes to your role;
- identify and focus on your achievements to date against targets set;
- make an action plan which will help you achieve your targets.

At the next appraisal the agreed objectives and targets set for the previous period are reviewed.

Fulfilling your job role

Image

Your image matters: it creates an expectation of the quality and standard that can be expected. The image must be consistent for all employees of the business.

A professional image creates confidence in clients, who learn to trust that they will be treated in a certain way – professionally, and with respect!

ACTIVITY
What is your organisation's dress code? Do you consistently achieve the required standard? If not, why is this?

Standards

Never use equipment or provide treatments for which you do not have the professional expertise. As well as the obvious potential health and safety hazards, you would not have the expertise to adapt the treatment to suit each client's treatment needs and you would not be able to provide the relevant treatment advice in order to obtain the optimum treatment results.

Targets

Targets to be achieved may be set by the employer either for individuals or for the team as a whole. These may review quality, efficiency and results. Computer systems are often able to provide relevant data about the salon.

You should set your *own* targets, however, to monitor your effectiveness and performance. In training situations trainees undergo a programme which states:

- what training activities will take place;
- what tasks need to be performed;
- what standards are expected to be reached;
- when assessment should be expected;
- when a review of progress towards the agreed targets is to take place.

In the same way, you can set targets for yourself.

Punctuality

It is important to the efficient working of the salon that you are punctual for work. This ensures that you are composed, that clients are not kept waiting, and that you are able to support other members of the team. You thereby minimise stress for yourself and for your colleagues.

Absence from work

If a member of staff is absent from work, other staff will need to review the daily work schedule to minimise disruption. In the case of holiday cover this can be planned for, but if a staff member is absent unexpectedly, teamwork will be needed to get the work done.

Performance Appraisal	
Name:	Jeanette Manners
Job Title:	Trainee beauty therapist
Date of Appraisal:	30 October 1999
Objectives:	To obtain competence within:- Improving facial skin condition across the range.
Notes on Achievement:	Competence has been achieved for most facial skin condition range requirements.
Training Requirements:	Further training and practice is needed within the area of facial massage.
Any Other Comments on Performance by Appraiser:	Jeanette has achieved most of the objectives set out during the last appraisal.
Any Comments on the Appraisal by the Staff Member Appraised:	I feel that this has been a fair appraisal of my progress although I did not achieve all of my performance targets. J Manners
Action Plan:	- To achieve occupational competence across the range for facial skin condition (i.e. facial massage). - To undergo training and practice in hair removal techniques. - To take assessment for hair removal.
Date of Next Appraisal:	28 April 1999

An appraisal form

When you learn that someone is to be absent, find out, if you can, for approximately how long. Then:

- Check the work schedule of the person who is absent.
- Reschedule clients, but without affecting the quality of the salon's service.
- Determine whether any clients' appointments can or must be cancelled, especially if there are double bookings or if the client must be treated specifically by the person who is away. Contact clients as soon as possible, so that they can reschedule their own time.

If you yourself are ill, to minimise disruption you must report your sickness as early as possible to the relevant person. You may be required to complete sickness forms.

Developing within the job role

There will be many opportunities to develop your skills and experience, and your understanding of your work:

- by attending trade seminars;
- by subscribing to professional trade magazines;
- by watching colleagues who have more advanced qualifications or experience;
- by developing your portfolio to include evidence and examples of experience gained;
- by using time effectively, and by practising – all tasks take time to master: the more you practise, the more skilled and efficient you will become.

ACTIVITY: EVALUATING CUSTOMER CARE

Visit a local salon. Beforehand, think of questions you would like to ask in relation to treatments. Then evaluate the customer care and services you received. Were the staff:

- friendly?
- dressed smartly?
- knowledgeable?
- efficient and eager to assist you?
- helpful?

If you answered 'no' to any of these questions, discuss your reasons with your colleagues. What have you learnt from this experience?

KNOWLEDGE REVIEW

Working relationshps

1. Why are positive working relationships with
   (a) your clients
   (b) colleagues
   important?
2. What is the function and what are the benefits of an induction process?
3. What should be included in a job description?
4. Why is it necessary to have a job description?
5. List examples of effective teamwork.
6. How should you speak to your colleagues when a client is present?
7. If a beauty therapist had an unprofessional attitude, how would this affect the reputation of the salon?
8. What might be included in a general code of conduct for staff?
9. If you had a grievance, how and to whom would you report this?
10. What are the main methods of communication? Give examples of both positive and negative approaches.
11. How can you make a client feel welcome?
12. List the main causes of client dissatisfaction and how these may be avoided.
13. What are the key objectives of appraisal?
14. What steps need to be taken if a customer makes a complaint?
15. What steps need to be taken when you are notified of a staff member's absence?

CHAPTER 4

Anatomy and physiology

ANATOMY AND PHYSIOLOGY OF THE SKIN

CARING FOR THE SKIN

The skin varies in appearance, according to our race, sex and age. It also alters from season to season and from year to year, and reflects our general health, lifestyle and diet.

At puberty the chemical substances (**hormones**) that control many of our bodies' activities become very active. Amongst other effects, this activity causes the skin to become more oily, and often blemishes appear on the skin's surface. Seven out of ten teenagers find that their skin becomes blemished with blackheads, inflamed angry spots and even scars at this time: a skin disorder called **acne vulgaris**.

During the twenties the skin should look its best; any hormonal imbalance that occurred at puberty should by now have stabilised. As we grow older, the skin ages too. In our late twenties and early thirties we will see fine lines appearing on the skin's surface, especially around the eyes where the skin is thinner and the skin gradually becomes drier.

At around the age of forty, hormone activity in the body becomes slower and the skin begins to lose its strength and elasticity. The skin becomes increasingly drier, and lines and wrinkles appear on the surface. In the late fifties brown patches of discoloured skin (**lentigines**) may appear: these are commonly seen at the temple region of the face and on the backs of the hands.

Fortunately help is at hand: there is an ever-increasing number of skin-care products from a vast and highly profitable cosmetics industry, and there are the skill and expertise of the qualified beauty professional.

If then it is *your* intention to become a qualified beauty professional – whether make-up artist or beauty therapist – you need to learn about skin: its construction, its function, and how and why it is changed by both internal and external influences.

BASIC STRUCTURES

The human body consists of many millions of microscopic **cells**. Each cell contains a chemical substance called **protoplasm**, which contains various specialised structures whose activities are essential to our health. If cells are unable to function properly, a disorder results.

Surrounding the cell is the **cell membrane**: this forms a boundary between the cell contents and their environment. The membrane has a porous surface which permits food to enter and waste materials to leave.

In the centre of the cell is the **nucleus**, which contains the **chromosomes**. On these are the **genes** we have inherited from our parents. The genes are ultimately responsible for cell reproduction and cell functioning.

The liquid within the cell membrane and surrounding the nucleus is called **cytoplasm**. Scattered throughout this are other small bodies, the **organelles**; each has a specific function within the cell.

Cells in the body tend to specialise in carrying out particular functions. Groups of cells which share function, shape, size or structure are called **tissues**. Tissues, in turn, may be grouped to form the larger functional and structural units we know as **organs**.

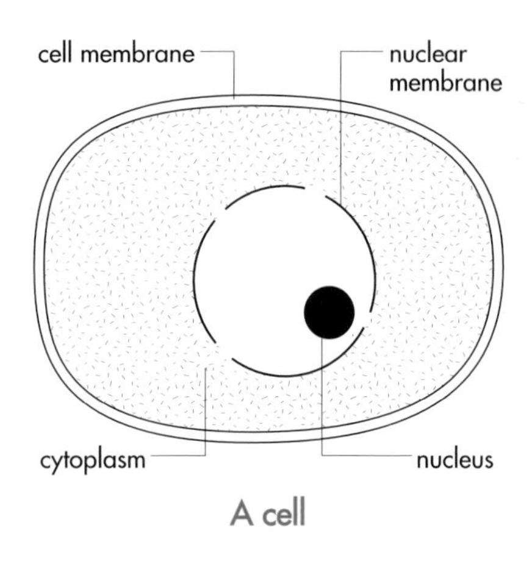

A cell

The skin

The human skin is an organ – the largest of the body. It provides a tough, flexible covering, with many different important functions.

FUNCTIONS OF THE SKIN

Protection

The skin protects the body from potentially harmful substances and conditions.

- The outer surface is **bactericidal**, helping to prevent the multiplication of harmful micro-organisms. It also prevents the absorption of many substances (unless the surface is broken), because of the construction of the cells on its outer surface which form a chemical and physical **barrier**.
- The skin cushions the underlying structures from physical injury.
- The skin provides a **waterproof coating.** Its natural oil, **sebum**, prevents the skin from losing vital water, and thus prevents skin dehydration.
- The skin contains a pigment called **melanin**. This absorbs harmful rays of ultra-violet light.

TIP

The skin accounts for one-eighth of the body's total weight. It measures approximately 1.5 m^2 in total, depending on body size.

It is thinnest on the eyelids (0.05 mm), and thickest on the soles of the feet (approximately 5 mm).

HEALTH & SAFETY: SKIN PROTECTION

Although the skin is structured to avoid penetration of harmful substances by absorption, certain chemicals *can* be absorbed through the skin. Always protect the skin when using potentially harmful substances.

Heat regulation

Body **temperature** is controlled in part by heat loss through the skin and by sweating.

Excretion

Small amounts of certain **waste products**, such as water and salt, are removed from the body by excretion through the surface of the skin.

Warning

The skin affords a warning system against outside invasion. **Redness** and **irritation** of the skin indicate that the skin is intolerant to something, either external or internal.

Sensitivity

The skin allows the feelings of **touch**, **pressure**, **pain**, **heat** and **cold**, and allows us to recognise objects by their feel and shape.

Nutrition

The skin provides storage for **fat**, which provides an energy reserve. It is also responsible for producing a significant proportion of our **vitamin D**, which is created by a chemical action when sunlight is in contact with the skin.

Moisture control

The skin controls the movement of moisture from within the deeper layers of the skin.

THE STRUCTURE OF THE SKIN

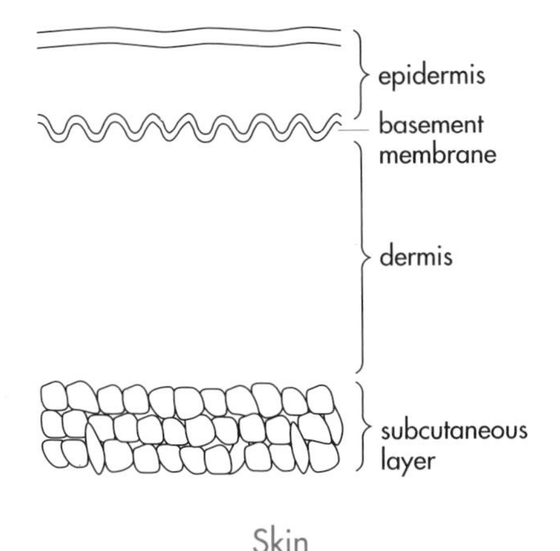

Skin

If we looked within the skin, using a microscope, we would be able to see two distinct layers: the **epidermis** and the **dermis**. Between these layers is a specialised layer which acts like a 'glue', sticking the two layers together: this is the **basement membrane**. If the epidermis and dermis become separated, body fluids fill the space, creating a **blister**.

Situated below the epidermis and dermis is a further layer, the **subcutaneous layer** or **fat layer**.

THE EPIDERMIS

The epidermis is located directly above the dermis. It is composed of five layers, with the surface layer forming the outer skin – what we can see and touch. The main function of the epidermis is to protect the deeper living structures from invasion and harm from the external environment.

Nourishment of the epidermis, essential for growth, is received from a liquid called the **interstitial fluid**.

Each layer of the epidermis can be recognised by its shape and by the function of its cells. The main type of cell found in the epidermis is the **keratinocyte**, which produces the protein **keratin**. It is keratin that makes the skin tough and that reduces the passage of substances into or out of the body.

Over a period of about four weeks, cells move from the bottom layer of the epidermis to the top layer, the skin's surface, changing in shape and structure as they progress. The process of cellular change takes place in stages.

- *The cell is formed* – by division of an earlier cell.
- *The cell matures* – it changes structure and moves upwards and outwards.
- *The cell dies* – it moves upwards and becomes an empty shell, which is eventually shed.

TIP

The epidermis is the most significant layer of the skin with regard to the external application of skin-care cosmetics and make-up.

TIP

Every five days we shed a complete surface layer. About 80 per cent of household dust is composed of dead skin cells.

The layers of the epidermis

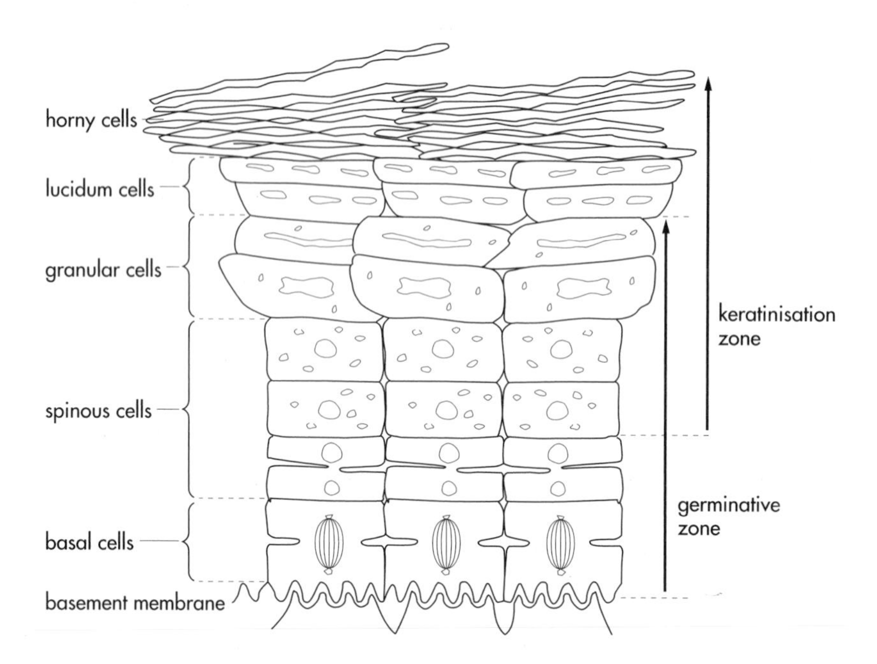

HEALTH & SAFETY: PSORIASIS

With the skin disorder *psoriasis*, cell division occurs much more quickly, resulting in clusters of dead skin cells appearing on the skin's surface.

The germinative zone

In the **germinative zone** the cells of the epidermis are living cells.

Stratum germinativum

The **stratum germinativum**, or **basal layer**, is the lowermost layer of the epidermis. It is formed from a single layer of column-shaped cells joined to the basement membrane. These cells

HEALTH & SAFETY: ALLERGIC REACTIONS

Most people will be unaware that a foreign body has invaded the skin. Sometimes, however, the skin's intolerance to a substance is apparent. It shows as an *allergic reaction*, in which the skin becomes red, itchy and swollen.

divide continuously and produce new epidermal cells (keratinocytes). This process of cell division is known as **mitosis**.

Stratum spinosum

The **stratum spinosum**, or **prickle-cell layer**, is formed from two to six rows of elongated cells; these have a surface of spiky spines which connect to surrounding cells. Each cell has a large nucleus and is filled with fluid.

Langerhan cells and melanocyte cells

Two other important cells are found in the germinative zone of the epidermis.

Langerhan cells

Langerhan cells absorb and remove foreign bodies that enter the skin. They then move from the epidermis to the dermis below, and finally enter the lymph system (the body's waste-transport system).

Melanocytes

Melanocytes are the cells that produce the skin pigment **melanin**, which contributes to our skin colour. About one in every ten germinative cells is a melanocyte. Melanocytes are stimulated to produce melanin by ultra-violet rays, and their main function is to protect the other epidermal cells in this way from the harmful effects of ultra-violet.

The quantity and distribution of melanocytes differs according to race. In a white Caucasian person the melanin tends to be destroyed when it reaches the granular layer (see below). With stimulation from artificial or natural ultra-violet light, however, melanin will also be present in the upper epidermis.

In contrast a black skin has melanin present in larger quantities throughout *all* the epidermal layers, a level of protection that has evolved to deal with bright ultra-violet light. This increased protection allows less ultra-violet to penetrate the dermis below, reducing the possibility of premature ageing from exposure to ultra-violet light. The more even quality and distribution of melanin also means that people with dark skins are less at risk of developing skin cancer.

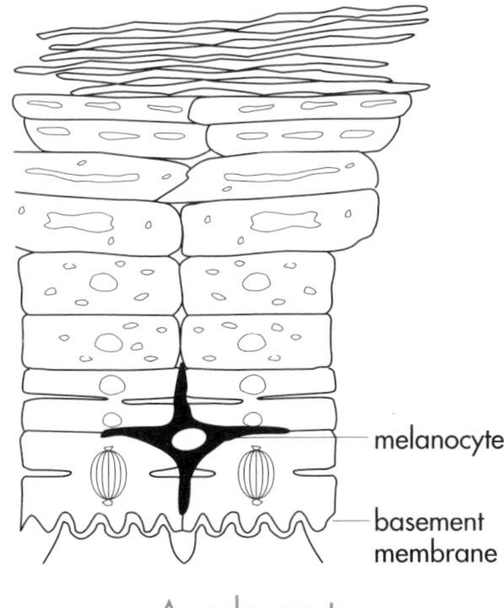

A melanocyte

HEALTH & SAFETY: VITILIGO

Lack of skin pigment is called *vitiligo* or *leucoderma.* It can occur with any skin colour, but is more obvious on a dark skin. Avoid exposing such skin to ultra-violet as it does not have the melanin protection.

Another pigment, **carotene**, which is yellowish, also occurs in epidermal cells. Its contribution to skin colour lessens in importance as the amount of melanin in the skin increases.

Skin colour also increases when the skin becomes warm. This is because the **blood capillaries** at the surface dilate, bringing blood nearer to the surface so that heat can be lost.

HEALTH & SAFETY: SUNBURN

If the skin becomes red on exposure to sunlight, this indicates that the skin has been over-exposed to ultra-violet. It will often blister and shed itself.

TIP

Many skin-care products and cosmetics, including lipsticks and mascaras, now contain *sunscreens*. This is because research has shown that ultra-violet exposure is the principal cause of skin ageing.

Stratum granulosum

The **stratum granulosum**, or **granular layer**, is composed of one, two or three layers of cells which have become much flatter. The nucleus of the cell has begun to break up, creating what appear to be granules within the cell cytoplasm. These are known as **keratohyaline granules** and later form keratin. At this stage the cells form a new, combined layer.

The keratinisation zone

The **keratinisation zone**, or **cornified zone**, is where the cells begin to die and where finally they will be shed from the skin. The cells at this stage become progressively flatter, and the cell cytoplasm is replaced with the hard protein keratin.

Stratum lucidum

The **stratum lucidum**, **clear layer** or **lucid layer**, is only seen in non-hairy areas of the skin such as the palms of the hands and the soles of the feet. The cells here lack a nucleus and are filled with a clear substance called **eledin**.

Stratum corneum

The **stratum corneum**, or **cornified layer**, is formed from several layers of flattened, scale-like overlapping cells, composed mainly of keratin. These help to reflect ultra-violet light from the skin's surface; black skin, which evolved to withstand strong ultra-violet, has a thicker stratum corneum than does Caucasian skin.

It takes about three weeks for the epidermal cells to reach the stratum corneum from the stratum germinativum. The cells are then shed, a process called **desquamation**.

HEALTH & SAFETY: CALLUSES

Constant friction causes the skin to thicken as a form of protection, developing calluses.

TIP

The skin will become much thicker in response to friction. A client with a manual occupation may therefore develop hard skin (calluses) on her hands. This skin condition can be treated with an *emollient* preparation, which will moisturise and soften the dry skin.

The dermis

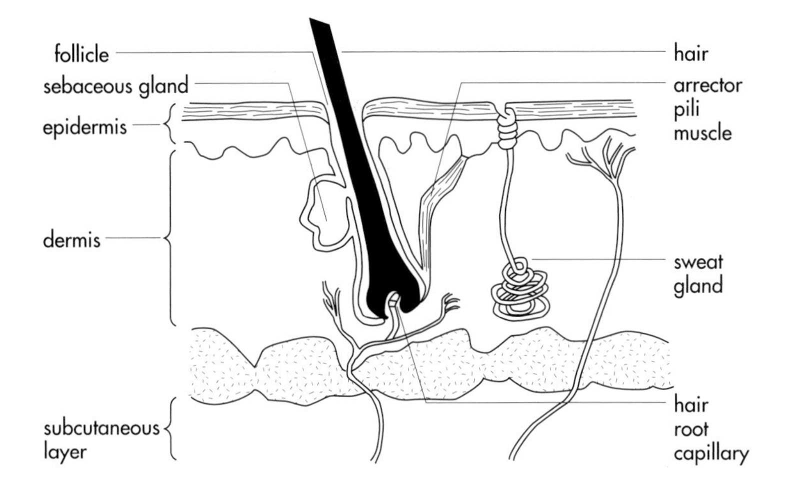

THE DERMIS

The dermis is the inner portion of the skin, situated underneath the epidermis and composed of dense **connective tissue**. It is much thicker than the epidermis.

The reticular layer

The dermis contains a network of protein fibres called the **reticular layer**. These fibres allow the skin to expand, to contract, and to perform intricate, supple movements.

This network is composed of two sorts of protein fibre, yellow **elastin** fibres and white **collagen** fibres: elastin fibres give the skin its elasticity, and collagen fibres give it its strength. The fibres are produced by specialised cells called **fibroblasts**, and are held in a gel called the **ground substance**.

Hugh Rushton

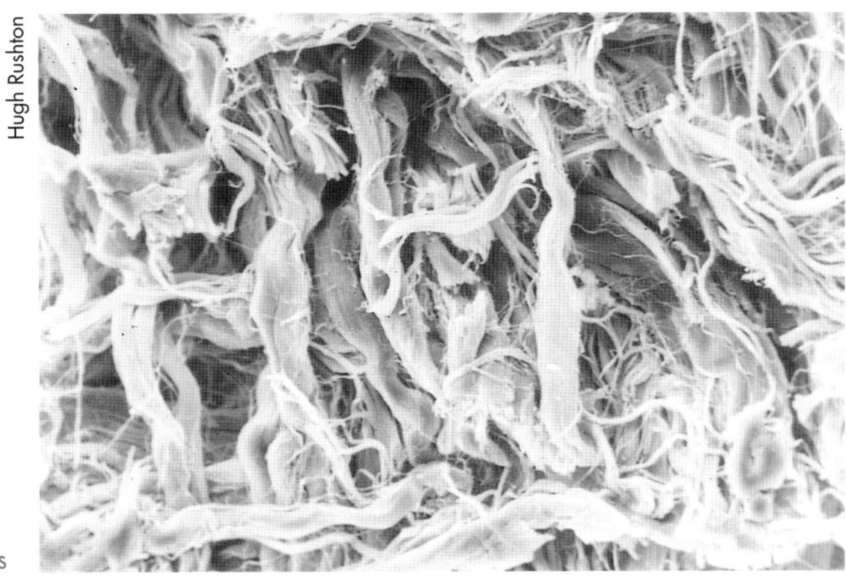

Collagen and elastin fibres

While this network is strong, the skin will appear youthful and firm. As the fibres harden and fragment, however, the network begins to collapse, losing its elasticity. The skin then begins to show visible signs of ageing.

A major cause of damage to this network is unprotected exposure of the skin to ultra-violet light and to weather. Sometimes, too, the skin loses its elasticity because of a sudden increase in body weight, for example at puberty or pregnancy. This results in the appearance of **stretch marks**, streaks of thin skin which is a different colour from the surrounding skin: on a white skin they appear as thin reddish streaks; on a black skin they appear slightly lighter than the surrounding skin. The lost elasticity cannot be restored.

HEALTH & SAFETY: SUNBATHING
When sunbathing, always protect the skin with an appropriate protective sun-care product, and always use an emollient after-sun preparation to minimise the cumulative effects of premature ageing.

Nerve endings

The dermis contains different types of sensory **nerve endings**, which register touch, pressure, pain and temperature. These send messages to the **central nervous system** and the **brain**, informing us about the outside world and what is happening on the skin's surface. The appearance of each of these nerve endings is quite varied.

TIP
Sensory nerve endings are most numerous in sensitive parts of the skin, such as the fingertips and the lips.

The papillary layer

Near the surface of the dermis are tiny projections called **papillae**; these contain both nerve endings and blood capillaries. This part of the dermis is known as the **papillary layer**, and it also supplies the upper epidermis with its nutrition.

Growth and repair

The body's blood system of arteries and veins continually brings blood to the capillary networks in the skin and takes it away again. The blood carries the nutrients and oxygen essential for the skin's health, maintenance and growth, and takes away waste products.

TIP
Appropriate external massage movements can be used to increase the blood supply within the dermis, bringing extra nutrients and oxygen to the skin and to the underlying muscle. At the same time, the lymphatic circulation is increased, improving the removal of waste products that may have accumulated.

HEALTH & SAFETY: SCARS
When the surface has been broken, the skin at the site of the injury is replaced but may leave a scar. This initially appears red, due to the increased blood supply to the area, required while the skin heals. When healed, the redness will fade.

Defence

Within the dermis are the structures responsible for protecting the skin from harmful foreign bodies and irritants.

One set of cells, the **mast cells**, burst when stimulated during inflammation or allergic reactions, and release **histamine**. This causes the blood vessels nearby to enlarge, thereby bringing more blood to the site of the irritation.

In the blood, and also in the lymph and the connective tissue, are another group of cells: the **macrophages**. These destroy micro-organisms and engulf dead cells and other unwanted particles.

Waste products

Lymph vessels in the skin carry a fluid (**lymph**) containing waste products such as used blood cells. The waste products are eliminated, and usable protein is recycled for further use by the body.

ACTIVITY: THE EMOTIONS
Emotions can affect blood supply to the skin: blood vessels in the dermis may enlarge or constrict. Think of different emotions, and how these might affect the appearance of the skin.

Control of functioning

Hormones are chemical messengers transported in the blood. They control the activity of many organs in the body, including the cells and glands in the skin. These include **melanosomes**, which produce skin pigment, and the **sweat glands** and **sebaceous glands**.

Hormone imbalance at different times of our life may disturb the normal functioning of these cells and structures, causing various **skin disorders**.

Skin appendages

Within the dermis are structures called **skin appendages**. These include:

- sweat glands;
- hair follicles, which produce hair;
- sebaceous glands;
- nails.

Sweat glands

TIP
We each have approximately 2–5 million sweat glands.

Sweat glands or **sudoriferous glands** are composed of **epithelial tissue**, which extends from the epidermis into the dermis. These glands are found all over the body, but are particularly abundant on the palms of the hands and the soles of

HEALTH & SAFETY: MOISTURE BALANCE
Excessive sweating, which can occur through exposure to high temperatures or during illness, can lead to *skin dehydration* – insufficient water content. Fluid intake must be increased to rebalance the body fluids.

TIP

Pores allow the absorption of some facial cosmetics into the skin. Many facial treatments are therefore aimed at cleansing the pores, some with a particularly deep cleansing action, as with *cosmetic cleansers* and *facial masks.*

The pores may become enlarged, because of congestion caused by dirt, dead skin cells and cosmetics. The application of an *astringent* skin-care preparation creates a tightening effect upon the skin's surface, slightly reducing the size of the pores.

the feet. Their function is to regulate body temperature through the evaporation of sweat from the surface of the skin. Fluid loss and control of body temperature are important to prevent the body overheating, especially in hot, humid climates. For this reason, perhaps, sweat glands are larger and more abundant in black skins than white skins.

There are two types of sweat glands: *eccrine glands* and *apocrine glands.* **Eccrine glands** are found all over the body, appearing as tiny tubes (**ducts**). These are straight in the epidermis, and coiled in the dermis. The duct opens directly onto the surface of the skin through an opening called a **pore**.

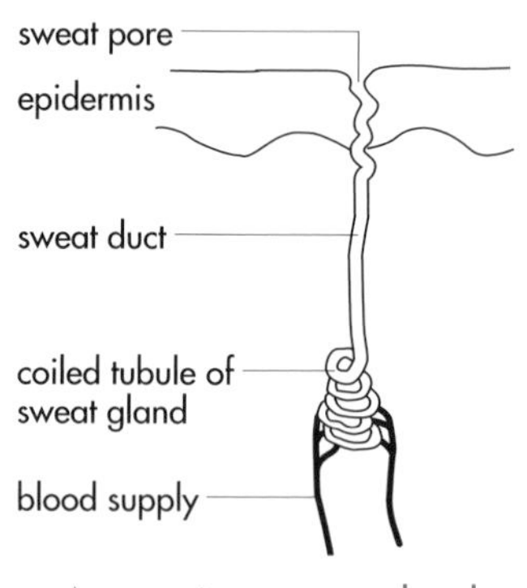

An eccrine sweat gland

Eccrine glands continuously secrete small amounts of sweat, even when we appear not to be perspiring. In this way they maintain the body temperature at a constant 36.8 °C.

Apocrine glands are found in the armpit, the nipples, and the groin area. This kind of gland is larger than the eccrine gland, and is attached to a hair follicle. Apocrine glands are controlled by hormones, becoming active at puberty. They also increase in activity when we are excited, nervous or stressed. The fluid they secrete is thicker than that from the eccrine glands, and may contain urea, fats, sugars and small amounts of protein. Also present are traces of aromatic molecules called **pheromones**

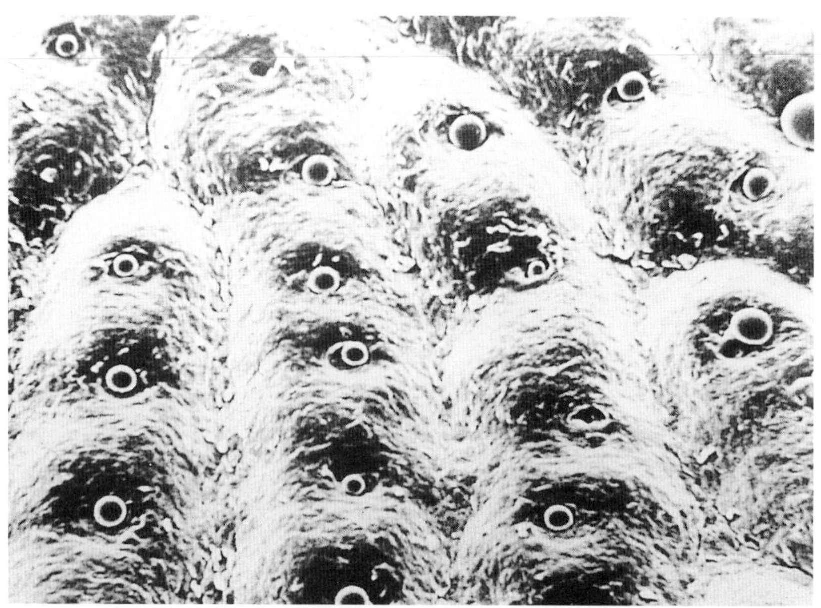

Unilever

Sweat pores

ACTIVITY: PREVENTING BODY ODOUR

Produce a checklist which should be followed daily to reduce the possibility of body odour.

which are thought to cause sexual attraction between individuals.

An unpleasant smell – **body odour** – develops when apocrine sweat is broken down by skin bacteria. Good habits of personal hygiene will prevent this.

Cosmetic perspiration control

To extend hygiene protection during the day, apply either a deodorant or an anti-perspirant. **Anti-perspirants** reduce the amount of sweat that reaches the skin's surface: they have an astringent action which closes the pores. **Deodorants** contain an active antiseptic ingredient which reduces the skin's bacterial activity, thereby reducing the risk of odour from stale sweat.

HEALTH & SAFETY: ANTI-PERSPIRANTS

The active ingredient in most anti-perspirants products is *aluminium chlorhydrate*. This is known to cause contact dermatitis in some people, especially if the skin has been damaged by recent removal of unwanted hair. Bear this in mind if you are performing an underarm depilatory wax treatment. (See the aftercare instructions, page 282.)

Hair follicles

Hair follicles are downward growths into the dermis of epidermal tissue. They are found all over the body, except on the palms of the hands, the soles of the feet, and the lips.

At the base of each hair follicle is a cluster of cells called the **germinal matrix**. These reproduce to form the lower part of the hair, the **bulb**. The bottom of the follicle is supplied with nerves and blood vessels, the **hair papillae**, which nourish the cells in this area. Cells move up the hair follicle from the hair bulb, changing in their structure and forming a **hair**.

A hair follicle

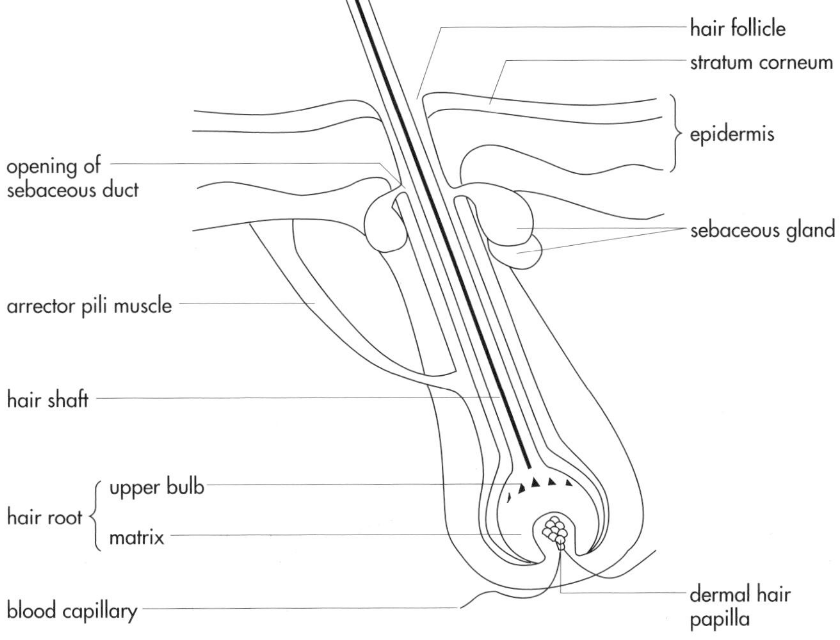

The hair follicle is attached to the base of the epidermis by a small muscle, the **arrector pili**. When this muscle contracts it causes the hair to stand upright in the hair follicle, temporarily raising the surrounding skin. This effect is often referred to as **goose pimples**.

Sebaceous glands

The **sebaceous gland** appears as a minute sac-like organ. Usually it is associated with the hair follicle with which it forms the **pilosebaceous unit**, but the two can appear independently.

Sebaceous glands are found all over the body, except on the palms of the hands and the soles of the feet. They are particularly numerous on the scalp, the forehead, and in the back and chest region. The cells of the glands decompose, producing the skin's natural oil, **sebum**. This empties directly into the hair follicle.

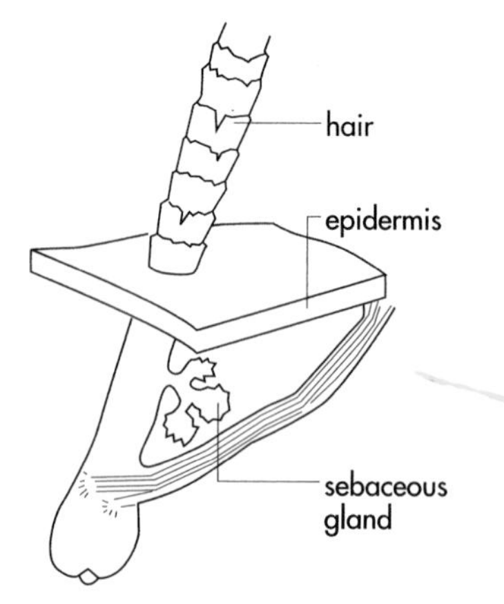

A sebaceous gland

> **HEALTH & SAFETY: THE LIPS**
> Sebaceous glands are not present on the surface of the lips. For this reason the lips should be protected with a lip emollient preparation, to prevent them from becoming dry and chapped.

The activity of the sebaceous gland increases at puberty when stimulated by the male hormone, **androgen**. In adults activity of the sebaceous gland gradually decreases again. Men secrete slightly more sebum than women; and on black skin the sebaceous glands are larger and more numerous than on white skin.

Sebum is composed of fatty acids and waxes. These have **bactericidal** and **fungicidal** properties, and so discourage the multiplication of micro-organisms on the surface of the skin. Sebum also reduces the evaporation of moisture from the skin, and so prevents the skin from drying out.

> **TIP**
> Cosmetic moisturisers mimic sebum in providing an oily covering for the skin's surface to reduce moisture loss.

Acid mantle

Sweat and sebum combine on the skin's surface, creating an acid film. This is known as the **acid mantle**, and discourages the growth of bacteria and fungi.

Acidity and alkalinity are measured by a number called the pH. An *acidic solution* has a pH of 0–7; a *neutral solution* has a pH of 7; and an *alkaline solution* has a pH of 7–14. The acid mantle of the skin has a pH of 5.5–5.6.

> **HEALTH & SAFETY: USING ALKALINE PRODUCTS**
> Because the skin has an acid pH, if alkaline products are used on it the acid mantle will be disturbed. It will take several hours for this protective film to be restored: during this time, the skin will be irritated and sensitive.

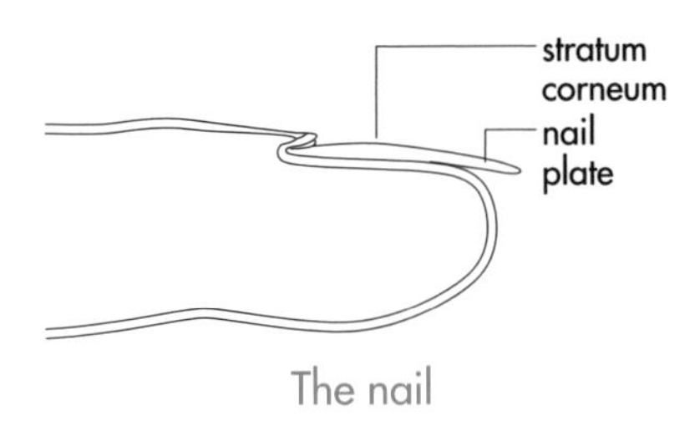

The nail

TIP

Fingernails grow more quickly than toenails. Fingernails grow about 0.1 mm each day, and grow faster in summer than in winter.

The nails

Nails are formed from a group of hard, horny, keratinised epidermal cells called the **nail plate**. This protects the living **nail bed** of the fingers and toes.

The nails are an extension of the skin. The nail plate – which we can cut with scissors without experiencing any pain – is composed entirely of dead epidermal cells.

KNOWLEDGE REVIEW

Anatomy and physiology of the skin

1 Name, and briefly describe, the structures shown in the cross-section of the skin (1–9).

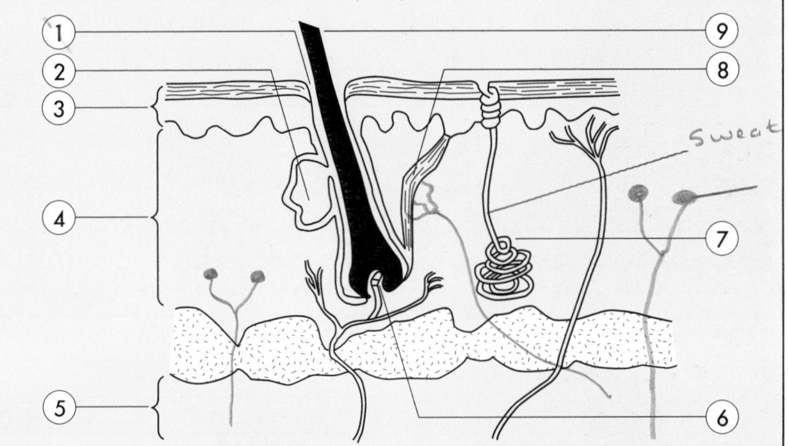

2 Name *six* functions of the skin.

3 Name the layers of the epidermis (1–5).

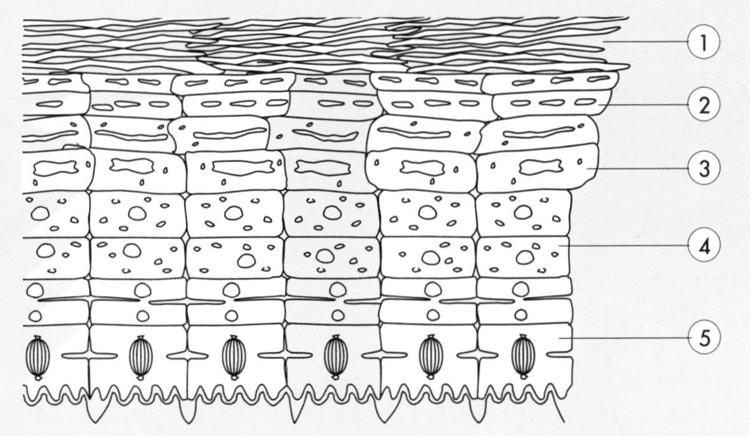

4 Which layer of the epidermis is continuously being shed, and is composed completely of dead skin cells?

5 What are melanocytes? What is their function?

6 What is the difference between the apocrine and eccrine sudoriferous (sweat) glands?

7 Which protein fibres found in the dermis give the skin its strength and elasticity?

8 Where is the skin thickest on the body?

9 What is the acid mantle? What is its approximate pH?

10 What is the name of the protein found in skin cells and hair?

ANATOMY AND PHYSIOLOGY FOR FACIAL MASSAGE

STRUCTURES AFFECTED BY MASSAGE

Manual massage is the external manipulation of the soft tissues of the body using the hands. The beneficial effects of massage can improve the appearance of the client's skin.

Before learning the practical techniques of facial massage you need to be aware of the structures you are working on. These are:

- the skin;
- the subdermal muscles;
- the bones of the head, neck and shoulders;
- the blood, the lymph and the nerve supply of the head and neck.

THE SKIN

Massage affects each of the skin's layers and all its structures, improving their general functioning and the skin's appearance.

- Dead epidermal cells are loosened and shed, which improves the appearance of the skin, exposing fresh new cells.
- The skin temperature increases, causing the pores and follicles to relax and so aiding the absorption of massage products, which soften the skin.
- Sweat and sebum production is increased, by stimulation of the sebaceous and sudoriferous glands.
- Blood circulation is increased, which nourishes the tissues of the skin.

THE MUSCLES

Muscles are responsible for the movement of body parts. Each is made up of a bundle of elastic fibres bound together in a sheath, the **fascia**. Movement occurs when these fibres contract.

A muscle is usually anchored by a strong **tendon** to one bone: the point of attachment is known as the muscle's **origin**. The

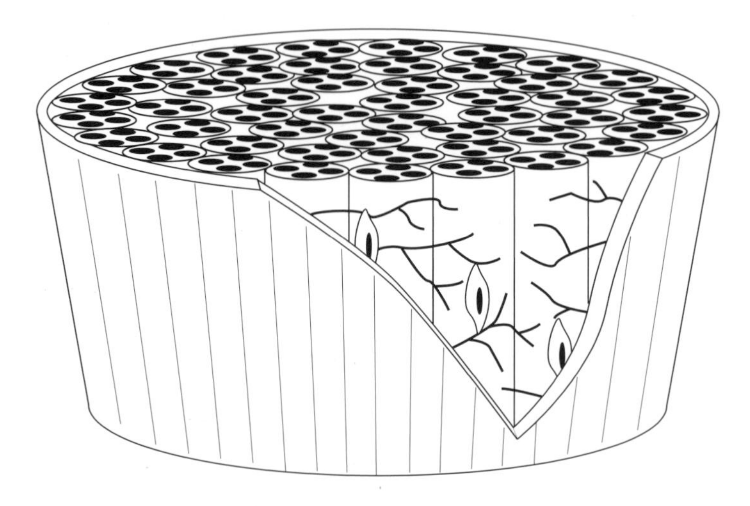

Muscle

muscle is likewise joined to a second bone: this attachment in this case is called the muscle's **insertion**. It is this second bone that is moved: the muscle contracts, pulling the two bones towards each other. (A different muscle, on the other side of the bone, has the contrary effect.) Not all muscles attach to bones, however: some insert into an adjacent muscle, or into the skin itself. The muscles with which we are concerned here are those of the face, the neck and the shoulders.

Facial muscles

Many of the muscles located in the face are very small and are attached to ('insert into') another small muscle or the facial skin. When the muscles contract, they pull the facial skin in a particular way: it is this that creates the facial expression.

With age, the facial expressions that we make every day produce lines on the skin – frown lines. The amount of tension, or **tone**, also decreases with age. When performing facial massage, the aim is to improve the general tone of the facial muscles.

Muscles of facial expression

Muscle	*Expression*	*Location*	*Action*
Frontalis	Surprise	The forehead	Raises the eyebrows
Corrugator	Frown	Between the eyebrows	Draws the eyebrows together
Orbicularis oculi	Winking	Surrounds the eyes	Closes the eyelid
Risorius	Smiling	Extends diagonally, from the corners of the mouth	Draws mouth corners outwards
Buccinator	Blowing	Inside the cheeks	Compresses the cheeks
Zygomaticus	Smiling, laughing	Extends diagonally from the corners of the mouth	Lifts the mouth corners upwards and outwards
Orbicularis oris	Pout, kiss, doubt	Surrounds the mouth	Purses the lip, closes the mouth
Triangularis	Sadness	The corner of the lower lip extends over the chin	Draws down the mouth's corners
Mentalis	Doubt	Chin	Raises the lower lip, causing the chin to wrinkle
Depressor labii	Sulking	The lower lip extends over the chin	Draws down the mouth's corners
Platysma	Fear, horror	The sides of the neck and chin	Draws the mouth's corners downwards and backwards

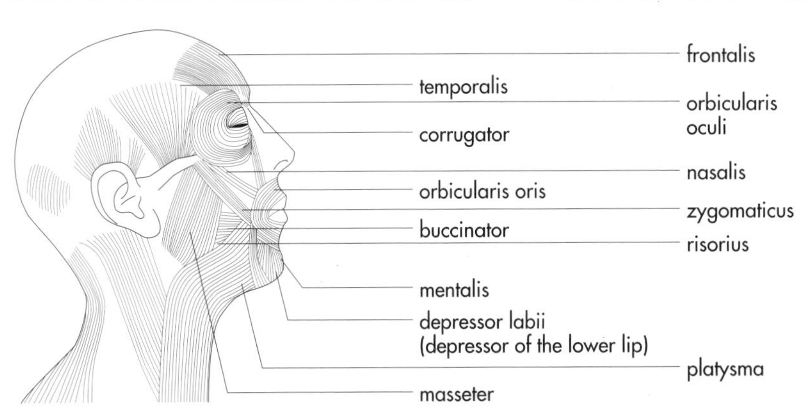

Muscles of the face

Muscles of mastication

The muscles responsible for the movement of the lower jawbone (the **mandible**) when chewing are called the **muscles of mastication**.

Muscle	*Location*	*Action*
Masseter	The cheek area: extends from the zygomatic bone to the mandible	Clenches the teeth; closes and raises the lower jaw
Temporalis	Extends from the temple region at the side of the head to the mandible	Raises the jaw and draws it backwards, as in chewing

Muscles that move the head

Muscle	*Location*	*Action*
Sternomastoid	Runs from the sternum to the clavical bone and the temporal bone	Flexes the neck; rotates and bows the head
Trapezius	A large triangular muscle, covering the back of the neck and the upper back	Draws the head backwards and allows movement at the shoulder

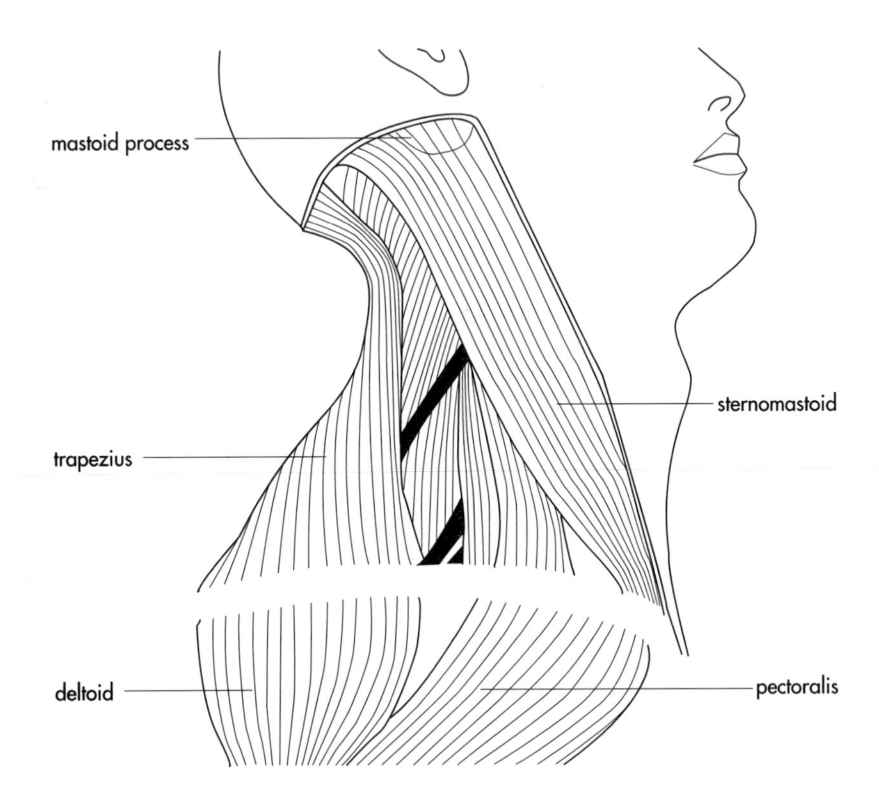

Muscles that move the head

Muscles of the upper body

When massaging the shoulder area you will cover the following muscles of the upper body.

Muscle	*Location*	*Action*
Pectoralis major	The front of the chest	Moves the arm towards the upper body
Deltoid	A thick triangular muscle, covering the shoulder	Takes the arm away from the side of the body

THE BONES

When carrying out a facial massage you will feel below your hands the underlying bones. **Bone** is the hardest structure in the body: it protects the underlying structures, gives shape to the body, and provides an attachment point for our muscles, thereby allowing movement.

Bones have different shapes, according to their function. Some have surfaces so that one can move against another. The point where two or more bones meet is known as a **joint**.

Kinds of bones

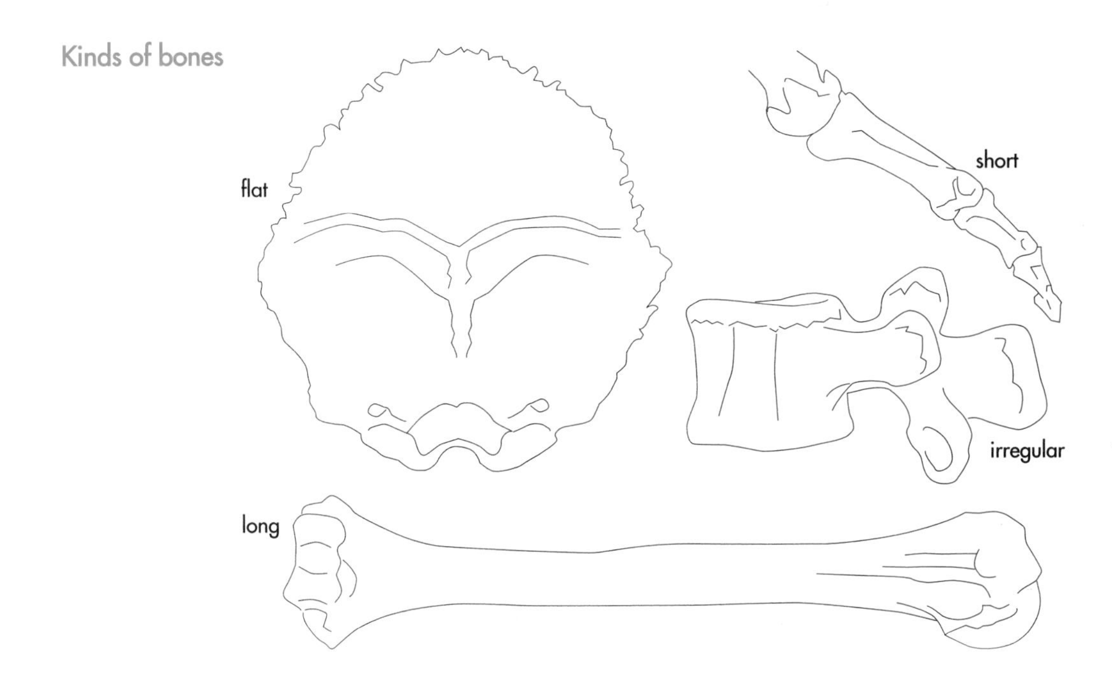

A ball-and-socket joint, as in the shoulder

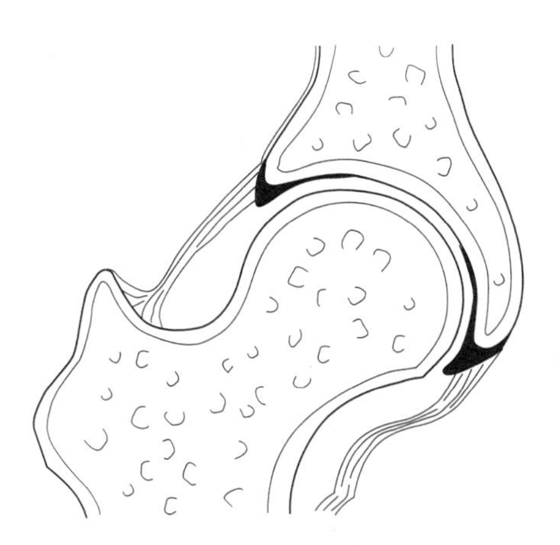

Bones of the head

The bones which form the head are collectively known as the **skull**. The skull can be divided into two parts, the face and the cranium, which together are made up of 22 bones:

- the 14 facial bones form the face;
- the 8 cranial bones form the rest of the head.

As well as forming our facial features, the facial bones support other structures such as the eyes and the teeth. Some of these bones, such as the nasal bone, are made from **cartilage**, a softer tissue than bone.

The cranium surrounds and protects the brain. The bones are thin and slightly curved, and are held together by connective tissue. After childhood, the joints become immovable, and are called **sutures**.

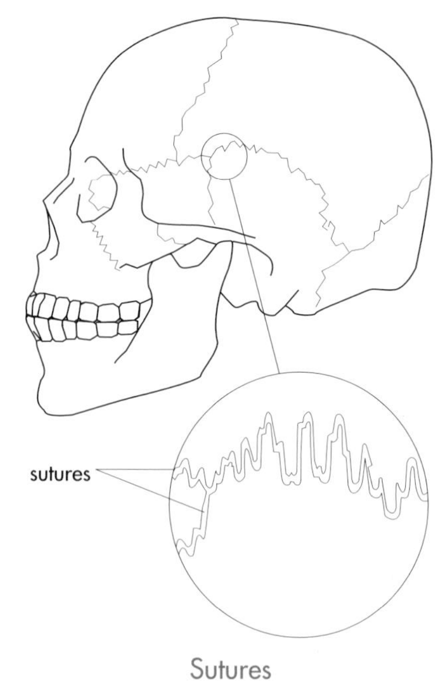

Sutures

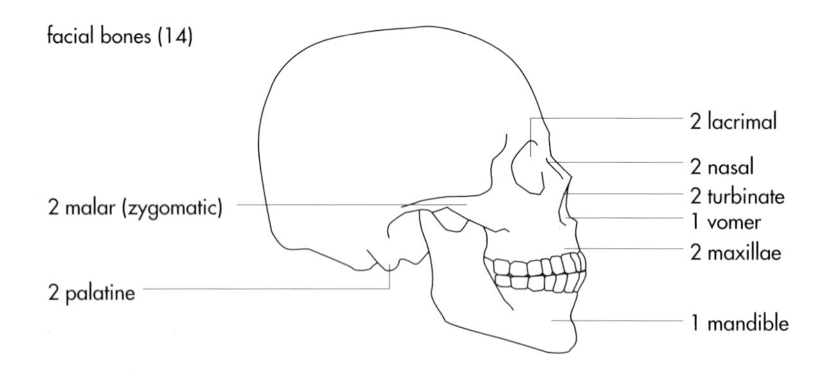

Bones of the face

Facial bones

Bone	*Number*	*Location*	*Function*
Nasal	2	The nose	Form the bridge of the nose
Vomer	1	The nose	Forms the dividing bony wall of the nose
Palatine	2	The nose	Form the floor and wall of the nose and the roof of the mouth
Turbinate	2	The nose	Form the outer walls of the nose
Lacrimal	2	The eye sockets	Form the inner walls of the eye sockets; contain a small groove for the tear duct
Malar (zygomatic)	2	The cheek	Form the cheekbones
Maxillae	2	The upper jaw	Fused together, to form the upper jaw, which holds the upper teeth
Mandible	1	The lower jaw	The largest and strongest of the facial bones; holds the lower teeth

ACTIVITY: FACIAL EXPRESSIONS

In front of a mirror, move the muscles of your face to create the expressions that you might form each day.

What expressions can you make? Which part or parts of the face are moving? Which facial muscles do you think have contracted to create these expressions?

HEALTH & SAFETY: CROW'S FEET

To avoid the premature formation of 'crow's feet':

- avoid squinting in bright sunlight – wear sunglasses;
- have your eyes tested regularly;
- if you use a visual display unit, ensure that you take regular breaks, and have a protective filter screen to remove glare.

Bones of the cranium

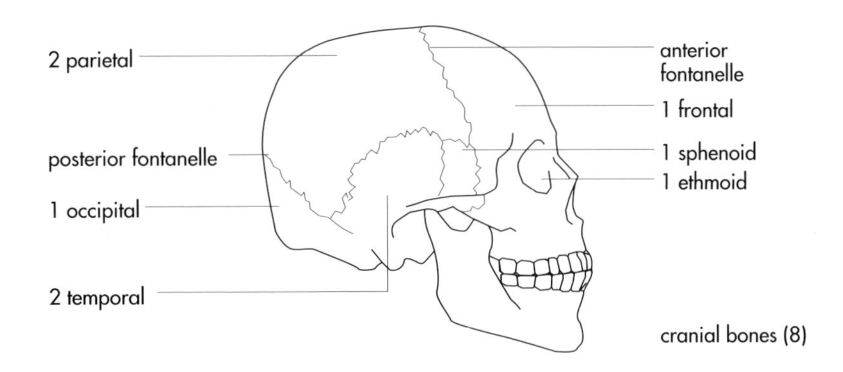

Cranial bones

Bone	*Number*	*Location*	*Function*
Occipital	1	The lower back of the cranium	Contains a large hole called the *foramen magnum:* through this pass the spinal cord, the nerves and blood vessels
Parietal	2	The sides of the cranium	Fused together to form the sides and top of the head (the 'crown')
Frontal	1	The forehead	Forms the forehead and the upper walls of the eye sockets
Temporal	2	The sides of the head	Provides two muscle attachment points: the mastoid process and the zygomatic process
Ethmoid	1	Between the eye sockets	Forms part of the nasal cavities
Sphenoid	1	The base of the cranium the back of the eye sockets	A bat-shaped bone which joins together all the bones of the cranium

Bones of the neck, chest and shoulder

Bone	*Number*	*Location*	*Function*
Cervical vertebra	7	The neck	These vertebrae form the top of the spinal column: the *atlas* is the first vertebra, which supports the skull; the *axis* is the second vertebra, which allows rotation of the head
Hyoid	1	A U-shaped bone at the front of the neck	Supports the tongue
Clavicle	2	Slender long bones at the base of the neck	Commonly called the *collar bone*: these form a joint with the sternum and the scapula bones, allowing movement at the shoulder
Scapula	2	Triangular bones in the upper back	Commonly called the *shoulder blade:* the scapulae provide attachment for muscles which move the arm. The *shoulder girdle,* which allows movement at the shoulder, is composed of the clavicles and the scapulae
Humerus	2	The upper bones of the arms	Form ball-and-socket joints with the scapulae: these joints allow movement in any direction
Sternum	1	The breastbone	Protects the inner organs; provides a surface for muscle attachment; and supports muscle movement

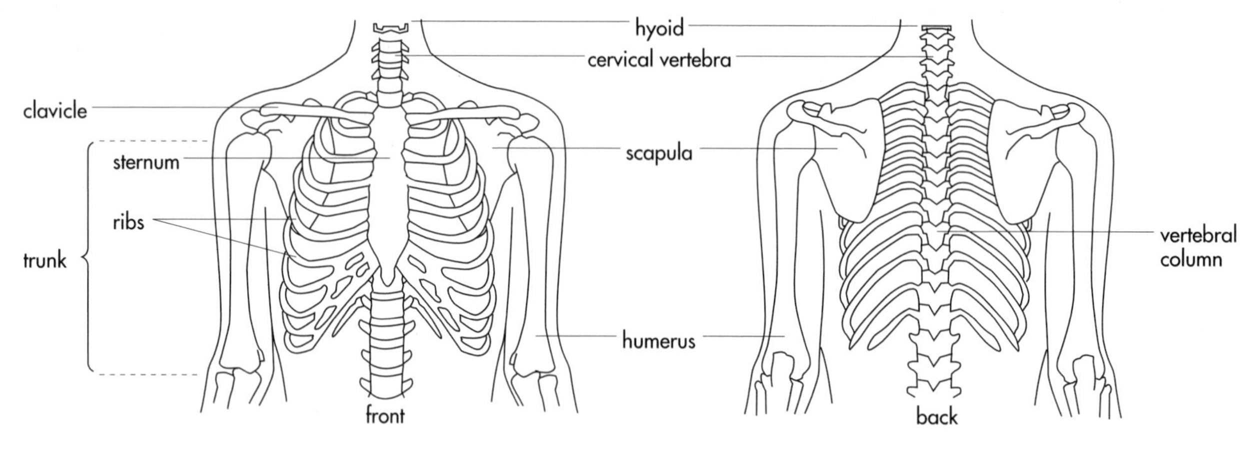

Bones of the neck, chest and shoulder

THE BLOOD SUPPLY

A healthy muscle is activated by a nerve supply, which effects movement; the necessary oxygen and nutrients are brought by the blood.

Blood transports various substances around the body:

- It carries oxygen from our lungs, and nutrients from our digested food to supply energy – these allow the cells to develop and divide, and the muscles to function.
- It carries waste products and carbon dioxide away for elimination from the body.
- It carries various cells and substances which allow the body to prevent or fight disease.

> **TIP**
>
> Blood helps to maintain the body temperature at 36.8 °C: varying blood flow near to the skin surface increases or diminishes heat loss.

The main constituents of blood

Blood consists of the following:

- **Plasma** A straw-coloured liquid: mainly water, with foods and carbon dioxide.
- **Red blood cells (erythrocytes)** These cells appear red because they contain **haemoglobin**; it is this that carries oxygen from the lungs to the body cells.
- **White blood cells (leucocytes)** There are several types of white blood cells: their main role is to protect the body, destroying foreign bodies and dead cells, and carrying away the debris (a process known as **phagocytosis**).
- **Plateletes (thrombocytes)** When blood is exposed to air, as happens when the skin is injured, these cells bind together to form a clot.
- **Other chemicals** Hormones also are transported in the blood.

The circulation

The circulation of blood is under the control of the **heart**, a muscular organ which pumps the blood around the body.

> **TIP**
> The pumping of the blood under pressure through the carotid arteries can be felt as a pulse in the neck. Press gently on the neck just inside the position of the sternomastoid muscle.

Blood leaving the heart is carried in large, elastic tubes called **arteries**. The blood to the head arrives via the **carotid arteries**, which are connected via other main arteries to the heart. There are two main carotid arteries, one on each side of the neck.

These arteries divide into smaller branches, the *internal carotid* and the *external carotid*. The **internal carotid artery** passes the temporal bone and enters the head, taking blood to the brain. The **external carotid artery** stays outside the skull, and divides into branches:

- the **occipital branch** supplies the back of the head and the scalp;
- the **temporal branch** supplies the sides of the face, the head, the scalp and the skin;
- the **facial branch** supplies the muscles and tissues of the face.

The blood supply to and from the head

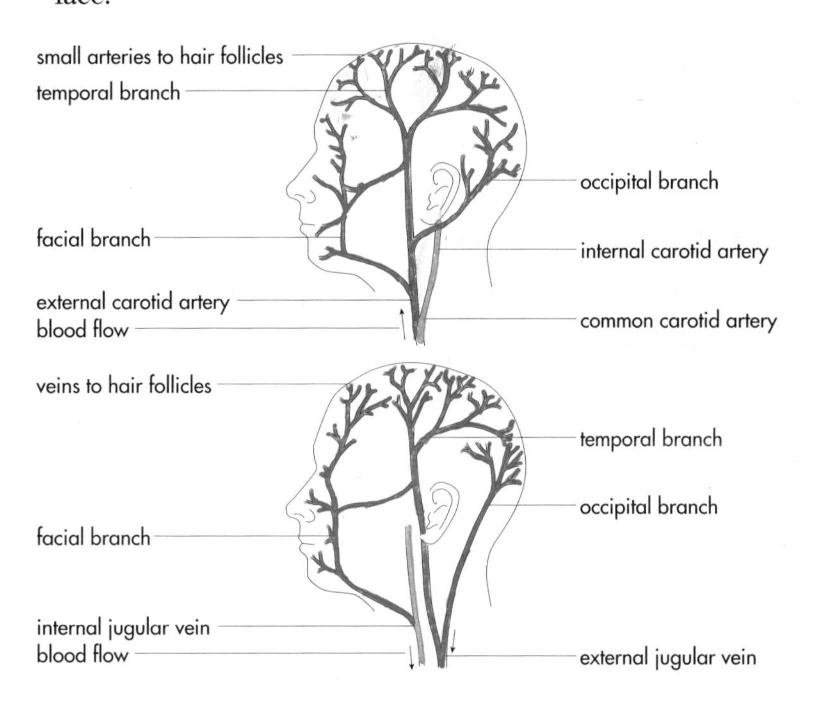

> **HEALTH & SAFETY: CAPILLARIES**
> The strength and elasticity of the capillary walls can be damaged, for example by a blow to the tissues.

These arteries also divide repeatedly, successive vessels becoming smaller and smaller until they form tiny blood **capillaries**. These vessels are just one cell thick, allowing substances carried in the blood to pass through them into the **tissue fluid** which bathes and nourishes the cells of the various body tissues.

The blood capillaries begin to join up again, forming first small vessels called **venules**, then larger vessels called **veins**. These return the blood to the heart.

Veins are less elastic than arteries, and are closer to the skin's surface. Along their course are **valves** which prevent the backflow of blood.

The main veins are the external and internal jugular veins. The **internal jugular vein** and its main branch, the **facial vein**, carry blood from the face and head. The **external jugular vein** carries blood from the scalp and has two branches: the **occipital branch** and the **temporal branch**. The jugular veins join to enter the **subclavian vein**, which lies above the clavicle.

Blood returns to the heart, which pumps it to the **lungs**, where the red blood cells take on fresh oxygen, and where carbon dioxide is expelled from the blood. The blood returns to the heart, and begins its next journey round the body.

THE NERVE SUPPLY

All muscles are made to work by electrical stimulation via the **nerves**. These, together with the **brain** and the **spinal cord**, form the nervous system.

Kinds of nerve

There are two types of nerve: *sensory nerves* and *motor nerves*. Both are composed of white fibres enclosed in a sheath.

- **Sensory nerves** These receive information and relay it to the brain. They are found near to the skin's surface and respond to touch, pressure, temperature and pain.
- **Motor nerves** These are situated in muscle tissue and act on information received from the brain, causing a particular response, typically muscle movement.

Appropriate massage manipulations, though applied to the skin, produce a stimulating or relaxing effect on nerves.

HEALTH & SAFETY: NERVE DAMAGE
Nerve cells do not reproduce; when damaged, only a limited repair occurs.

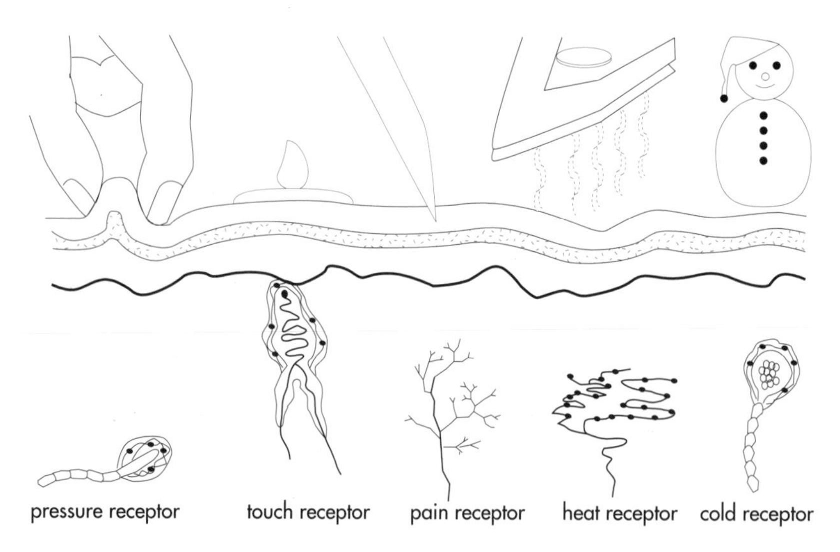

Sensory nerves

Nerves of the face and neck

These nerves link the brain with the muscles of the head, face and neck.

There are 12 pairs of **cranial nerves**. Those of concern to the beauty therapist when performing a facial massage are as follows:

- the 5th cranial nerve, or **trigeminal**;
- the 7th cranial nerve, or **facial**;
- the 11th cranial nerve, or **accessory**.

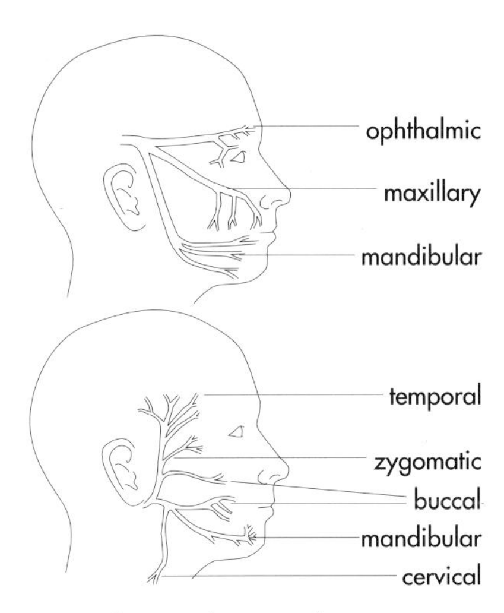

above 5th cranial nerve
below 7th cranial nerve

5th cranial nerve

This nerve carries messages to the brain from the sensory nerves of the skin, the teeth, the nose and the mouth. It also stimulates the motor nerve to create the chewing action when eating.

The 5th cranial nerve has three branches:

- the **ophthalmic nerve** serves the tear glands, the skin of the forehead, and the upper cheeks;
- the **maxillary nerve** serves the upper jaw and the mouth;
- the **mandibular nerve** serves the lower jaw muscle, the teeth and the muscle involved with chewing.

7th cranial nerve

This nerve passes through the temporal bone and behind the ear, and then divides. It serves the ear muscle and the muscles of facial expression, the tongue and the palate.

The 7th cranial nerve has five branches:

- the **temporal nerve** serves the orbicularis oculi and the frontalis muscles;
- the **zygomatic nerve** serves the eye muscles;
- the **buccal nerve** serves the upper lip and the sides of the nose;
- the **mandibular nerve** serves the lower lip and the mentalis muscle of the chin;
- the **cervical nerve** serves the platysma muscle of the neck.

11th cranial nerve

This nerve serves the sternomastoid and trapezius muscles of the neck.

THE LYMPHATIC SYSTEM

The **lymphatic system** is closely connected to the blood system, and can be considered as supplementing it. Its primary function is defensive: to remove bacteria and foreign materials, thereby preventing infection. It also drains away excess fluids for elimination from the body.

The lymphatic system consists of the fluid **lymph**, the **lymph vessels**, and the **lymph nodes** (or glands). You may have experienced swelling of the lymph nodes in the neck when you have been ill.

Unlike the blood circulation, the lymphatic system has no muscular pump equivalent to the heart. Instead, the lymph moves through the vessels and around the body because of movements such as contractions of large muscles. Facial massage can play an important part in assisting this flow of lymph fluid, thereby encouraging the improved removal of the waste products transported in the lymph.

Lymph

Lymph is a straw-coloured fluid, derived from blood plasma, which has filtered through the walls of the capillaries. The composition of lymph is similar to that of blood, though less oxygen and fewer nutrients are available. In the spaces between the cells where there are no blood capillaries, lymph provides nourishment. It also carries **lymphocytes** (a type of white blood cell).

Lymph travels only in one direction: from body tissues back towards the heart.

Lymph vessels

Lymph vessels often run very close to veins, forming an extensive network throughout the body. The lymph moves quite slowly, and there are valves along the lymph vessels to prevent backflow of the lymph.

The lymph vessels join to form larger lymph vessels, which eventually flow into one or other of two large lymphatic vessels: the **thoracic duct** (or **left lymphatic duct**) and the **right lymphatic duct**. The thoracic duct receives lymph from the left side of the head, neck, chest, abdomen and lower body; the right lymphatic duct receives lymph from the right side of the head and upper body.

These principal lymphatic vessels then empty their contents into a vein at the base of the neck, which in turn empties into the **vena cava**. The lymph is mixed into the venous blood as it is returned to the heart.

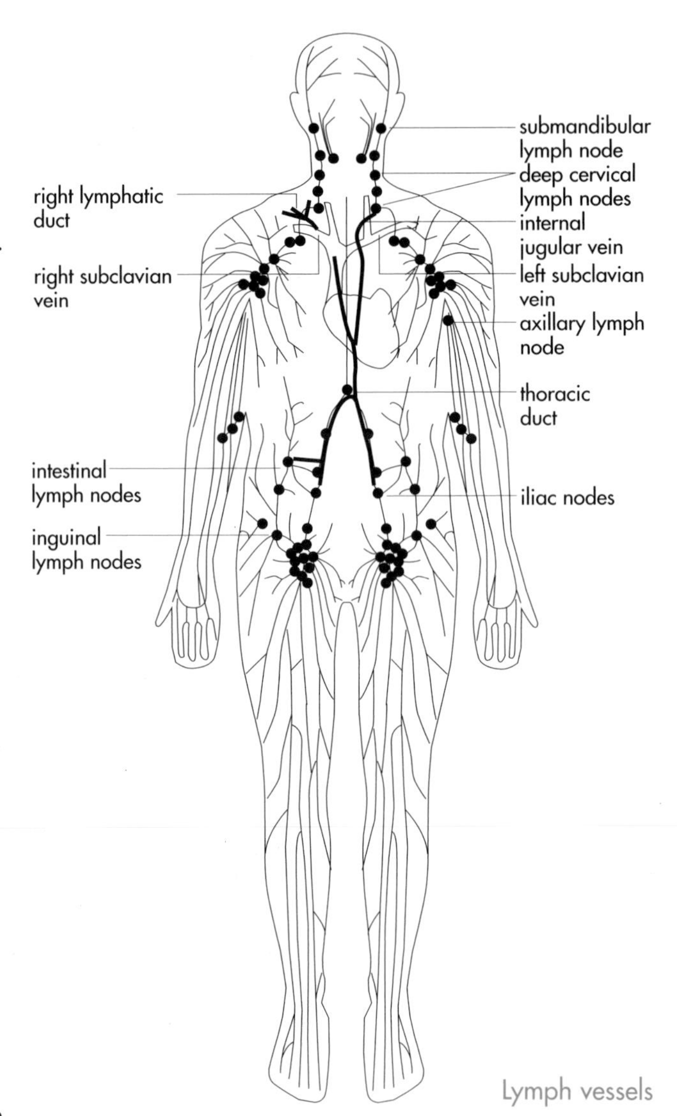

Lymph vessels

Lymph nodes

Lymph nodes or **glands** are tiny oval structures which filter the lymph, extracting poisons, pus and bacteria, and thus defending the body against infection by destroying harmful organisms. **Lymphocytes**, found in the lymph glands, are special cells which produce **antibodies** which enable us to resist invasion by micro-organisms.

When performing massage, the hands should be used to apply pressure to direct the lymph towards the nearest lymph node: this encourages the speedy removal of waste products. Various groups of lymph nodes drain the lymph of the head and neck.

Lymph nodes of the head

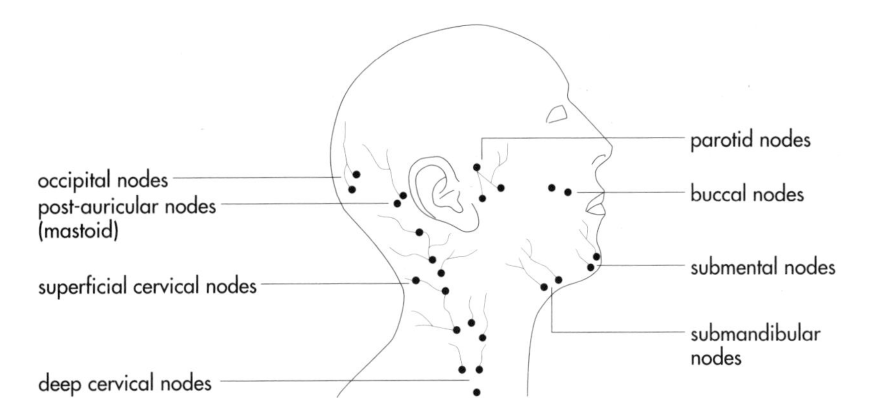

Lymph nodes of the head

- The **buccal group** drain the eyelids, the nose and the skin of the face.
- The **mandibular group** drain the chin, the lips, the nose and the cheeks.
- The **mastoid group** drain the skin of the ear and the temple area.
- The **occipital group** drain the back of the scalp and the upper neck.
- The **submental group** drain the chin and the lower lip.

Lymph nodes of the neck

- The **external cervical group** drain the neck below the ear.
- The **upper deep cervical group** drain the back of the head and the neck.
- The **lower deep cervical group** drain the back area of the scalp and the neck.

Lymph nodes of the chest and arms

- The nodes of the armpit area drain various regions of the arms and chest.

TIP

Frontal–Temporal–Parietal–Occipital–Mandible (mandibular)–Cervical
Learn and remember these names of the main regions of the head and neck. Not only will this assist you in recalling the names and locations of the bones, it will also help you greatly with the names and locations of muscles, arteries, veins, nerves and lymph nodes.

KNOWLEDGE REVIEW

Anatomy and physiology for facial massage

1. What is the difference in function between the blood and the lymph in the body?
2. What is the difference between sensory nerves and motor nerves?

3 To what are the main sensory nerve endings in the skin receptive?

4 Name the lymph nodes of the head (1–5).

5 Name the labelled parts of the skull (1–11). How many bones form:
(a) the cranium?
(b) the face?

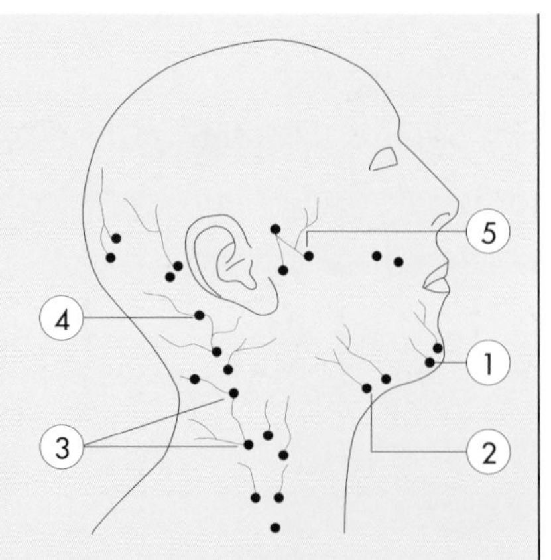

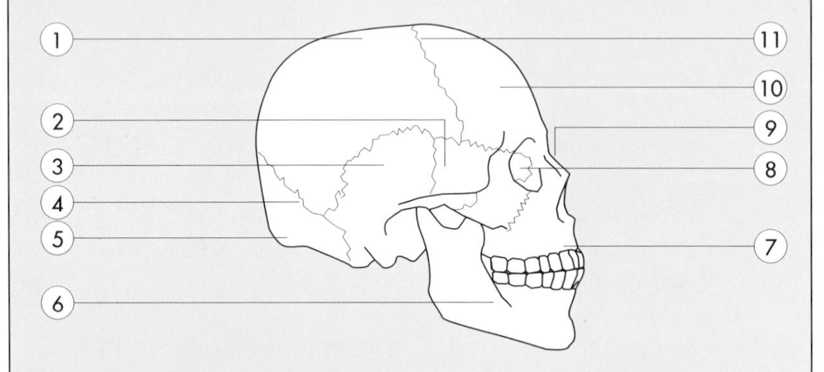

6 Which of the bones form the face? Which bone or bones form:
(a) the forehead?
(b) the cheekbones?
(c) the jawbone?

7 Name the muscles (1–6). What is the action of each of these muscles?

8 What is the general action of the following muscles:
(a) the deltoid muscle?
(b) the trapezius?
(c) the pectoralis major?

9 Name the main branches of the 7th cranial facial nerve (1–5).

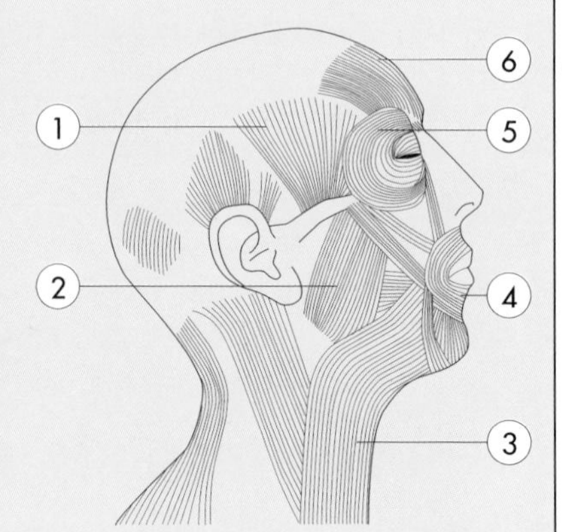

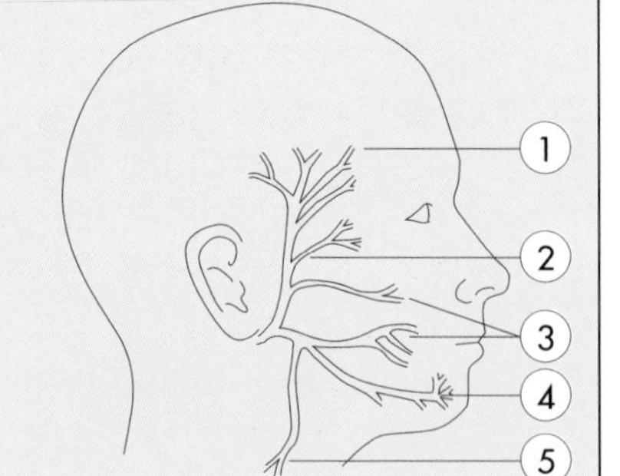

CHAPTER 5

Skin care and facial massage

Cereal Partners UK

BASICS OF SKIN CARE

Nutrition

If the skin is to function efficiently, the skin must be cared for both internally and externally.

Internally, a nutritionally balanced diet is vital to the health and appearance of the skin. A number of skin allergies and disorders are in part the result of a poorly balanced diet.

Foods contain the chemical substances we need for health and growth, the **nutrients:** a healthy diet contains all the essential nutrients. The nutrients are carried to the skin in the blood, where they nourish the cells in the processes of growth and repair.

There are six principal groups of nutrients.

Carbohydrates

Carbohydrates provide energy quickly. They are either simple sugars or starches that the body can turn into simple sugars.

Food sources: Carbohydrates are found in fruit, vegetables, milk, grains, and honey.

Fats

Fats provide a concentrated source of energy, and are also used in carrying certain vitamins (see below) around the body. Fat is stored in the body around organs and muscles and under the skin. However, if too much fat is deposited under the skin, the elastic fibres there may be damaged by the expansion of the 'adipose' tissue. Fat is also used in the formation of sebum, the skin's natural lubricant.

Food sources: Although this is not always evident, fats are present in almost all foods, from plants and from animals.

Proteins

Proteins provide material for the growth and repair of body tissue, and are also a source of energy. Severe protein deficiency in children gives the skin a yellowish appearance, known as **jaundice**.

Food sources: Proteins are found in meat, fish, eggs, dairy products, grains, and nuts.

> **HEALTH & SAFETY: VEGAN DIETS**
> Proteins are composed of many smaller units called *amino acids*. Animal protein sources contain all the amino acids essential to health. A *vegetarian* consuming dairy food will likewise obtain all the essential amino acids. A *vegan*, however, must be careful to eat an adequate quantity and variety of vegetables and other foods, in order to be sure that she receives all the amino acids she needs.

Minerals

Minerals provide materials for growth and repair and for regulation of the body processes. The major minerals are calcium, iron, phosphorus, sulphur, sodium, potassium, chlorine, and magnesium. Of these, the most important to the skin is iron. A pale, dry skin may indicate **anaemia**, caused by a shortage of iron.

Food sources: Fruit and vegetables; iron is found in liver, egg yolks, and green vegetables.

Vitamins

Vitamins regulate the body's processes and contribute to its resistance to disease. Vitamins are divided into two groups, according to whether they are soluble in water or in fat:

- the fat-soluble vitamins are A, D, E, and K;
- the water-soluble vitamins are B and C.

The vitamins most important to the condition of the skin are vitamins A, B_2 and B_3.

> **HEALTH & SAFETY: VEGAN DIETS**
> Vegans must take vitamin B_{12} as a vitamin supplement as this vitamin occurs naturally only in animal-derived foods.

Vitamin A

Insufficient vitamin A in the diet leads to **hyperkeratinisation** (production of too much keratin). This causes blockages in the skin tissue. The skin becomes rough and dry, and eye disorders such as styes may occur.

Food sources: Vitamin A is found in red, yellow and green vegetables, and in egg yolk, butter and cheese.

Vitamin B_2

Vitamin B_2 (also called **riboflavin**) helps to break down other foodstuffs, releasing energy needed by cells to function efficiently.

> **HEALTH & SAFETY: SPECIAL DIETS**
> Warn your clients about the danger of very low-fat, low-carbohydrate or low-protein diets: these can deprive the body of the nutrients it needs for growth, repair, and energy.

A deficiency of vitamin B_2 causes the skin at the corners of the mouth to crack.

Food sources: Vitamin B_2 is found in brewer's yeast, milk products, leafy vegetables, liver, and whole grains.

Vitamin B_3

Vitamin B_3 (also called **niacin**) has the same function as vitamin B_2, but is also vital in the maintenance of the tissues of the skin.

Food sources: Vitamin B_3 is found in meat, brewer's yeast, nuts, and seeds.

Vitamin C

Vitamin C (also called **ascorbic acid**) maintains healthy skin. A lack of vitamin C causes the capillaries to become fragile, and haemorrhages of the skin, such as bruising, may occur. Severe deficiency results in **scurvy**.

Food sources: Vitamin C is found in fruit and vegetables.

Water

HEALTH & SAFETY: WEIGHT LOSS

If you lose weight too quickly, your skin will sag and wrinkle.

Water forms about two-thirds of the body's weight, and is an important component both inside and outside the body cells. At least one litre of natural water should be drunk every day, to avoid dehydration of the body and the skin.

Food sources: Water is also a constituent of many foods, including fruits and vegetables.

Fibre

Fibre is not broken down into nutrients, but it is very important for effective digestion.

Food sources: Fibre is found in fruit, vegetables and cereals.

Threats to the skin

Internal

Alcohol

Alcohol deprives the body of its vitamin reserves, especially vitamin B and C, which are necessary for a healthy skin. Alcohol also tends to dehydrate the body, including the skin.

Caffeine

TIP

Advise your clients to replace tea with herbal infusions and to drink decaffeinated coffee (in moderation).

Coffee, tea, cocoa and soft fizzy drinks contain a mild drug called **caffeine**. In moderate doses, such as two or three cups of coffee per day, caffeine is safe. If you drink too much, however, caffeine can cause nervousness, interfere with digestion, block the absorption of vitamins and minerals, and spoil the appearance of the skin.

Smoking

Smoking interferes with cell respiration and slows down the circulation. This makes it harder for nutrients to reach the skin cells and for waste products to be eliminated. Cigarette smoking also releases a chemical that destroys vitamin C. This interferes with the production of collagen, and thereby contributes to premature wrinkling. Nicotine is a **toxic** substance – a poison!

Medication

Certain **medicines** taken by mouth can cause skin dehydration, oedema or swelling of the tissues (this may for example be caused by steroids) or irregular skin pigmentation (sometimes caused by the contraceptive pill). During the initial consultation with the client, find out whether she is taking any medication – and take this into account in your diagnosis and treatment plan.

Stress

Stress is shown in the face as tension lines where the facial muscles are tight. Because blood and lymph cannot circulate properly, this causes a 'sluggish' skin condition and poor facial nutrition. A person suffering from stress usually experiences disturbed sleep, sometimes sleeplessness (**insomnia**). Lack of sleep causes the skin to become dull and puffy, especially the tissue beneath the eyes, where dark circles also appear. Too *much* sleep also can cause the facial tissue to become puffy – because the circulation is less active, body fluids collect in the tissues.

If someone is suffering from stress, she may drink more tea, coffee or alcohol, or smoke more cigarettes: this too damages the skin.

Stress and anxiety are often the underlying cause of certain skin disorders. Some skin conditions, such as boils and styes, appear at times of stress; others, such as psoriasis and eczema, may become much worse. At the consultation, try to determine whether the client is suffering from stress: if she is, make sure that the salon treatments promote relaxation.

TIP

Some cosmetic companies offer bath preparations that aid relaxation by the inhalation of aromatic ingredients. These ingredients, combined with the warm water, may help a client who has difficulty in sleeping.

ACTIVITY: CAUSES OF STRESS

1 Can you think of everyday situations that may trigger stress?
2 How do *you* react physically when put in a stressful situation?
3 How can you create a relaxing environment in the salon?

External

As well as looking after the skin from the *inside*, by diet, it needs care from the *outside* – it must be kept clean, and it must be nourished.

With normal physiological functioning, the skin becomes greasy, and sweat is deposited on its surface. The skin's natural oil (**sebum**) can easily build up and block the natural openings, the hair **follicles** and **pores**: this may lead to infection. Facial cosmetics too affect the health of the skin; if not regularly removed, they may cause congestion. Skin-care treatments help to maintain and improve the functioning of the skin.

Ultra-violet light

Although recently **ultra-violet (UV)** has been identified as a hazard to skin, it also has some *positive* effects. One of these is its ability to stimulate the production of **vitamin D**, which is absorbed into the bloodstream and nourishes and helps to maintain bone tissue. Secondly, UV light activates the pigment **melanin** in the skin, and thereby creates a **tan**. Many people feel better when they have a tan, as it gives a healthy appearance.

Ultra-violet light is divided into different bands. The most important to skin tanning are UVA and UVB. **UVA** stimulates the melanin in the skin to produce a rapid tan, which does not last very long. UVA penetrates deep into the dermis where it can cause premature ageing of the skin. **Free radicals** – highly reactive molecules which cause skin cells to degenerate – are also formed. These molecules disrupt production of collagen and elastin, the fibres that give skin its strength and elasticity. Reduced elasticity leads to wrinkling.

UVB stimulates the production of vitamin D. Melanin activation by UVB produces a longer-lasting tan than that produced by UVA. UVB is partially absorbed by the atmosphere – it has a shorter wavelength than UVA – and only 10 per cent reaches the dermis. UVB causes thickening of the stratum corneum layer, which reflects ultra-violet away from the skin's surface.

UVB causes **sunburn**: the skin becomes red as the cells are damaged, and the skin may blister. UVB is also implicated in skin cancers, especially malignant melanoma.

The relaxing, warming effect of the sun is caused by the **infra-red (IR)** light. This penetrates the skin to the subcutaneous layer and is thought to speed skin ageing and possibly to cause a cancer called squamous cell carcinoma. The tan is actually a sign of skin damage, therefore, and both UV and IR probably contribute to **photo-ageing** – the premature ageing of the skin by light.

Although black skin has a high melanin content, which absorbs more ultra-violet and allows less to reach the dermis, it is not fully protected against the UV and still requires additional protection.

Chemical skin protection, or **sunscreens**, are designed to absorb ultra-violet light (UVA and UVB), reducing the rate of skin ageing in all skin types. Various sunscreens are available, classified by numbers accordingly to their **sun-protection factor (SPF)**. This is the amount of protection that the sunscreen gives

TIP

UVB is most intense between 11 a.m. and 3 p.m, when the sun is at its highest. Advise clients to avoid the sun between these times.

HEALTH & SAFETY: COLD SORES

A client who suffers from cold sores (herpes simplex) should avoid UV light as it can stimulate production of the sores.

ACTIVITY: THE EFFECTS OF UV

To see evidence of the damaging effects of ultra-violet on the skin, compare the skin on the back of your hands to skin on parts of the body that are not normally uncovered.

HEALTH & SAFETY: SUNBATHING
Never wear perfume, cosmetic products or deodorants when sunbathing, either in natural sunlight or in artificially produced (sun-canopy) ultra-violet. The chemicals in these products can sensitise the skin, causing an allergic skin reaction.

TIP
Pigmentation marks caused by repeated exposure to UV light eventually remain, even without the exposure to UV. Some clients dislike such marks, and you could advise them on camouflage make-up.

you from the sun. The application of the sunscreen extends your natural skin protection, allowing you to stay in the sun for longer without burning. For example, if normally you can be in the sun for 10 minutes before the skin begins to go red, a sunscreen with an SPF of 10 will allow you 10 × 10 minutes' – 100 minutes' – safe exposure in the sun.

Artificial UV light produced by sunbeds, used for cosmetic skin tanning, also causes premature ageing. Most sunbeds use concentrated UVA, which causes dermal tissue damage resulting in lines and wrinkles.

It may take years to see the effects of the dermal damage caused by unprotected UV exposure, but once they have occurred the effects are irreversible. UVA rays are present all year round, so to prevent premature ageing cosmetic preparations containing sunscreens should be worn at all times.

TIP
Ultra-violet can penetrate water to a depth of 1 metre, so even when swimming you need to wear a sunscreen. Special waterproof products are designed for this purpose.

Climate

Sebum, the skin's natural grease, provides an oily protective film over the surface of the skin which reduces evaporation. Despite this, unprotected exposure of the skin to the environment allows evaporation from the epidermis which results in a dry, dehydrated skin condition.

The climate has several effects on the skin:

- *Sebum production* When the skin is exposed to the cold, less sebum is produced. The skin has reduced protection, allowing moisture to evaporate.
- *Perspiration* In very hot weather more moisture is lost as **perspiration**: perspiration increases, to cool the skin and regulate the body's temperature.
- *Humidity* Moisture loss from the skin is also affected by the **humidity** (water content) of the surrounding air. In hot, dry weather humidity will be low, so water loss will be high. In temperate, damp conditions humidity will be high, so water loss will be low.
- *Extremes of temperature* Alternating heat and cold often leads to the formation of **broken capillaries**. These appear as fine

TIP
To acquire a tan without the damaging effects of the sun or sunbeds, use a *fake tanning* preparation. This treatment is becoming popular in the beauty therapist's salon.

red lines on Caucasian skin, and as discoloration on black skin.

- *Stratum corneum* The cells of the stratum corneum multiply with repeated unprotected exposure to the climate, as the body's natural defence.

The damaging effects caused by the climate can be reduced by using protective skin-care preparations such as moisturisers. These spread a layer of oil over the skin's surface, reducing evaporation.

TIP

At the consultation, ask the client her occupation: this will guide you as to her likely skin-care requirements. The client who works outdoors, for example, will have different treatment needs from the one who works indoors.

HEALTH & SAFETY: ASHINESS

A black skin can sometimes appear ashen, with patches of skin becoming grey and flaky. This is caused by sudden changes in temperature combined with low humidity, which can cause the skin to lose moisture. Moisturisers help in alleviating this problem.

ACTIVITY: GEOGRAPHICAL VARIATIONS

Consider the effects of climate on the skin. Name *four* different geographical locations, and state the humidity level and climate you would expect at each location.

Think of the probable effects on the skin. What treatments by a beauty therapist might be needed in each country?

Environmental stress and pollution

Further causes of moisture loss include harsh alkaline chemicals such as detergents and soaps – which remove sebum from the skin's surface – and air conditioning and central heating.

Environmental **pollutants** such as lead, mercury, cadmium and aluminium can accumulate in the body. One result is the formation of dangerous chemicals that attack proteins in the cells. Such pollutants find their way into food through polluted waters, rain and dust. To protect the body, always wash vegetables thoroughly, and eat a diet rich in vitamins C and E.

Air pollution, involving carbon from smoke, chemical discharges from factories, and fumes from car exhausts, should

HEALTH & SAFETY: EFFECTS OF WATER LOSS

Water loss can sometimes be sufficient to disrupt living cells in the skin to the extent that some actually die. This results in skin irritation and reddening.

TIP

Central heating creates an environment with humidity similar to that of the desert. Unless you use a moisturiser, the loss of moisture from the skin will cause a dry, dehydrated skin condition.

ACTIVITY: ENVIRONMENTAL FACTORS

It is important that you are aware of all the factors that can affect the health of a client's skin: this knowledge allows you to assess each client's treatment needs. Summarise the different factors you have learnt about.

be removed from the skin by effective cleansing. Absorption of these pollutants is reduced by the application of moisturiser: this forms a barrier over the skin's surface.

SKIN-CARE TREATMENTS

The beauty therapist has the professional expertise to help each client improve the appearance and condition of her skin by the application of appropriate cosmetic treatments and preparations. The facial skin services she offers include:

- consultation and skin analysis;
- skin cleansing;
- specialised skin-care treatments;
- the application of face masks or face packs;
- manual massage of the face, neck and shoulders;
- facial cosmetic make-up.

The beauty therapist cannot change the underlying skin type, which is genetically determined, but she can keep the physiological characteristics of each skin type in check.

Bellitas

Skin-care products

SKIN TYPES

The basic structure of the skin does not vary from person to person, but the physiological functioning of its different features does; it is this that gives us **skin types**.

The first and most important part of a facial treatment is the correct diagnosis of the skin type. This is carried out at the beginning of each facial treatment. The beauty therapist must choose the correct skin-care products and facial treatments for the client's skin type. This assessment is called a **skin analysis**.

Basic types

There are four main skin types:

- normal;
- dry;
- greasy;
- combination.

Normal skin

Normal skin is often referred to as **balanced**, because it is neither too oily nor too dry. Because when young this skin type seldom has any problems, such as blemishes, it is often neglected. Neglect causes the skin to become dry, especially around the eyes, cheeks and neck, where the skin is thinner.

A normal skin type in adults is very rare. It has these characteristics:

- the pore size is small or medium;
- the moisture content is good;

- the skin texture is even, neither too thick nor too thin;
- the colour is healthy (because of good blood circulation);
- the skin elasticity is good, when young;
- the skin feels firm to the touch;
- the skin pigmentation is even-coloured;
- the skin is usually free from blemishes.

Dry skin

Dry skin is lacking in either sebum or moisture, or both. Because sebum limits moisture loss by evaporation from the skin, skin with insufficient sebum rapidly loses moisture. The resulting dry skin is often described as **dehydrated**.

Dry skin has these characteristics:

- the pores are small and tight;
- the moisture content is poor;
- the skin texture is coarse and thin, with patches of visibly flaking skin;
- there is a tendency towards sensitivity (broken capillaries often accompany this skin type);
- premature ageing is common, resulting in the appearance of wrinkles, seen especially around the eyes, mouth and neck;
- skin pigmentation may be uneven, and disorders such as ephelides (freckles) usually accompany this skin type;
- milia are often found around the cheek and eye area.

Greasy skin

In **greasy skin** the sebaceous glands become very active at puberty, when stimulated by the male hormone **androgen**. An increase in sebum production often causes the appearance of skin blemishes. Sebaceous gland activity begins to decrease when the person is in her twenties.

Greasy skin has these characteristics:

- the pores are enlarged;
- the moisture content is high;
- the skin is coarse and thick;
- the skin is sallow in colour, as a result of the excess sebum production, dead skin cells having become embedded in the sebum, and the skin having sluggish blood and lymph circulation;
- the skin tone is good, due to the protective effect of the sebum;
- the skin is prone to shininess, due to excess sebum production;
- there may be uneven pigmentation;
- certain skin disorders may be apparent – comedones, pustules, papules, milia, or sebaceous cysts.

Acne vulgaris and **seborrhoea** are skin disorders that occur when the skin becomes excessively greasy due to the influence of hormones. Treatment of these skin disorders should be carried out to control sebum flow (see pages 19–20).

Combination skin

Combination skin is partly oily and partly dry. The oily parts are generally the chin, nose and forehead, known as the **T-zone**. The upper cheeks may show signs of oiliness, but the rest of the face and neck area is dry.

Combination skin is the commonest skin type. It has these characteristics:

- the pores in the T-zone are enlarged, while in the cheek area they are small to medium;
- the moisture content is high in the oily areas, but poor in the dry areas;
- the skin is coarse and thick in the oily areas, but thin in the dry areas;
- the skin is sallow in the oily areas, but shows sensitivity and high colour in the dry areas;
- the skin tone is good in the oily areas, but poor in the dry areas;
- there is uneven pigmentation, usually seen as ephelides and lentigines;
- there may be blemishes such as pustules and comedones on the oily skin at the T-zone;
- milia and broken capillaries may appear in the dry areas, commonly on the cheeks and near the eyes.

ACTIVITY: RECOGNISING SKIN TYPES

Refer to the illustration on the right. What skin characteristics would you expect to see in each of the numbered areas? Consider each of the skin types discussed

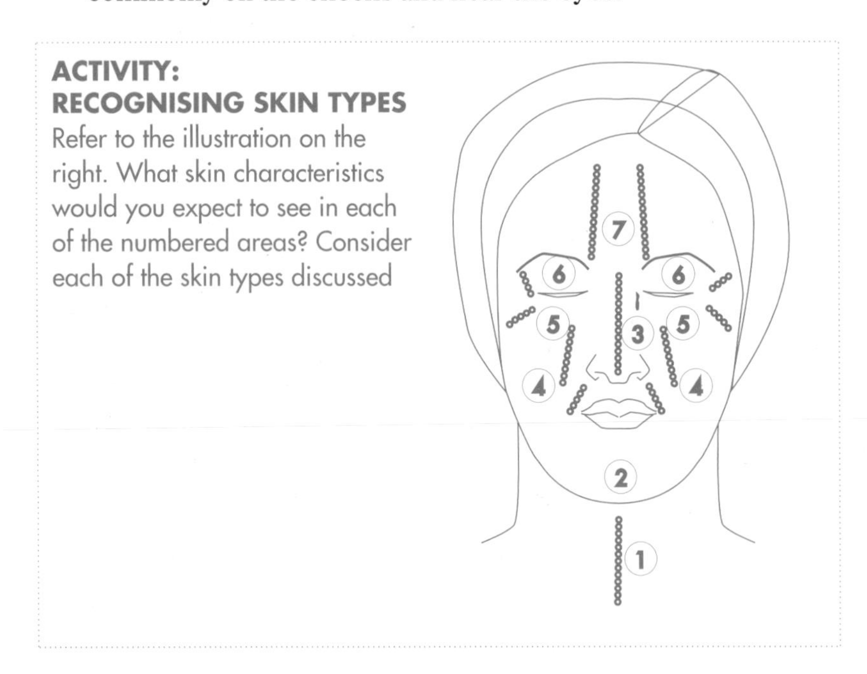

Additional characteristics

Whilst looking closely at the skin, further skin characteristics may become obvious. The skin may be:

- sensitive;
- dehydrated;
- moist;
- oedematous.

HEALTH & SAFETY: SENSITIVE SKIN
Use hypoallergenic products – these do not contain any of the known common skin sensitisers.
A skin that is sensitive may also be allergic to certain substances.

Sensitive skin

Sensitive skin usually accompanies a dry skin type, but not always.

The characteristics of sensitive skin are these:

- the skin may show high colouring;
- there are usually broken capillaries in the cheek area;
- the skin feels warm to the touch;
- there is superficial flaking of the skin;
- the skin may show high colouring after skin cleansing, if it is sensitive to pressure.

In a black skin, instead of the redness shown by a Caucasian skin, irritation shows up as a darker patch.

Allergic skin

Allergic skin is irritated by external **allergens**, including chemicals in some cosmetics. The allergens inflame the skin and may damage its protective function. At the consultation, always try to discover whether the client has any allergies, and if so, to what.

The allergies of most concern to the beauty therapist are those caused by substances applied to the skin. The therapist must be aware of such substances and avoid their use. Contact with an allergen, especially if repeated, may cause skin disorders such as eczema or dermatitis (see page 26).

TIP
Encourage clients with dehydrated skin to drink 6–8 glasses of water a day to replenish the moisture in the body.

Dehydrated skin

Dehydrated skin is skin that has lost water from the skin tissues. The condition can affect any skin type, but most commonly accompanies dry or combination skin types. The problem may be related to the client's general health. If she has recently been ill with a fever, for example, the skin will have lost fluid through sweating. If she is taking medication, this too may cause dehydration, as may drastic dieting. In many cases the dehydration is caused by working in an environment with a low humidity, or in one that is air-conditioned. You must try to discover the cause, and provide both corrective treatment and advice.

The characteristics of dehydrated skin are as follows:

- the skin has a fine orange-peel effect, caused by its lack of moisture;
- there is superficial flaking;
- fine, superficial lines are evident on the skin;
- broken capillaries are common.

Moist skin

Moist skin appears moist and feels damp: this is due to the over-secretion of sweat. The beauty therapist cannot correct this skin condition, which is often caused by some internal physiological disturbance such as a hormonal or metabolic imbalance.

Advise the client to use lightweight cleansing preparations. She should avoid skin-toning preparations with a high alcohol content; these would stimulate the skin, causing yet further perspiration and skin sensitivity. She should avoid highly spiced food, and be aware that alcoholic or hot drinks will cause dilation of the skin capillaries, thereby increasing the skin's temperature.

Oedematous skin

Oedematous skin is, and appears, swollen and puffy: this is because the tissues are retaining excess water. The condition may be caused by a medical disorder, or may be a side-effect of medication. Hot weather can cause temporary swelling of the tissues, as can local injury to the tissues. Poor blood circulation and lymphatic flow may cause puffy skin, too; this is often seen around the eyes. In this case the condition may benefit from gentle massage around the eye area. Tissue-fluid retention in the facial skin may be caused by an incorrect diet, such as one that includes too much salt or the drinking of too much alcohol, tea or coffee.

Unless you are quite sure about the cause of the oedema, always seek permission from your client's doctor before treating the skin.

TIP

Whilst completing the record card you may be able to recognise aspects of your client's lifestyle that probably contribute to the oedematous skin. If so, you can advise the client accordingly.

The age of the skin

Having identified the skin type, the beauty therapist must classify the age of the skin.

Often the age of a client will relate to skin problems that are evident. A young client, for example, may have skin blemishes such as comedones, pustules and papules. These disorders are caused by overactivity of the sebaceous gland at puberty, when the body is developing its secondary sexual characteristics. It is at this time that acne vulgaris is most likely to occur, due to the hormonal imbalance. The skin of clients aged over 25 years, however, is generally termed **mature skin**.

The beauty therapist should also consider the client's skin tone and muscle tone in relation to her age. A young skin will probably have good skin tone, and the skin will be supple and elastic. This is because the collagen and elastin fibres in the skin are strong. Poor skin tone, on the other hand, is recognised by the appearance of facial lines and wrinkles.

A healthy young skin will also have good muscle tone, and the facial contours will appear firm. With poor muscle tone, the muscles becomes slack and loose.

TIP

To test skin tone, gently lift the skin at the cheeks between two fingers and then let go. If the skin tone is good, the skin will spring back to its original shape.

Mature skin

The change in appearance of women's skin during ageing is closely related to the altered production of the hormones oestrogen, progesterone and androgen at the menopause.

Mature skin has the following characteristics:

ACTIVITY: THE AGEING PROCESS

Cut out photographs from magazines or newspapers showing men and women of different cultures and various ages.

1. Can you identify the visible characteristics of ageing?
2. Does ageing occur at the same rate in men and women and in different cultures?
3. Discuss your findings with your tutor.

- The skin becomes dry, as the sebaceous and sudoriferous glands become less active.
- The skin loses its elasticity as the elastin fibres harden, and wrinkles appear due to the cross-linking and hardening of collagen fibres.
- The epidermis grows more slowly and the skin appears thinner, becoming almost transparent in some areas such as around the eyes, where small veins and capillaries show through the skin.
- Broken capillaries appear, especially on the cheek area and around the nose.
- The facial contours become slack as muscle tone is reduced.
- The underlying bone structure becomes more obvious, as the fatty layer and the supportive tissue beneath the skin grow thinner.
- Blood circulation becomes poor, which interferes with skin nutrition, and the skin may appear sallow.
- Due to the decrease in metabolic rate, waste products are not removed so quickly, and this leads to puffiness of the skin.
- Patches of irregular pigmentation appear on the surface of the skin, such as lentigines and chloasmata.

HEALTH & SAFETY: UV LIGHT AND AGEING

The ageing process is accelerated when the skin is regularly exposed to ultra-violet light.

The skin may also exhibit the following skin conditions, although these are not truly *characteristic* of an ageing skin:

- Dermal naevi may be enlarged.
- Sebhorrheic warts may appear on the epidermal layer of the skin.
- Verruca filiformis warts may increase in number.
- Hair growth on the upper lip or chin, or both, may become darker or coarser, due to hormonal imbalance in the body.
- Dark circles and puffiness may occur under the eyes.

RECEPTION

Before carrying out *any* facial treatments, the beauty therapist must consult with the client to determine her treatment needs and to discuss the services that are available. **Consultation** is a service that should be offered separately: there should be no pressure on the client to book a treatment following the consultation.

Making the appointment

When a new client telephones to make an appointment for a service, always allocate extra time for the consultation beforehand. Explain to the client how long she should allow for the appointment. For example, if a client is seeking a basic skin cleansing and mask treatment, allow 30 minutes; if she is a *new* client allow 45 minutes so that there is time for the consultation and filling in the record card. If it is necessary to remove facial blockages and to carry out other specialised treatments, allow 45 minutes to 1 hour.

On arrival

When the client arrives for the treatment, the **record card** is completed. Record the client's personal details, such as her name and address. Check that there are no contra-indications to facial treatment.

The beauty therapist will add further information to the record card at the consultation and during treatment.

The consultation

The consultation is the time when the beauty therapist can assess whether the client is actually suited to treatment. The therapist must look for contra-indications, and must give no treatment if there are any – this is to safeguard the therapist, the client herself and, in the case of a client with a contagious skin disorder, other clients who would be at risk of cross-infection.

The beauty therapist's knowledge of facial treatments and advice on skin care will instil confidence in the client. Explain what is involved with each treatment, how long it takes, the aftercare involved and any home care that is required. This will demonstrate your professional expertise.

The client is likely to ask which is the most suitable treatment for her, and you must advise her as to which would best meet her needs.

Make the client aware of the cost of the individual treatment – or treatment programme, if necessary – so that she can decide whether or not to undertake the financial commitment involved.

Invite the client to ask questions during the consultation. By the end, she should understand fully what the proposed treatment involves.

She may receive the treatment immediately, following the consultation, or go away to consider the proposals.

During the consultation, details are noted on the client's record card. You can fill this in as you speak to her, without diverting your attention away from her.

TIP

Never make assumptions about what the client can afford. Usually the client will indicate to you what she is prepared to spend.

HEALTH & SAFETY: HERPES SIMPLEX
A small skin lesion due to herpes simplex may initially appear to be simply a pustule with a scab. At the consultation, always check whether the client knows that she suffers from some skin disease or disorder.

CONTRA-INDICATIONS

The consultation and skin analysis will draw your attention to any contra-indications or aspects that require special care and attention.

Remember that not all contra-indications are visible – a current bone fracture, for example, would not be. Refer to the checklist of contra-indications on the client's record card.

The following contra-indications are relevant to *all* facial treatments:

- *Skin disorder,* such as acne vulgaris (unless medical approval has been sought and given).
- *Skin disease,* such as impetigo.
- *Bruising* in the area.
- *Haemorrhage,* if recent – wait until the condition has healed.
- *Operation* in the area, if recent – wait for 6 months.
- *Fracture,* if recent – wait for 6 months.
- *Furuncle* (boil).
- *Inflammation or swelling* of the skin.
- *Scar tissue,* if recent – wait for 6 months.
- *Sebaceous cyst.*
- *Eye disorder*, such as conjunctivitis.

ACTIVITY: RECOGNISING CONTRA-INDICATIONS
Think of *six* skin disorders and *six* skin diseases. List them in a chart.
Briefly describe how you would recognise each skin condition. Why would it be inappropriate to treat each?

EQUIPMENT AND MATERIALS

Beauty salons differ in how much floor space is available and how much of that is allocated to each beauty service. There may be one or several facial-treatment cubicles. In the case of a salon having only one such cubicle, it is important that the range of facial services offered can all be delivered safely and hygienically.

Basic equipment

Each treatment cubicle should have the following basic equipment.

HEALTH & SAFETY: THE COUCH OR CHAIR
The couch should be wiped over with a disinfectant solution after each client, and should be cleaned thoroughly with hot water and detergent at the end of each day.

Treatment couch or beauty chair

The **couch** or **chair** should be covered with easy-to-clean upholstery: it must withstand daily cleaning with warm water and detergent. It must have an adjustable **back rest**, for the comfort of both the client and the therapist. If possible, purchase a couch that also has an adjustable **leg rest**, as this allows treatments such as pedicure to be carried out.

Equipment trolley

The **equipment trolley** should be large enough to accommodate all the necessary equipment and products; trolleys are usually of a two- or three-shelf design. Like the chair or couch, the trolley should be made of a material that will withstand regular cleaning. Some models have restraining bars to prevent objects sliding off the trolley. Drawers are useful in storing tools and small consumables. The trolley should have securely fixed easy-glide castors.

Beauty stool

The **stool** should be covered in a fabric similar to that covering the treatment couch. It may or may not have a back rest; in some designs the back rest is removable. For the comfort of the therapist, it should be adjustable in height; to allow mobility, it should be mounted on castors.

Sorisa

left A treatment couch (with a disposable paper roll)

below left An equipment trolley

below centre A beauty stool

below A magnifying lamp

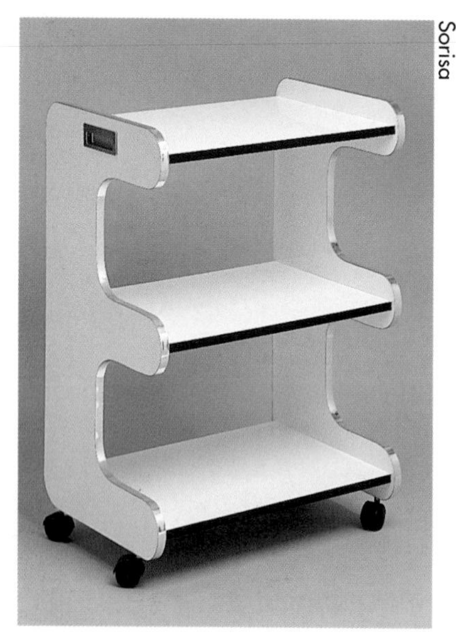

Sorisa

Sorisa

Depilex/RVB

Ellisons

Covered waste bins

TIP

Disposable paper roll may be placed over the surface of the treatment couch if desired. This should be changed for each client.

TIP

Wall-mounted towel racks save storage space.

HEALTH & SAFETY: CONTAINERS

Bottles and other containers should be clean and clearly labelled.

Magnifying lamp

The **magnifying lamp** is available in three models: floor-standing, wall-mounted, and trolley-mounted.

Covered waste bin

A **covered waste bin** should be placed unobtrusively within easy each. It should be lined with a disposable bin-liner. You should also have a 'sharps' box for the disposal of contaminated equipment.

Preparing the cubicle

The following guidelines describe the basic preparation of the **facial treatment cubicle.** Further equipment and materials relevant to other beauty services are discussed within the appropriate chapters.

- **Towel drapes** There should be a *large towel* to cover the client's body, and a *small hand towel* to drape across the client's chest and shoulders.
- **Headband** A clean material headband should be provided for each client.
- **Skin-cleansing preparations** The trolley should carry a display of facial skin-cleansing preparations to suit all skin types.
- **Trolley** The surface of each shelf can be protected with a sheet of 500 mm disposable bedroll.
- **Towel** A clean towel should be placed on the trolley for the therapist to wipe her hands on as necessary.
- **Gown** A clean gown should be provided for each client as necessary.

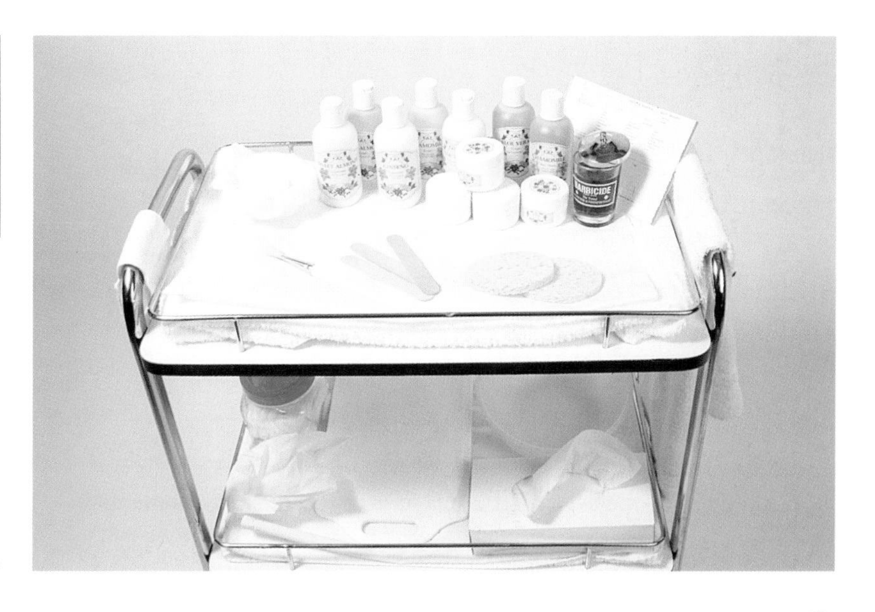

Equipment and materials

TIP

It is a bad practice to have to leave the client in order to fetch more cottonwool. It is also bad practice to prepare too much and be wasteful!

Pre-shaped cottonwool discs are ideal for facial treatments. Alternatively, cut high-quality cottonwool into squares (6 cm × 6 cm).

- **Cottonwool** There should be a plentiful supply of both damp and dry cottonwool, sufficient for the treatment to be carried out. *Dry* cottonwool should be stored in a covered container; *damp* cottonwool is usually placed in a clean bowl.
- **Tissues** Facial tissues should be large and of a high quality. They should be stored in a covered container.
- **Mirror** A clean hand mirror should be available, for use in consulting with the client before, during and after her treatment.
- **Waste bin** A covered container for waste may be placed on the bottom shelf of the trolley.
- **Container for jewellery** A container may be provided in which the client can place her jewellery if she needs to remove it prior to treatment.
- **Spatulas** Several clean spatulas (preferably disposable) should be provided for each client. One should be used in tucking any stray hair beneath the headband. Others will be used in removing products from their containers.

PREPARING THE CLIENT

By the time the client is shown through to the treatment cubicle, the record card will already have been partly filled in at reception. The card should be collected by the beauty therapist, who will add to it during and after the treatment.

In the privacy of the treatment cubicle, carry out the consultation. This takes place when the client first meets the therapist and again whenever a new treatment is to be carried out.

The consultation

The consultation enables the therapist:

- to assess whether the client is suited to treatment or whether treatment is contra-indicated in some way;
- to ask the client specific questions about her present skin-care routine and her general health.

HEALTH & SAFETY: POOR HEALTH

If the client is in poor health, or is taking medication that affects her skin condition, it may not be possible to treat her until the medical condition has been treated by her physician.

The client's answers will indicate to the therapist what is required, and what is achievable, from a skin-care programme.

During the consultation the therapist can also:

- explain what is involved in each treatment, how long it takes, and what aftercare and home care are required (if relevant);
- assess how much the client is willing to spend, and so design a treatment programme within her budget.

The consultation is also helpful to the client. She can:

- discover what the beauty therapist can offer;
- ask questions, and receive honest professional advice concerning the most appropriate choice of skin-care treatment;
- decide how much she is willing to spend.

At the end of the consultation the client should understand fully what the proposed treatment involves.

ACTIVITY: ANSWERING QUESTIONS

You should be able to answer honestly, competently and tactfully any question related to the beauty services you offer. Below are examples of the questions you may be asked at consultation.

- 'I have always used soap and water upon my face. Is this a satisfactory way to cleanse my face?'
- 'Why do I need to use a separate night cream as well as a day moisturiser?'
- 'How often should I have a facial?'
- 'I have extremely oily skin. Why do I need to wear a moisturiser?'
- 'What can I do to treat these fine lines around my eyes?'

What answers would *you* give? Think of further questions you might be asked with regard to skin care. An experienced beauty therapist will be able to advise you.

TIP

Always give clear instructions to your client. This will help to ensure that she is not embarrassed or uncomfortable at any time.

TIP

Check that the headband is comfortable. A tight headband will cause tension and eventually a headache.

Prior to treatment

Depending on the treatment to be carried out, the client may need to remove some clothing. Offer her a **gown** to wear.

1. Position the client on the couch according to the treatment to be given. Cover the client with the large bath towel. If necessary, drape a small hand towel across her shoulders.
2. If facial, neck and shoulder massage is to be given, ask the client politely to remove her arms from her bra straps in preparation: this avoids disturbance later.
3. Fasten a clean headband around the client's hairline. Position the headband so that it does not cover the skin of the face. If using steam, cover the hair to stop it getting damp.
4. After preparing the client, wash your hands: this demonstrates to the client your concern to work hygienically.

CLEANSING

Skin cleansing is essential in promoting and maintaining a healthy complexion. There are various cleansing preparations to choose from; basically their action is the same in each case:

- to exfoliate gently dead skin cells from the stratum corneum, exposing younger cells and improving the skin's appearance;
- to remove make-up, dirt and pollutants from the skin's surface, reducing the possibility of blemishes and skin irritation;
- to remove excess sweat and sebum from the skin's surface, reducing congestion of the skin and the subsequent formation of comedones and pustules;
- to prepare the skin for further treatments.

Cleansing preparations

A cleanser is required that will remove both oil-soluble and water-soluble substances. Oil is capable of dissolving grease; water will dissolve other substances. Usually, therefore, a cleanser is a combination of both oil and water.

Oil and water do not combine: if you simply mix the two together they separate again, with the oil floating on the top of the water. If the two substances are shaken together vigorously, however, one substance will break up and become suspended in the other. The result is known as an **emulsion**.

Emulsions are used in many cosmetic preparations. They are either:

- **oil-in water** (O/W) – minute droplets of oil, surrounded by water;
- **water-in-oil** (W/O) – minute droplets of water, surrounded by oil.

To give the emulsion stability, and to stop it separating out again, an **emulsifier** is added.

Various cleansing preparations are available to the beauty therapist, with formulations designed to suit the different skin types. They include:

- cleansing milks;
- cleansing creams;
- cleansing lotions;
- facial foaming cleansers;
- cleansing bars;
- eye make-up removers.

Whichever cleanser is chosen, it should have the following qualities:

- it should cleanse the skin effectively, without causing irritation;
- it should remove all traces of make-up and grease;
- it should feel pleasant to use;

TIP

Seasonal changes affect the skin. You may need to alter the client's basic skin-care routine through the year.

- it should be easy to remove from the skin;
- ideally, it should be pH-balanced.

The **pH scale** is used to measure the **acidity** or **alkalinity** of a substance. Using a numbered scale of 1–14, acids have a pH less than 7; alkalis have a pH greater than 7. Substances with a pH of 7 are **neutral**.

The skin is naturally slightly acidic: it has an **acid mantle**. Alkalis strip the skin of its protective film of sebum, making it feel dry and taut. To avoid skin irritation it is preferable to use a product that matches the acid mantle, a product whose pH is 5.5–5.6.

The pH scale

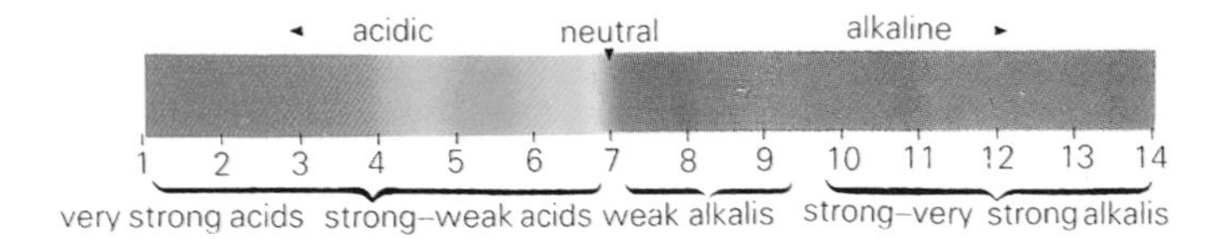

ACTIVITY: ACIDIC OR ALKALINE?

To discover whether a liquid product is acidic or alkaline, carry out a simple test using litmus paper. Litmus changes colour according to the pH:

- litmus paper turns *blue* if an *alkali* is present;
- litmus paper turns *red* if an *acid* is present.

Test the pH of various cleansing preparations.

HEALTH & SAFETY: SENSITIVE SKIN

When treating a sensitive skin, choose a cleansing product that does not contain common known allergens such as mineral oil, alcohol and lanolin. Such products are usually referred to as *hypoallergenic* or *dermatologically tested.*

Cleansing milks

Cleansing milks are usually oil-in-water emulsions, with a relative high proportion of water to oil, making the milk quite fluid in its consistency.

Cleansing milks have these specific treatment uses:

- treating dry skin that is prone to sensitivity;
- treating sensitive skin.

Cleansing creams

Cleansing creams have a relatively high proportion of oil to water, making the emulsion thicker and richer in its consistency than cleansing milks. The high oil content allows the product to be massaged over the skin surface without dragging the tissues. The cream is also more effective in removing grease and oil-based make-up from the skin.

Cleansing creams have these specific treatment uses:

- removing facial cosmetics;
- treating by deep cleansing massage;
- treating very dry skin.

Cleansing lotions

Cleansing lotions are solutions of detergents in water. They do not usually contain oil, and are therefore unsuitable for the removal of facial cosmetics.

Cleansing lotions have these specific treatment uses:

- cleansing a normal to combination skin type;
- treating greasy skin (where a high oil content could aggravate the skin, causing yet further sebum production).

Medicated ingredients may be included in a cleansing formulation: these are only suitable for greasy, congested, spotty skin types.

If the client has a mature, normal or combination skin, a cleansing lotion may not be effective because of the reduced oil content.

Facial foaming cleansers

Facial foaming cleansers usually contain a mild detergent which foams when mixed with water. Additional ingredients are selected for the treatment of different skin types. These cleansers are quick to use and afford a suitable alternative for the client who likes to cleanse her face with soap and water. If the client wears an oil-based make-up, advise her to use a cleansing cream first to remove make-up thoroughly before using this cleanser.

Facial foaming cleansers have general application:

- treating most skin types except very dry or sensitive skin.

TIP

If a client has a very dry skin, advise her to avoid the use of tap water on the face – tap water contains salts and chlorine, which can dry the skin.

Cleansing bars

Although it is efficient as a cleanser, **soap** is usually considered unsuitable for use on the skin. It has an *alkaline* pH, which disturbs the skin's natural *acidic* pH balance. Soap strips the skin of its protective acid mantle, leaving insoluble salts on the skin's surface. The skin may be left feeling itchy, taut and sensitive.

Cleansing bars are a milder alternative to soap, and are specially formulated to match the skin's acidic pH of 5.5–5.6. They are less likely to dry out the skin.

Cleansing bars have this specific application:

- treating greasy to normal skin which is not sensitive.

TIP

Cleansing rinse-off formulations are very popular with male clients.

Eye make-up remover

Eye tissue is a lot finer than the skin on the rest of the face. It readily puffs if aggravated by oil-based cleansing preparations, and becomes very dry if harsh cleansing preparations are used.

To remove make-up from this area, use an **eye make-up remover**. This product cleanses the eyelid and lashes gently emulsifying the make-up. It also conditions the delicate skin. Formulated as a lotion or a gel, it is designed to remove either water-based or oil-based products (or both) from the eye area.

Oily eye make-up removers have these specific treatment uses:

- treating clients who wear waterproof mascara;
- removing wax or oil-based eyeshadow.

Non-oily eye make-up removers have these applications:

- treating clients with sensitive skin around the eyes;
- treating clients who wear contact lenses;
- treating clients who wear individual false eyelashes.

Cleansing treatment

There are two manual processes involved in the cleansing routine: the *superficial* and the *deep* cleanse.

The superficial cleanse uses lightweight cleansing preparations to emulsify surface make-up, dirt and grease. This is followed by the more thorough deep cleanse, in which a heavier cleansing cream is applied to the face. The high percentage of oil contained in the cream formulation allows the cream to be massaged over the skin's surface without evaporation of the product.

TIP

Unlike the muscles of the body, which attach to bones, most of the facial muscles are attached to the facial skin itself. You should therefore avoid stretching the skin unnecessarily – if you do, you may also stretch the facial muscles and contribute to premature ageing.

Superficial cleansing

Each part of the face requires a special technique in the application and removal of the cleansing product. The face is cleansed in the following order:

- the eye tissue and lashes;
- the lips;
- the neck, chin, cheeks, and forehead.

HEALTH & SAFETY: EYE CARE

Never apply pressure over the eyeball when cleansing the eye area.

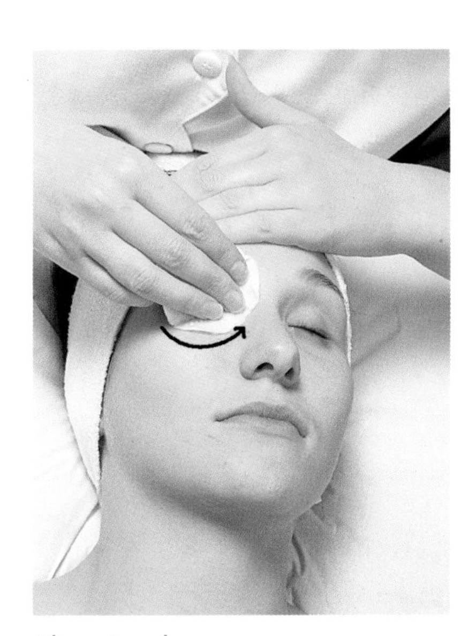

Cleansing the eye area

Sequence for superficial cleansing

1 Wash your hands.

2 Cleanse the eye area, using a suitable eye make-up remover. Each eye is cleansed separately. Your non-working hand lifts and supports the eye tissue whilst the working hand applies the eye make-up remover.

   If a water-based eye make-up remover is used, this is applied directly to a clean piece of cottonwool. Stroke down the length of the eyelashes, from base to points. Next, cleanse the eye tissue in a sweeping circle, outwards across the upper eyelid, circling beneath the lower lashes towards the nose. Repeat, regularly changing the cottonwool until the eye area and the cottonwool show clean.

Sometimes a cleansing milk is used to remove eye make-up. In this case, apply a little of the product to the back of one hand. The ring finger is then used to apply the cleansing milk to the lashes.

Use damp cottonwool to remove the emulsified product.

Repeat the cleansing process until the eye area is clean.

TIP

The ring finger is always used as it provides the least pressure.

TIP

If there is any make-up left at the base of the lower lashes after eye cleansing, this may be removed with a cotton bud or a thin piece of clean cottonwool. Ask the client to look upwards, and gently draw the cottonwool along the base of the lower lashes, towards the nose.

HEALTH & SAFETY: HYGIENE

Never use the reverse side of the cottonwool pad – this is unhygienic.

3 Cleanse the lips, preferably with a cleansing milk or lotion (as this readily emulsifies the oils or waxes contained in lipstick).

Apply a little of the product to the back of your non-working hand. Support the left side of the client's mouth with this hand. With the working hand, apply the product in small circular movements across the upper lip, from left to right; and then across the lower lip, from right to left.

Remove the cleanser from the lips. Support the corner of the mouth; using a clean damp piece of cottonwool wipe across the lips.

Repeat the cleansing process as necessary, until the lips and the cottonwool show clean.

TIP

When cleansing, be careful that cleanser does not enter the eyes or mouth.

Cleansing the lips

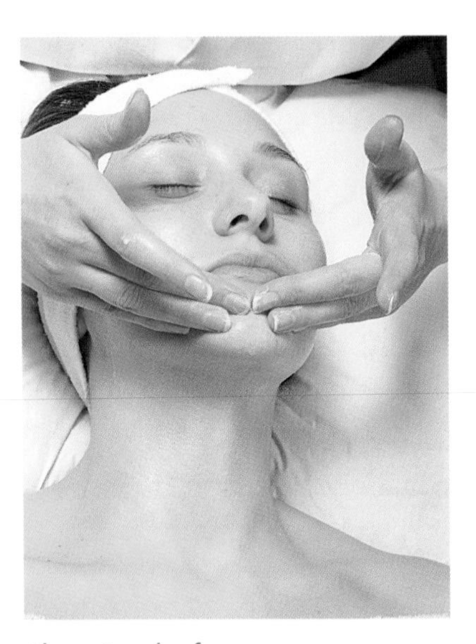

Cleansing the face

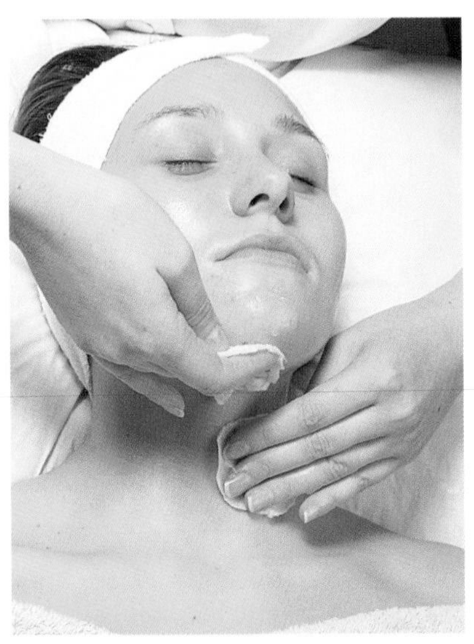

Removing cleanser

4 Select a cleansing milk or lotion to suit your client's skin type.

Pour the product into one hand – sufficient to cover the face and neck, and to massage gently over the surface of the skin. Massage the surface of the hands together: this warms the product (so that it isn't cold on the client's skin) and distributes it over your hands.

Clasp the fingers together at the base of the neck, and

HEALTH & SAFETY: SAFE PRESSURE

Reduce pressure when working over the *hyoid bone* and bony prominences such as the *zygomatic bone* (the cheekbone) and the *frontal bone* (the forehead).

Pressure must always be *upwards* and *outwards*, to avoid stretching the tissues.

HEALTH & SAFETY: CLIENT COMFORT

When cleansing around the nose, avoid restricting the client's nostrils.

TIP

If the client is wearing facial make-up, it is usual to perform the superficial cleanse twice.

unlink them as you move up the neck.

Clasp the fingers together again at the chin, drawing the fingers outwards to the angle of the jawbone.

Stroke up the face, towards the forehead, with your fingertips pointing downwards and your palms in contact with the skin.

Using a series of light circular movements of your fingertips, gently massage the product into the skin, beginning at the base of the neck and finishing at the forehead.

Remove the cleanser thoroughly with clean damp cottonwool, simultaneously stroking over the skin surface, upwards and outwards in a rolling motion. Repeat this process as necessary, using clean cottonwool each time.

Deep cleansing

The deep cleanse involves a series of massage manipulations which reinforce the cleansing achieved with the cleansing product. Blood circulation is increased to the area: this has a warming effect on the skin which relaxes the skin's natural openings, the hair follicles and pores. This aids the absorption of cleanser into the hair follicles and pores, where it can dissolve make-up and sebum.

There are various deep-cleansing sequences; all are acceptable if carried out in a safe, hygienic manner, and all can achieve the desired outcomes. Here is one sequence for deep cleansing.

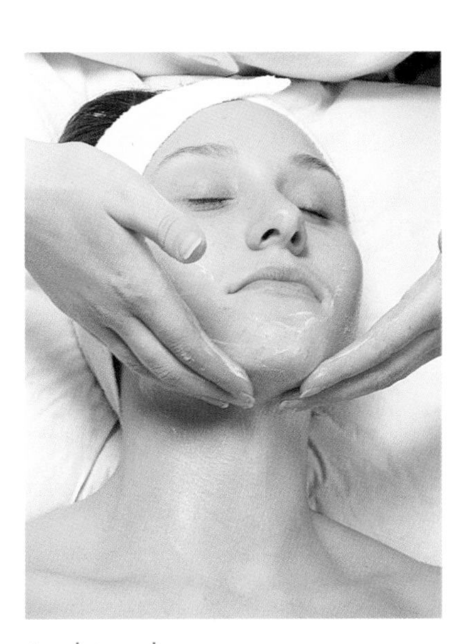

Applying cleanser

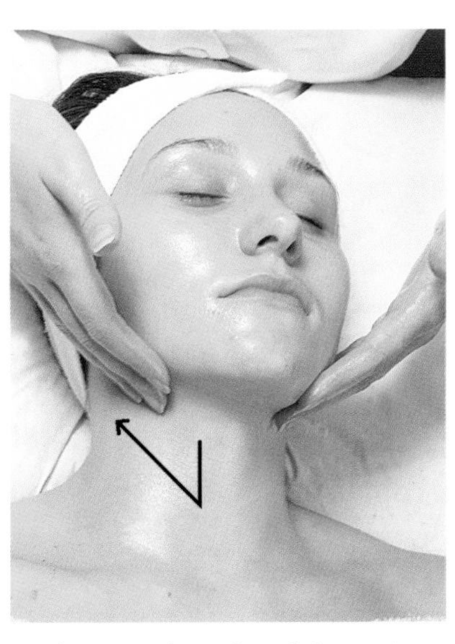

Stroking up the sides of the neck

1. Select a cleansing cream to suit your client's skin type. The procedure for application is the same as that for the superficial cleanse.
2. Stroke up either side of the neck, using your fingertips. At the chin, draw the fingers outwards to the angle of the jaw, and lightly stroke back down the neck to the starting position.
3. Apply small circular manipulations over the skin of the neck.

4 Draw the fingertips outwards to the angle of the jaw. Rest each index finger against the jawbone (you will be able to feel the lower teeth in the jaw). Place the middle finger beneath the jawbone. Move the right hand towards the chin where the index finger glides over the chin; return the fingers, beneath the jawbone, to the starting position. Repeat with the left hand.

Repeat step **4** *a further 5 times.*

5 Apply small circular manipulations, commencing at the chin working up towards the nose, and finishing at the temples. Slide the fingers from the temples back to the chin, and repeat.

Repeat step **5** *a further 5 times.*

6 Position the ring finger of the right hand at the bridge of the nose. Perform a running movement, sliding the ring, middle and index fingers off the end of the nose. Repeat immediately with the left hand.

Repeat step **6** *a further 5 times with each hand.*

7 With the ring fingers, trace a circle around the eye orbits. Begin at the inner corner of the upper browbone; slide to the outer corners of the browbone, around and under the eyes, and return to the starting position.

Repeat step **7** *a further 5 times.*

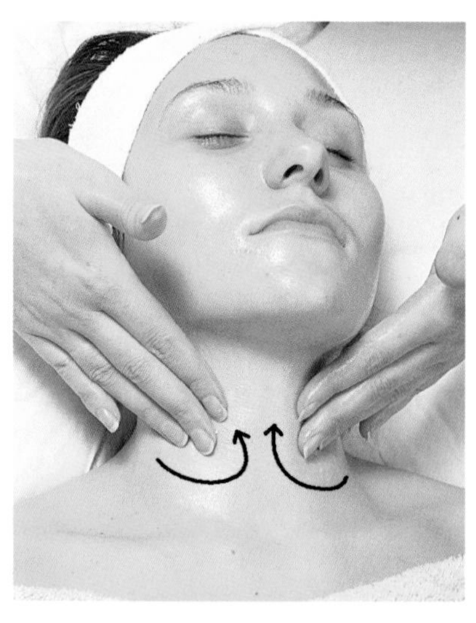

Circular manipulations of the neck

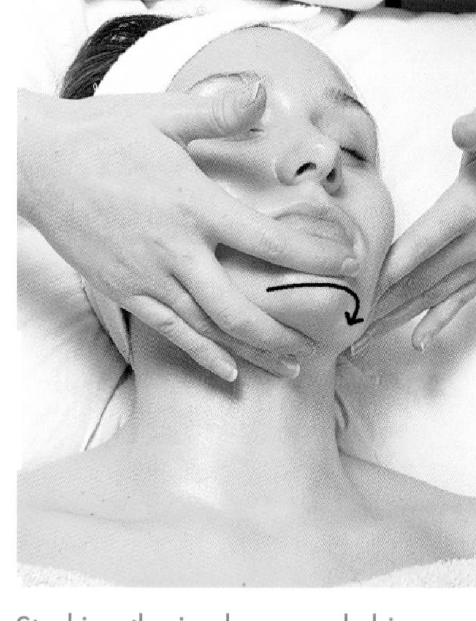

Stroking the jawbone and chin

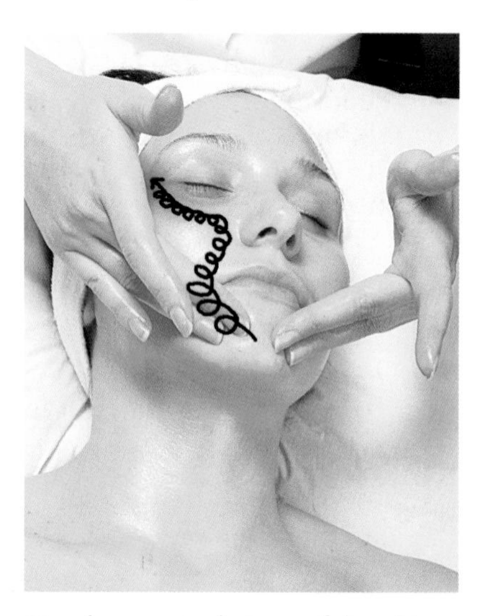

Circular manipulations of the chin, nose and temples

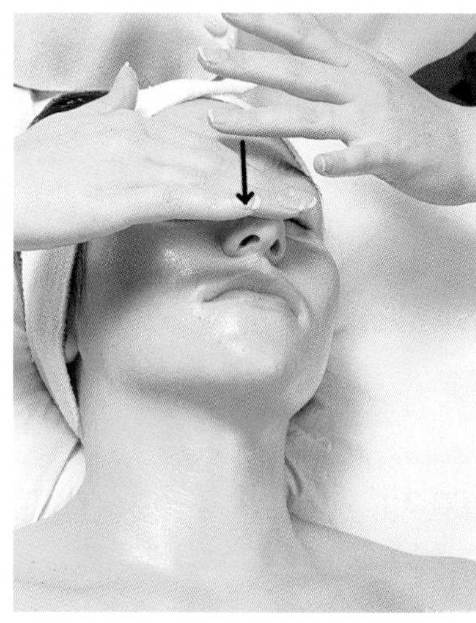

Running movement on the nose

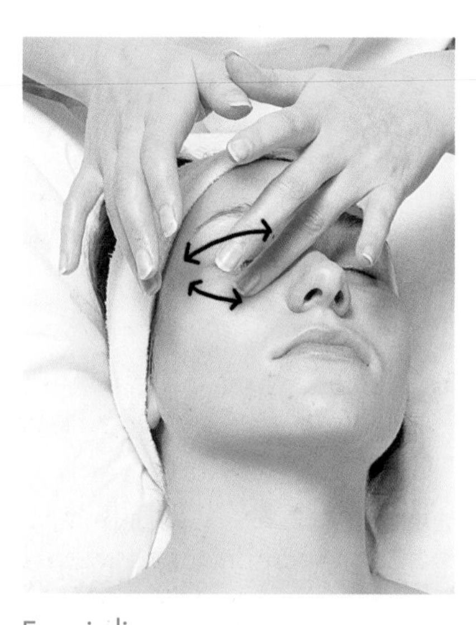

Eye circling

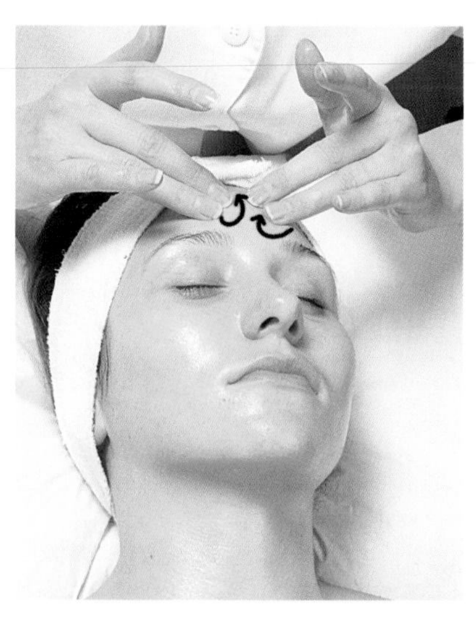

Circling manipulations of the forehead

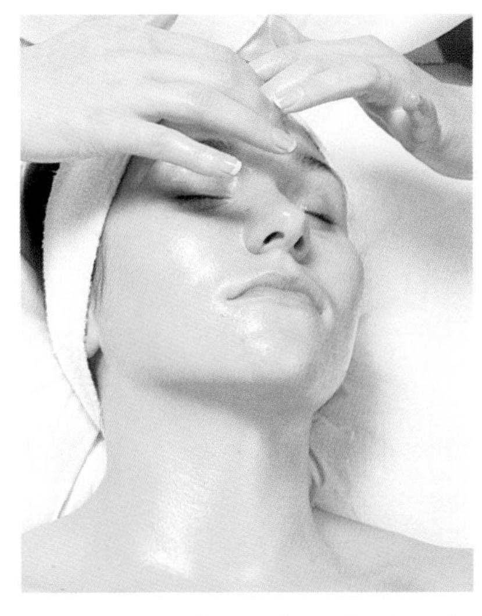

Crisscross stroking of the forehead

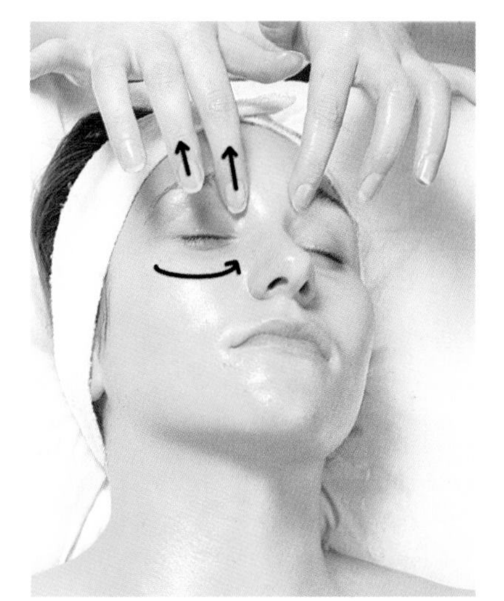

Lifting the eyebrows

8 Using both hands, apply small circular manipulations across the forehead.
Repeat step **8** *a further 5 times.*

9 Open the index and middle fingers of each hand and perform a crisscross stroking movement over the forehead.

10 Slide the index finger upwards slightly, lifting the inner eyebrow. Lift the centre of the eyebrow with the middle finger. Finally, lift the outer corner of the eyebrow with the ring finger. Slide the ring fingers around the outer corner and beneath the eye orbit.
Repeat step **10** *a further 5 times.*

11 With the pads of each hand, apply slight pressure at the temples. This indicates to the client that the cleansing sequence is complete.

12 Remove the cleansing cream from the skin, using damp cottonwool.

TIP

Check that the brow hair and the skin beneath the chin are free of grease: it is easy to overlook some cleansing product in these areas.

TONING

After the skin has been cleansed it is then toned with an appropriate lotion.

Toning preparations

Toning lotions remove from the skin all traces of cleanser, grease and skin-care preparations. The toning lotion's main action is as follows.

- It produces a cooling effect on the skin when the water or alcohol in the toner evaporates from the skin's surface. (When a liquid evaporates it changes to a gas, which takes energy. In the case of toner, the energy is taken from the skin, which therefore feels cooler.)

- It creates a tightening effect on the skin, because of a chemical within the toner called an astringent. This causes the pores to close, thereby reducing the flow of sebum and sweat onto the skin's surface.
- It helps to restore the acidic pH balance of the skin. Milder skin toners have a pH 4.5–4.6; stronger astringents disturb the pH more severely, and may cause skin irritation and sensitivity.

There are three main types of toning lotions, the difference being largely the amounts of alcohol they contain. They include:

- bracers and fresheners;
- tonics;
- astringents.

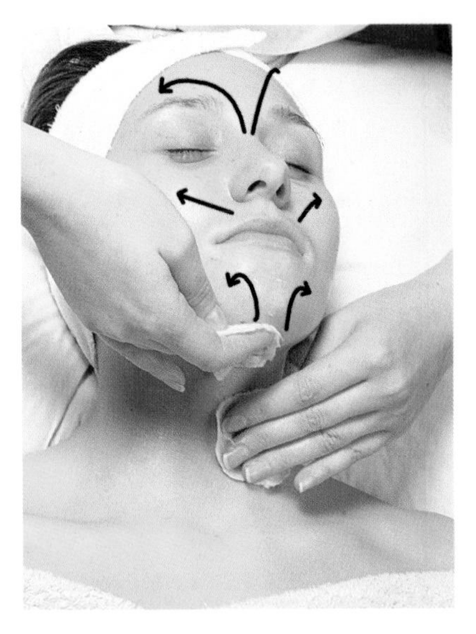

Toning

Skin bracers and fresheners

Skin bracers and **skin fresheners** are the mildest toning lotions: they contain little or no alcohol. They consist mainly of purified water, with floral extracts such as **rose water** for a mild toning effect.

Skin bracers and fresheners are recommended for:

- dry, delicate skin;
- sensitive skin;
- mature skin.

TIPS

All toning lotions have *some* astringent effect on the skin, but those that contain a relatively high alcohol content are actually marketed as astringents.

Do not use toning lotions that contain more than 20 per cent alcohol on a dry skin – they may cause skin irritation.

Avoid the excessive use of astringent on a greasy skin – the astringent will make the skin dry, and it will then produce more sebum.

Skin tonics

Skin tonics are slightly stronger toning lotions. Many contain a little of some astringent agent such as witchhazel.

Skin toners are recommended for:

- normal skin.

Astringents

Astringents are the strongest toning lotions; they have a high proportion of alcohol. They may contain antiseptic ingredients such as **hexachlorophene**; these are for use on a blemished skin, to promote skin healing.

Astringents are recommended for:

- oily skin with no skin sensitivity;
- mild acne in young skin.

Application

Toning lotion may be applied in several ways. Whichever method you choose, it should leave the skin thoroughly clean and free of grease.

The most popular method of application is to apply the toner directly to two pieces of clean damp cottonwool, which are wiped gently upwards and outwards over the neck and face.

Alternatively, the toner may be applied under pressure as a fine

TIP

When applying toning lotion to a combination skin, you may need to apply different toning lotions to treat separate skin conditions.

spray, using a vaporiser. This produces a fine mist of the toning lotion over the skin. If using this method, always protect the eye tissue with cottonwool pads and hold the vaporiser about 30 cm from the skin, directing the spray across the skin in a sweeping movement.

To produce a stimulating effect, the toning lotion can be applied to dampened cottonwool: hold this firmly at one corner, and gently tap it over the skin.

After applying toning lotion, immediately blot the skin dry with a soft facial tissue to prevent the toner evaporating from the skin's surface (which would stimulate the skin).

TIP

Before facial application it is a good idea to spray the mist onto the back of the client's hand: this helps her to relax.

TIP

For home use, advise the client to apply toning lotion using dampened cottonwool: this is more economical.

Blotting the skin

Make a small tear in the centre of a large facial tissue, for the client's nose. Place the tissue over her face and neck, and mould it into position to absorb excess moisture.

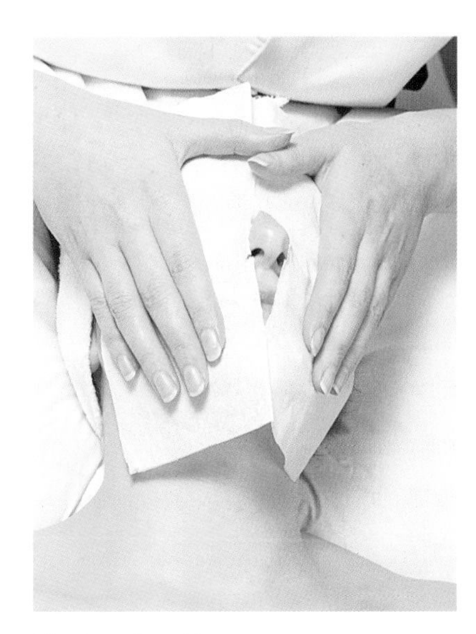

Blotting the skin

HEALTH & SAFETY: CLIENT COMFORT

To avoid claustrophobia and discomfort, tell the client before you start how and why facial blotting is carried out.

MOISTURISING

The skin depends on water to keep it soft, supple and resilient. Two-thirds of our body is composed of water and the skin is an important reservoir, containing about 20 per cent of the body's total water content. Most of the fluid is in the lower layers of the dermis, but it circulates to the top layer of the epidermis, where it evaporates.

The skin protects its water content in these ways:

- sebum keeps the skin lubricated, and reduces water loss from skin;
- the skin cells have **natural moisturising factors (NMFs)**, a complex mix of substances which are able to fix moisture inside the cells;
- a 'cement' of fats (lipids) between the skin cells forms a watertight barrier.

The natural moisture level is constantly being disturbed. The application of a cosmetic **moisturiser** helps to maintain the natural oil and moisture balanced by locking moisture into the tissues, offering protection and hydration.

The basic formulation of a moisturiser is oil and water to make an oil-in-water emulsion. The water content helps to return lost moisture to the surface layers; the oil content prevents moisture loss from the surface of the skin. Often a **humectant**, such as **glycerine** or **sorbitol**, is included: this attracts moisture to the skin from the surrounding air and stops the moisturiser from drying out. If a humectant is included, less oil is used in the formulation: this results in a lighter cream.

TIP

Even a greasy skin requires a moisturiser. This skin can become dehydrated by the over-use of harsh cleansers and astringents.

Moisturiser also has the following benefits:

- it protects the skin from external damage caused by the environment;
- it softens the skin and relieves skin tautness and sensitivity;

- it plumps the skin tissue with moisture, which minimises the appearance of fine lines;
- it provides a barrier between the skin and make-up cosmetics;
- it may contain additional ingredients which improve the condition of the skin (such as vitamin E, which has a humectant action and is an excellent skin conditioner);
- it may contain ultra-violet filters, which protect the skin against the age-accelerating sunlight.

Moisturisers are available for wear during the day or the night. These are available in different formulations, to treat all skin types and conditions.

TIP

If your client likes a natural look for the day, she may wish to wear a moisturiser that is tinted; this gives the skin a healthy appearance.

TIP

Even on an overcast day, as much as 80 per cent of the sun's age-accelerating UVA can penetrate the skin.

Moisturisers for daytime use

Moisturising lotions

Moisturising lotions contain up to 85–90 per cent water and 10–15 per cent oil. They have a light, liquid formulation, and are ideal for use under make-up.

Moisturising lotions have these specific applications:

- greasy skin;
- young combination skin;
- dehydrated skin;
- normal skin.

Moisturising creams

Moisturising creams contain up to 70–85 per cent water and 15–30 per cent oil. They have a thicker consistency, and cannot be poured.

Moisturising creams have these specific applications:

- mature skin;
- dry skin.

TIP

Some clients dislike the heavier cream, feeling that it is too heavy for their skin. Offer them a suitable moisturising lotion alternative.

HEALTH & SAFETY: ALLERGIES

Hypoallergenic moisturisers are available for clients with sensitive skin. These are screened from all common sensitising ingredients, such as lanolin and perfume, and they also have soothing properties.

ACTIVITY: MOISTURISERS

Collect information on different moisturisers from various skin-care suppliers. You could visit local beauty salons, retail stores or beauty wholesale suppliers, or write to professional skin-care companies.

Application

Moisturiser is applied after the final application of toning lotion. If the moisturiser is being applied before make-up, use a light formulation so that it does not interfere with the adherence of the foundation.

1 Remove some moisturiser from the jar, using a disposable spatula or sanitised plastic spatula. Place it on the back of the non-working hand, then take it on the fingertips of your working hand.
2 Apply the moisturiser in small dots to the neck, chin, cheeks, nose and forehead. Quickly and evenly spread it in a fine film over the face, using light upward and outward stroking movements.
3 Blot excess moisturiser from the skin using a facial tissue.

Applying moisturiser

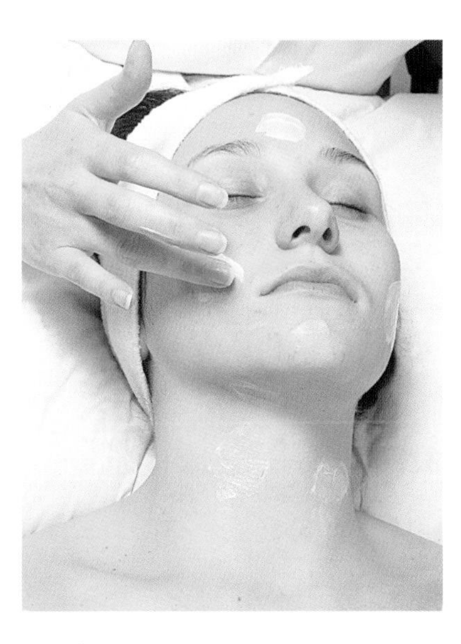

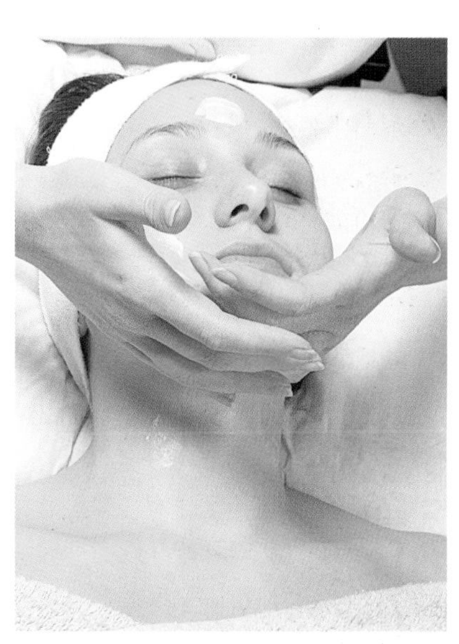

TIP

If the moisturiser is very fluid, apply it directly to the fingertips of the non-working hand – it would run if applied to the back of the hand.

Moisturisers for night-time use

Moisturisers are applied to the skin in the evening, after the skin has been cleansed, toned and blotted dry.

An emulsion **night cream** with a higher proportion of oil is the most effective for application in the evening: by this time the surrounding air is dry and warm, which encourages water loss from the skin; the oil seals the surface of the skin, preventing this water loss.

A small amount of **wax** (such as beeswax) may be included in the formulation: this improves the *'slip'* of the product, making it easier to apply and helping its skin-conditioning effect.

Throat creams

The neck can become dry as it is exposed to the weather, often without the protection of a moisturiser. As a client ages, the collagen molecules in the dermis become increasingly cross-linked; they are then unable to retain the same volume of water, and the skin loses its plump appearance.

The formulation for a **throat cream** is similar to that for a night cream; it also contains various skin conditioning supplements, such as collagen or vitamin E, which help maintain moisture in the stratum corneum.

Encourage your client to include the neck in her cleansing routine, applying facial moisturiser to the neck during the day and either a night cream or specially formulated throat cream in the evening. Recommend that she always applies the throat cream gently in an upward and outward direction, using her fingertips.

To improve the appearance of the neck, good posture is important. If the client is round-shouldered the head often drops forwards, putting strain on the muscles of the neck. This causes tension in the muscles, which become tight and painful. Correct the client's posture, and advise her to massage the neck when applying the throat cream to relieve tension.

TIP

Advise the client against losing weight quickly, especially when older – the neck tissue can look loose and very wrinkled as the underlying fat is lost.

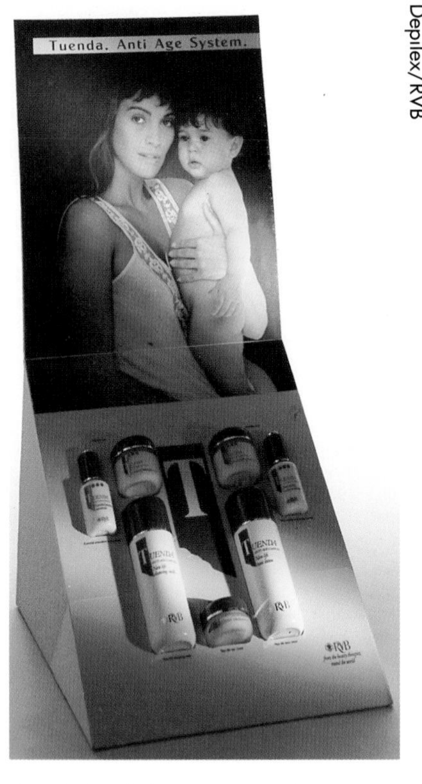

Depilex/RVB

Specialised skin-care products

Eyecreams

The eye tissue is very thin and readily becomes very dry, emphasising fine lines and wrinkles (**crow's feet**). Special care must be taken when applying products near this area: it contains a large number of **mast cells**, the cells that respond to contact to an irritant by causing an allergic reaction.

Eye cream is a fine cream formulated specifically for application to the eye area. A small quantity of the product is applied to the eye tissue using the ring finger of one hand, gently stroking around the eye, inwards and towards the nose. Support the eye tissue with the other hand. Do not apply the product too near to the inner eyelid or you will cause irritation to the eye.

Further skin-care preparations

Eye gel

Eye gel is usually applied in the morning: it has a cooling, soothing, slightly astringent effect. (This is caused partly by the evaporation of the water in the gel, and partly by the inclusion of plant extracts such as **cornflower** or **camomile**.)

Eye gel is recommended for all clients, but especially for those suffering with slightly puffy eye tissue. It may also be applied following a facial treatment, to normalise the pH of the skin.

Eye gel may be applied with a light tapping motion, using the pads of the fingers. This will mildly stimulate the lymphatic circulation in the area and help to reduce any slight swelling.

ACTIVITY: EYE CONDITIONS

Think of different *non-medical* eye conditions for which a client might seek your advice. Discuss with colleagues the possible cause of these conditions, and what you could recommend to improve the appearance in each case.

Depilex/RVB

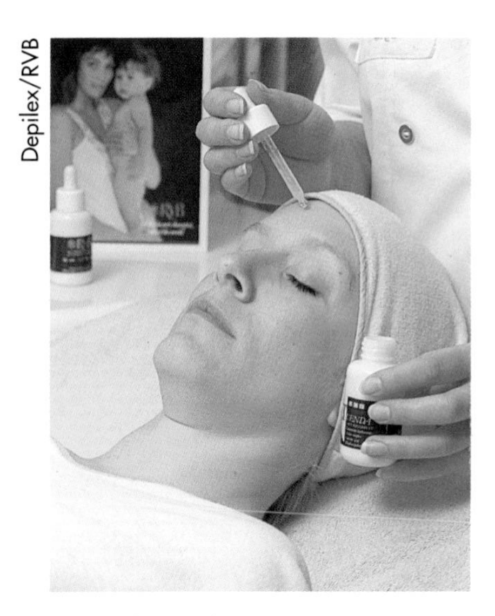

Ampoule application

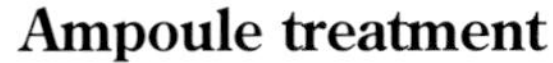

Ampoule treatment

Serums are chemicals used to revitalise the skin. They are supplied in **ampoules**, sealed glass or plastic phials which prevent the content from evaporating and losing their effectiveness. Serums are usually applied for 7–28 days as a skin-tonic course for the treatment of different skin types and conditions.

REMOVING SKIN BLOCKAGES

After the skin has been cleansed, you may wish to remove minor skin blemishes such as comedones (blackheads) and milia (whiteheads). It is preferable to warm the tissues first: this softens the skin and relaxes the openings of the skin which are blocked.

> **HEALTH & SAFETY: SKIN BLOCKAGES**
> Do not attempt to remove larger skin blockages such as sebaceous cysts – these should be treated by a general practitioner.

Warming the skin

Steam is the ideal means of producing the required warming effect on the skin to achieve both cleansing and stimulation. Skin warming is often incorporated into a facial treatment after the manual cleansing, so as to stimulate the skin and make it more receptive to subsequent treatments.

The effects are these:

- the pores are opened;
- locally the blood circulation and the lymphatic circulation are stimulated;
- the surface cells of the epidermis are softened, which helps desquamation;
- sebaceous gland activity is improved, which benefits a dry, mature skin type;
- skin colour is improved.

Steam is provided by an electric **vapour unit**. In this, distilled water is heated electrically.

Sorisa

A vapour unit

> **HEALTH & SAFETY: VAPOUR APPLICATION**
> Keep the vapour directed away from the client's face until a visible jet of steam can be seen. To avoid skin sensitisation, consider carefully where to position the steam so as to ensure even heat distribution.

The resulting steam is applied as a fine mist over the facial area. As the steam settles upon the skin it is absorbed by the surface epidermal cells. These cells are softened and can be gently loosened with exfoliation treatment.

Contra-indications

Although the treatment is suitable for most clients, do not use steam if you discover that the client has any of the following:

- *Respiratory problems*, such as asthma or a cold.
- *Vascular skin disorders*, such as acne rosacea.
- *Claustrophobia.*
- *Excessively dilated arterioles.*
- *Skin with reduced sensitivity.*
- *Diabetes*, unless the client's GP has given permission.

Application

The duration of the application and the distance differ according to the skin type.

> **HEALTH & SAFETY: STEAM APPLICATION**
> A greasy skin will tolerate a shorter application distance and a longer application time. For a sensitive skin type, increase the application distance and reduce the time. What are the manufacturer's guidelines for your equipment? Remember that these are only *guidelines* – observe the skin's reaction and check that the client is comfortable.

Before applying steam, protect the client's eyes with damp cottonwool. Areas of delicate skin should be protected with damp cottonwool and if necessary a barrier cream.

Explain to the client:

- for how long the treatment is to be applied;
- the sensations that will be experienced;
- the physical effect on the skin.

If you are using ozone, this is applied following the steam application for the final few minutes, as directed by the manufacturer. The steam will change in appearance to a bluish-white cloud.

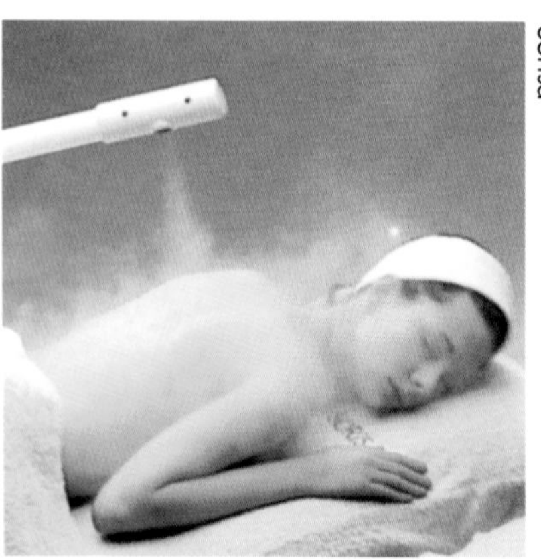

Sorisa

Applying steam to the back

> **HEALTH & SAFETY: OZONE**
> Most vapour units produce ozone when the oxygen in the steam is passed over a high-intensity quartz mercury arc tube. Ozone is thought to be beneficial in the treatment of a blemished skin, as it kills many bacteria, but in excess it may also be carcinogenic (liable to cause cancer) if inhaled. Use a vapour unit only in a well-ventilated room, and only for short periods of time.

After applying steam vapour, blot the skin dry with a soft facial tissue and proceed to remove any blockages.

At the end of the treatment, turn off the machine and unplug it. Check that you have tidied away the trailing lead so that there is no risk of it causing an accident.

Contra-actions

Contra-actions to steaming include the following:

- *over-stimulation of the skin*, caused by incorrect application distance and duration of the steam;
- *scalding*, caused by spitting from a faulty steam jet or by the vessel being over-filled;
- *discomfort*, caused by the steam being too near the skin, leading to breathing difficulties, or by the treatment being applied for too long.

ACTIVITY: VAPOUR UNITS

Collect literature on different vapour units. Compare their efficiencies and features. Which would be the best buy? Consider:

- Is the unit transportable, for marketing demonstrations?
- Is it height-adjustable, to suit the height of the treatment couch?
- Is it easy to clean?
- If floor-standing, does it move easily?
- Does it allow the addition of aromatic oils?
- Has it safety features to prevent overheating or the vessel running dry?

Towel steaming

Towel steaming is an alternative to steaming which can be used if an electrical vapour unit is not available to achieve similar beneficial effects. Several clean small towels are required: these are heated in a bowl of clean hot water and are then applied to the face.

TIP

Provided that you have received instruction in the use of aromatic oils, and provided that the client has no contra-indications to them, you may add oils to the water.

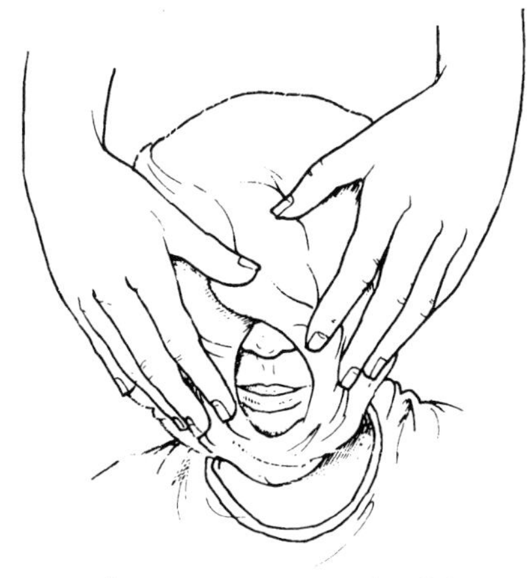

Towel steaming: one side of the towel should overlap the other

Application

Seat the client in a semi-reclined position. Neatly fold a small clean towel and immerse it in very warm water. Wring it out quickly and, standing behind the client, transfer it to her face – with the towel folded in half, place it over the lower half of the face, directly under the client's lower lip; then unfold it to cover the upper face, leaving the mouth and nostrils uncovered.

Press the towel gently against the face for 2 minutes; during this time the towel will begin to cool. Remove the towel and replace it with another heated towel. Continue in this way, heating and replacing the towels, for approximately 10 minutes.

HEALTH & SAFETY: TREATMENT PROGRAMME
If the client suffers from severe congestion, do not attempt to carry out all the removals in one session. This would sensitise the skin, making it appear very red, and would be most uncomfortable for the client. Instead she should visit the salon weekly for you to clear the skin gradually as part of an overall treatment programme.

Equipment and materials

You will need the following equipment and materials:

- *Disposable rubber gloves.*
- *Medical swabs*, impregnated with isopropyl alcohol or antiseptic lotion.
- *Facial tissues* – soft and white.
- *Stainless steel comedo extractor* and a *disposable individual milium extractor needle* (sterile), or combined *stainless steel comedo and milium extractor.*
- *'Sharps' disposal container* – for the contaminated needle.

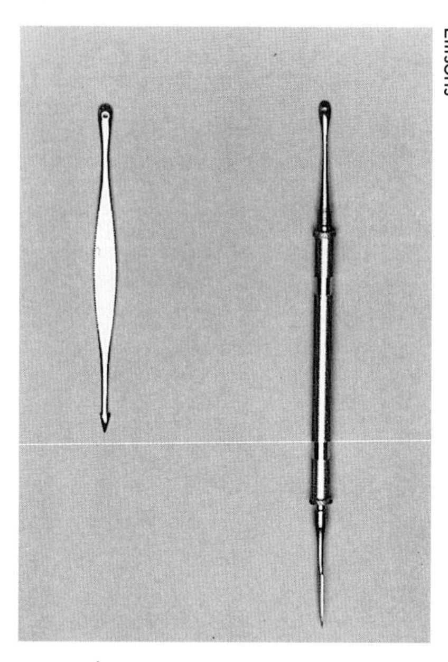

Ellisons

Comedo extractors

Sterilisation and sanitisation

You must put the disposable milium extractor in a 'sharps' container. All waste material from this treatment (such as facial tissues and gloves) should be disposed of in an identified waste container, as directed by your local health authority.

After use the stainless steel comedo extractor should be cleaned with an alcohol preparation and then sterilised in an autoclave.

Wear disposable rubber gloves whilst carrying out the treatment.

Treatment

Comedo removal

Using the loop end of the extractor, apply gentle pressure around the comedo. The comedo should leave the skin, apparent as a plug. You may need to apply gentle pressure with your fingers at the sides of the comedo to ensure that it is effectively removed; when doing this, wrap a tissue around the pads of the index fingers.

Contra-actions:

- Skin bruising could occur if too much pressure is applied.
- Capillary damage could result if too much force is used when squeezing the comedo. The surrounding blood capillaries can rupture, causing permanent skin damage.

Milium extraction

HEALTH & SAFETY: CLIENT COMFORT
- Never obstruct the client's nostrils when removing a comedo from the nose area.
- Never apply pressure on the soft cartilage of the nose.

HEALTH & SAFETY: AVOIDING INFECTION
Ensure that *all* contents of the skin blockage are removed, or infection may occur.

Milium extraction

Hold the point of the extractor parallel with the skin's surface, and *superficially* pierce the epidermis. This makes an opening through which the sebaceous matter can pass to the skin's surface. Using either the comedo extractor or tissue wrapped around the index fingers, apply gentle pressure. A mild antiseptic soothing lotion applied after extraction will help the skin to heal.

EXFOLIATION

The natural physical process of losing dead skin cells from the stratum corneum layer of the epidermis is called **desquamation**. **Exfoliation** is a salon technique used to accelerate this process. It is normally carried out after the skin has been cleansed and toned, and before further facial treatments.

Exfoliation has the following benefits:

- dead skin cells, grease and debris are removed from the surface of the skin;
- fresh new cells are exposed, improving the appearance of the skin;
- skin preparations such as moisturising lotions are more easily absorbed;
- the blood circulation in the area is mildly stimulated, bringing more oxygen and nutrients to the skin cells and improving the skin colour.

TIP
This treatment should be strongly recommended for the mature client. The removal of the surface dead cells has a rejuvenating effect on the skin's appearance.

TIP
A man who shaves exfoliates every day, as shaving removes the top layer of skin. You can recommend that he also use a cosmetic exfoliating product on areas such as the nose and forehead.

ACTIVITY: EXFOLIANTS
Apply an exfoliant to the back of your own – or a colleague's – hand or arm. Compare the appearance and the feeling of the skin before and after application.

This technique produces a particularly marked effect on black skin, as it reduces the greyish appearance of the skin. (This is due to the skin having a thicker stratum corneum.)

Exfoliants

Various exfoliants are available; they may be of chemical or vegetable origin. Alternatively, mechanical exfoliation may be used.

Biochemical skin peel

Natural acids (alpha-hydroxy acids – AHAs), derived from fruits, sugar cane and milk, are applied to the skin as a face mask. The natural acids dissolve dead surface cells and stimulate circulation in the underlying skin. These masks are available to suit all skin types.

When you apply this type of face mask, warn the client that there will be a stinging sensation and then a tightening effect as the mask sets.

Pore grains

Pore grains are the most popular exfoliants: a base of cream or liquid containing tiny spheres of polished plastic or crushed nuts is gently massaged over the skin's surface.

TIP

Steaming before exfoliation softens the outer skin cells, making them easier to remove.

Clay exfoliants

Gentler **clay exfoliants** have a clay base which is applied like a face mask. As it dries, the clay absorbs dead skin cells and sebum. The mask is then gently stroked away, using the pads of the fingers.

HEALTH & SAFETY: EXFOLIANTS

Avoid exfoliating products that contain sharp grains of nut shells: these can scrape, split and damage the epidermis. Before purchasing, test the exfoliating product on the back of your own hand to feel its action.

Mechanical exfoliation

Mechanical exfoliation, or 'facial brushing', softens and cleanses the skin. Dead skin cells and excess sebum are removed as the soft hair bristles rotate over the skin's surface. The rotary action also increases the cleansing action of exfoliation.

If steam is applied before mechanical exfoliation, this will soften the dead skin cells, and **skin peeling cream** may be applied: together these will maximise the result of exfoliation.

Be careful to avoid over-stimulation, sensitising the skin's surface, or disturbing the skin's natural protective qualities.

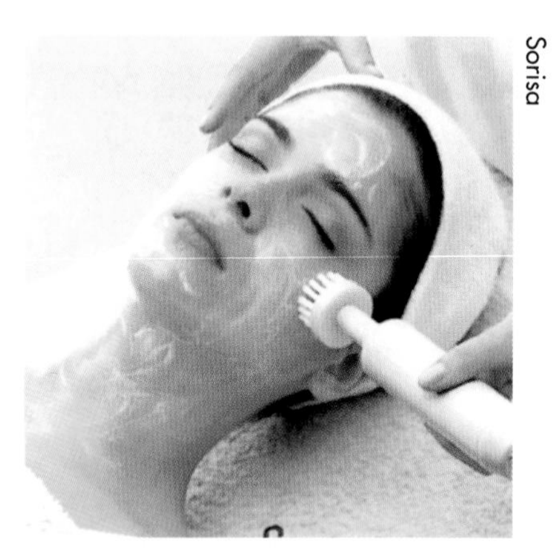
Sorisa

Mechanical exfoliation

Advice on home use

The client can be advised to use exfoliants at home as a specialised cleansing treatment after normal cleansing and toning. Exfoliants should be applied once a week for all skin types except greasy, for which it may be applied twice a week.

Advise the client to massage the product gently over the skin using her fingertips. Application should always be upwards and outwards.

After application, any product residue should be thoroughly rinsed from the face using clean, tepid water. The client may then apply a face mask or tone the skin and apply a nourishing skin moisturiser.

TIP

Point out to clients that exfoliants designed for use on the *body* are unsuitable for use on the *face* – their action is not as gentle.

HEALTH & SAFETY: EXFOLIANTS

- Tell the client how the skin should look after exfoliation treatment. The skin's colour should be *slightly* heightened; but too vigorous a massage application may cause the formation of broken capillaries.
- Tell the client to avoid contact with the delicate eye tissue.
- A client with a pustular skin should not use exfoliant products – they would probably cause discomfort, and any lesions present might burst.

TIP

To encourage sales of this skin-care product, demonstrate the product on the back of one of the client's hands. She will be amazed when she compares the appearance of her hands.

TIP

The term *face pack* is sometimes used instead of *face mask*: a face pack does not set, but remains soft; a face mask sets and becomes firm.

MASK TREATMENT

The **face mask** is a skin-cleansing preparation which may contain a variety of different ingredients selected to have a deep cleansing, toning, nourishing or refreshing effect on the skin. The mask achieves this through the following actions:

- if it contains *absorbent* materials, dead skin cells, sebum and debris will adhere to it when it is removed;
- if it contains *astringent* ingredients, the pores and the skin will tighten;
- if it contains *emollient* ingredients, the skin will be softened and nourished;
- if it contains *soothing* ingredients, the skin can be desensitised to reduce skin irritation.

Depilex/RVB

A face mask

Mask preparations

There are basically two types of mask: setting and non-setting.

Setting masks

Setting masks are applied in a thin layer over the skin and then allowed to dry. The mask need not necessarily set solid – a solid mask can become uncomfortable, and be difficult to remove.

Setting masks come in these varieties:

- clay packs;
- peel-off masks – gel, latex, or paraffin wax;
- thermal masks.

Clay masks

The **clay mask** absorbs sebum and debris from the skin surface, leaving it cleansed. It can also stimulate or soothe the skin, according to the ingredients chosen. Various clay powders are available – select from these according to the physiological effects you require:

- **Calamine** A light pink powder which soothes surface blood capillaries. *Uses in treatment:* For sensitive or delicate skin.
- **Magnesium carbonate** A very light, white powder which creates a temporary astringent and toning effect. *Use in treatment:* For open pores on dry and normal skins.

HEALTH & SAFETY: AVOIDING INFECTION

Clay mask ingredients must always be made from sterilised materials, because of the danger from tetanus spores.

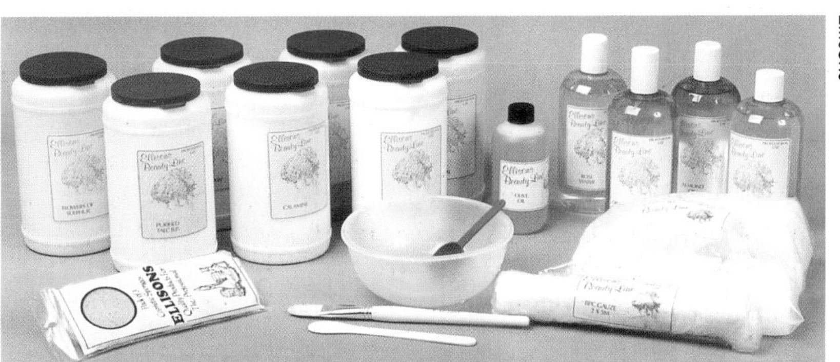

Ellisons

Clay mask ingredients

- **Kaolin** A cream-coloured powder which has a very stimulating effect on the skin's surface capillaries, thereby helping the skin to remove impurities and waste products. *Use in treatment:* For congested, greasy skin.
- **Fuller's earth** A green, heavy clay powder. It has a very stimulating effect, such that the skin will show slight reddening. It also produces a whitening, brightening effect. *Use in treatment:* For greasy skin with a sluggish circulation. Due to its strong effect, it is *not* suitable for a client with a sensitive skin.
- **Flowers of sulphur** A light, yellow clay powder, which has a drying action on pustules and papules. *Use in treatment:* Applied only to specific blemishes (pustules).

To activate these masks it is necessary to add a liquid – an **active lotion** – which turns the powder to a liquid paste. Active lotions are selected according to the skin type of the client and the mask to be used; they reinforce the action of the mask.

- **Rose water and orange-flower water** These are very popular; they have a very mild stimulating and toning effect.
- **Witchhazel** This has a soothing effect on blemished skin; it is also an astringent and is suitable for use on greasy skin.
- **Distilled water** This is used on highly sensitive skins.
- **Almond oil** This is mildly stimulating. Because it is an oil, it does not allow the mask to dry: it is therefore recommended for highly sensitive skin or dehydrated skin.
- **Glycerol** A humectant, which prevents the mask drying and is suitable for a dry, mature skin.

Clay masks have the disadvantage when treating a black skin that they tend on removal to leave streaks of white residue. Choose a mask that does not have this effect.

The mask should be kept in place for about 10–15 minutes.

HEALTH & SAFETY: ALLERGIES

When treating a sensitive skin, remove the mask before it has set to avoid the stimulating action of the mask sensitising the skin.

ACTIVITY: CHOOSING FACE MASKS

Which clay powder and which active lotion would you mix for clients with the following skin types?

1. A mature, sensitive skin type.
2. A young, normal skin type.
3. A combination skin type: cheeks, neck area dry; forehead, nose and chin area greasy.

TIP

Ensure that you position the gauze correctly so that the nose and eye holes are properly placed.

Peel-off masks

Peel-off masks may be made from gel, latex or paraffin wax. Because perspiration cannot escape from the skin's surface,

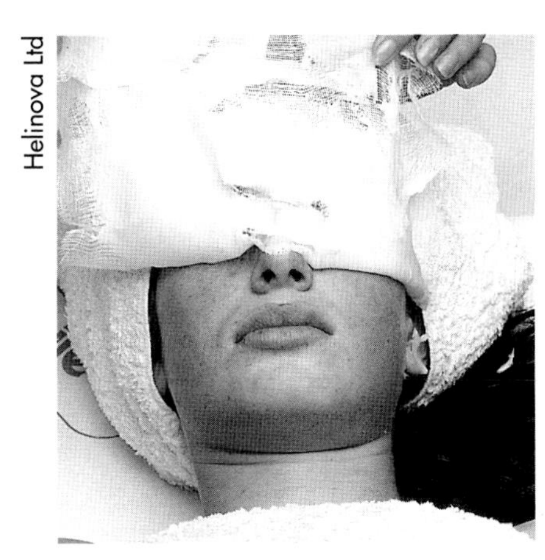

Helinova Ltd

A paraffin-wax mask

HEALTH & SAFETY: CLIENT COMFORT

Before applying the wax to the client's skin always test the temperature of the wax on your own inner wrist.

moisture is forced into the stratum corneum. The mask also insulates the skin, causing an increase in temperature.

The **gel mask** is either a suspension of biological ingredients, such as starches, gums or gelatin, or a mixture of synthetic non-biological resin ingredients. The mask is applied over the skin; on contact with the skin it begins to dry. When dry it is peeled off the face in one piece. *Uses in treatment:* Depending on the biological ingredients added, the gel mask can be used to treat all skin types. (If the client has excessive facial hair, such as at the sides of the face, this mask may cause discomfort on removal. To avoid this place a lubricant under the mask, or use a different sort of mask.)

The **latex mask** is an emulsion of latex and water: when applied to the skin, the water evaporates to leave a rubber film over the face. This produces a rise in temperature, thereby stimulating the skin. In alternative peel-off masks, latex is replaced by a synthetic resin emulsion such as **polyvinyl acetate** (PVA) resin. *Uses in treatment:* Latex masks tighten the skin temporarily, and are suitable for a mature skin; they can also be used with dry skin.

The **paraffin-wax mask** is stimulating in its action. The paraffin wax is blended with petroleum jelly or acetyl alcohol which improve its spreading properties. The wax is heated to approximately 37 °C and is then applied to the skin as a liquid. It sets on contact, so speed is essential if the mask is to be effective. The wax mask is loosened at the sides and removed in one piece after 15–20 minutes. *Use in treatment:* The paraffin-wax mask is suitable for dry skin. Because of its stimulating action, it is unsuitable for greasy skin or highly sensitive skin.

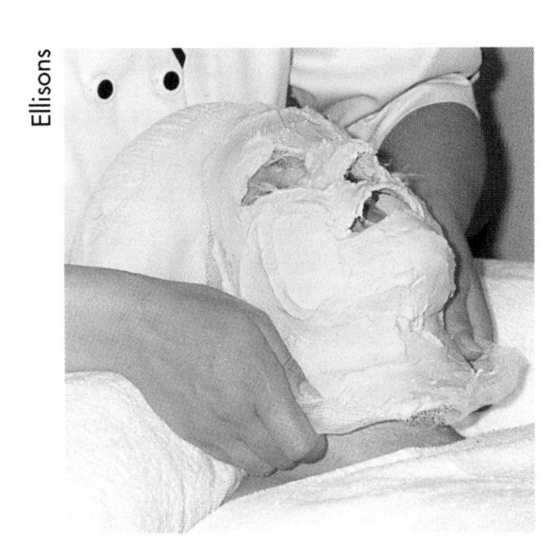

Ellisons

A thermal mask

HEALTH & SAFETY: CONTRA-INDICATIONS

Do not use thermal masks on a client with a circulatory disorder or one who has lost tactile sensation.

Thermal masks

The **thermal mask** contains various minerals. The ingredients are mixed and applied to the face and neck, avoiding the mouth and eye tissue. The mask warms on contact with the skin: this causes the pores to enlarge, thereby cleansing the skin. As the mask cools it sets, and the pores constrict slightly. The mask is removed from the face in one piece. *Uses in treatment:* Thermal masks have a stimulating, cleansing action, suitable for a normal skin or for a congested, greasy skin with open pores.

Non-setting masks

Some **non-setting masks** stay soft on application; others become firm, but they do not tighten like a setting mask. For this reason they do not tone the skin as effectively as setting masks. Non-setting masks include:

- warm oil;
- natural masks – fruits, plant and herbal;
- cream.

HEALTH & SAFETY: CLIENT COMFORT
If a client is particularly nervous, choose an effective non-setting mask. Some clients feel claustrophobic when wearing a setting mask.

Warm-oil masks

A plant oil, typically **olive oil** or **almond oil**, is warmed and then applied to the skin. It softens the skin and helps to restore the skin's natural moisture balance.

A **gauze mask** is cut to cover the face and neck, with holes for the eyes, nostrils and lips. This is then soaked in warm oil. A dampened cottonwool eye pad is placed over each eye. The gauze is then placed over the face and neck. It is usually left in place for 10–20 minutes. *Uses in treatment:* Warm-oil masks are recommended for mature skin and dry or dehydrated skin.

HEALTH & SAFETY: ALLERGIES
When using biological masks, always check first whether the client has any food allergies.

Natural masks

Natural masks are made from natural ingredients rich in vitamins and minerals. Fresh **fruit** and **vegetables** have a mildly astringent and stimulating effect. Usually the fruit is crushed to a pulp and placed between layers of gauze which are laid over the face.

Honey is used for its toning, tightening, antiseptic and hydrating effect. **Egg white** has a tightening effect and is said to clear impurities from the skin. **Avocados** have a nourishing effect; **bananas** soften the skin, and are for sensitive skins.

HEALTH & SAFETY: ACIDIC FRUIT
Lemon and grapefruit are generally considered too acidic for use on the face.

ACTIVITY: CREATING NATURAL MASKS
Create some masks, listing the ingredients to suit each of the following skin types:

- dry;
- greasy;
- mature, with superficial wrinkling;
- sensitive.

If possible, arrange to carry out one of the masks on a suitable client in the workplace.

Evaluate the natural face mask. Consider: cost, preparation, application, removal, and effectiveness. Remember to ask your client for *her* opinion!

HEALTH & SAFETY: NATURAL MASKS
Because natural masks are prepared from natural foods, they must be prepared *immediately* before use – they very quickly deteriorate.

Cream masks

Cream masks are prepared for you. They have a softening and moisturising effect on the skin. Each mask contains various biological extracts or chemical substances to treat different skin types or conditions. Instructions will be provided with the mask, stating how the product is to be used professionally.

These masks are popular in the beauty salon: they often complement a particular facial treatment range used by the salon, and they are available for retail sale to clients.

Contra-indications

The contra-indications to general skin care apply also to face-mask application. In addition, observe the following:

- *Allergies* Check whether your client knows that she has allergies. If so, avoid all contact with known allergens.
- *Claustrophobia* Do not use a setting mask on a particularly nervous client. Some clients feel claustrophobic under its tightening effect.
- *Sensitive skins* Do not use stimulating masks on clients with highly sensitive skin.

Equipment and materials

TIP

When purchasing mask brushes, note that a plastic-handled brush is preferable to one with a painted wooden handle – the painted one would be likely to peel on immersion in water, which spoils the professional image!

When applying masks you will need the following equipment and materials:

- *Disposable tissue roll* – such as bedroll.
- *Towels (2)* – freshly laundered for each client.
- *Flat mask brushes (3)* – sanitised.
- *Trolley.*
- *Client's record card.*
- *Dampened clean cottonwool.*
- *Cottonwool eye pads (2)* – pre-shaped round and dampened.
- *Scissors* – to cut cottonwool eye pads (if cottonwool discs are not to be used).
- *Facial tissues (white)*– to blot the skin dry after applying toner following mask removal.
- *Protective headband (clean).*
- *Clean spatulas (several)* – to mix individual masks (if required).
- *Facial toning lotions (a selection)* – to suit various types of skin.
- *Sterilised mask-removal sponges (2)* – for use when removing the mask using clean warm water.
- *Large bowl* – to hold warm water during removal of the mask.
- *Lukewarm water* – if required for mask removal.
- *Moisturisers (a range)* – to suit different skin types for use after mask removal.
- *Hand mirror (clean)* – to show the client her skin following the facial treatment.
- *Face-mask ingredients.*
- *Gauze* – used in applying certain masks.
- *Waste bin (covered and lined)* – for waste consumables.

TIP

Place a clean facial tissue under the edge of the headband at the forehead, so that it overlaps the headband. This will protect the headband from staining.

TIP

Many eye make-up removers are also eye treatments. If your workplace uses a commercial product in this way apply it to the dampened eye pads used during mask treatment. This may also assist your retail sales, as the client may wish to buy some for home use.

TIP

Ensure that you have plenty of dampened cottonwool – it is used to apply toner to the skin before mask application, to remove the mask or mask residue left on the skin, and to apply toner to the skin after mask removal.

Ideally, buy cottonwool discs to use for the eye pads. This will reduce preparation time and ensure an evenly-shaped protective shield for the eye area.

Sterilisation and sanitisation

After applying the mask, clean the mask brush thoroughly in warm water and detergent. Next, place it in a chemical sanitising agent; rinse it in clean water; allow it to dry; and then store it in the ultra-violet cabinet.

When you use a paraffin-wax mask, remove as much mask residue as possible from the brush, then place the brush in boiling water to completely remove all the wax. Sanitise the brush as usual before use.

If you use sponges to remove the mask, place them in warm water and detergent. After rinsing them in clean water, place them ready for sterilisation in an autoclave. (With repeated sanitisation, sponges will begin to break up.)

A large high-quality cottonwool disc may be purchased to use in mask removal.

HEALTH & SAFETY: MAINTAINING HYGIENE

You need several mask brushes and mask sponges to allow effective sterilisation of the tools, and so that you can provide freshly sterilised tools for each client.

Preparing the cubicle

Check that you have all the materials you need to carry out the treatment. You may like to place a paper roll at the head of the couch, underneath the client's head, to collect any mask residue on mask removal.

The head of the couch should be flat or slightly elevated. Don't have it in a semi-reclined position during the mask application, as some masks are liquid in consistency and may run into the client's eyes and behind her neck.

TIP

If you are using paraffin wax, remember to heat it in advance.

Treatment

Preparing the client

For maximum effect, the mask must be applied on a clean, grease-free surface. If the mask application follows a facial massage, ensure that the massage medium has been thoroughly removed.

Select the appropriate mask ingredients to treat the skin type and the facial conditions that require attention.

HEALTH & SAFETY: ALLERGIES

When using a commercial mask, try to find out *exactly* what it contains, so that you don't apply a sensitising ingredient to an allergic skin type.

Applying the mask

The mask is usually applied as the *final* facial treatment, because of its cleansing, refining and soothing effects upon the skin. The methods of preparation, application and removal are different for the various face-mask types, so the guidelines below are a general outline of effective treatment technique.

> **HEALTH & SAFETY: CLIENT COMFORT**
> It is important to check that your client is comfortable whilst the mask is on her face. (She will suffer discomfort if her skin is intolerant to a particular mask.)

1. Having determined the client's treatment requirements, select the appropriate mask ingredients. If you use a commercial mask, always read the manufacturer's instructions first.
2. Discuss the treatment procedure with the client. Tell her:
   - what the mask will feel like on application;
   - what sensation, if any, she will experience;
   - how long the mask will be left on the skin.

   Generally the mask will be left in place for 10–20 minutes, but the exact time depends on the type and effect required.

> **TIP**
> Don't mix the mask with the mask brush – if you do, the solid contents will tend to collect in the bristles, the mask won't be mixed effectively, and the brush won't spread the mask evenly.

3. Prepare the mask ingredients for application.
4. Using the sterilised mask brush or spatula, begin to apply the mask. The usual sequence of mask application is neck, chin, cheeks, nose, and forehead.

   If you are using more than one mask to treat different skin conditions, apply first the one that will need to be on longest.

   Apply the mask quickly and evenly so that it has maximum effect on the whole face. Don't apply it too thickly; as well as making mask removal difficult, this is wasteful as only the part that is in contact with the skin has any effect.

   Keep the mask clear of the nostrils, the lips, the eyebrows and the hairline.

> **HEALTH & SAFETY: SETTING MASKS**
> When using a setting mask, ensure that the mask is evenly applied or it will dry unevenly.

> **HEALTH & SAFETY: PEEL-OFF MASKS**
> When using a peel-off mask, make sure that the border of the mask is thick enough – if it isn't, it will be difficult to remove and the client may experience discomfort.

5. To relax the client, apply cottonwool eye pads dampened with clean water.
6. Leave the mask for the recommended time, and according to the effect required. Take account also of the the sensitivity of the skin and your client's comfort.

> **TIP**
> While the mask is on the face you can tidy the working area and collect together the materials required for mask removal. Do not disturb the client, who will be relaxing at this time.

> **TIP**
> When using water to remove a mask you may need to renew the water as you work.

7. Wash your hands.
8. When the mask is ready for removal, remove the eyepads.

   Explain to the client that you are going to remove the mask. Briefly describe the process, according to whether this is a setting or a non-setting mask.

   Remove the mask. Mask sponges, if used, should be damp, not wet, so that water doesn't run into the client's eyes, nose or mouth.

9 When the mask has been completely removed, apply the appropriate toning lotion using dampened cottonwool. Blot the skin dry with a facial tissue.

10 Apply an appropriate moisturiser to the skin.

11 Remove the headband, and tidy the client's hair.

12 With a mirror, show the client her skin. Evaluate the treatment.

13 Record the results on her record card.

Contra-actions

Before you apply the mask, explain to the client what the action of the mask will feel like on the skin. This will enable her to identify any undesirable skin reaction, evident to her as skin irritation – a burning sensation.

Ask the client initially whether she is comfortable: this will give her the opportunity to tell you if she is experiencing any discomfort. Should there be a contra-action to the mask, remove the mask immediately and apply a soothing skin-care product.

If on removal of the mask you can see that there has been an unwanted skin reaction (that is, if you see inflammation), apply a soothing skin-care product. In either case, note the skin reaction on the record card, and choose a different mask next time.

Advice on home use

The client may be given a sample of the face mask for use at home. Explain to her the procedure for application and removal, so that she achieves maximum benefit from the mask. Encourage her to apply a mask once or twice a week depending on her skin type, to dislodge dead skin cells and to cleanse and stimulate the skin.

Advise the client not to apply the mask directly before a special occasion, in case it causes blemishes, as sometimes happens.

The skin should be toned and moisturised after removing the mask.

FACIAL MASSAGE

Manual massage is the external manipulation using the hands, of the soft tissues of the face, neck and upper chest. Massage can improve the appearance of the skin and promote a sensation of stimulation or relaxation.

Each massage performed is adapted to the client's physiological and psychological needs. The skin's physiological needs are observed during the skin analysis; the client's psychological needs are usually discovered during the consultation.

The benefits of the facial massage include the following:

- Dead epidermal cells are loosened and shed. This improves the appearance of the skin, exposing fresh, younger cells.

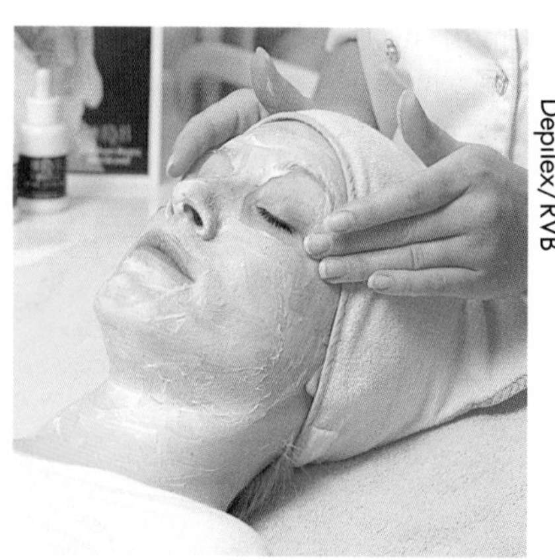
Depilex/RVB

Facial massage

- The muscles receive an improved supply of oxygenated blood, essential for cell growth. The tone and strength of the muscles are improved, firming the facial contour.
- The increased blood circulation in the area warms the tissues. This induces a feeling of relaxation which is particularly beneficial when treating tense muscles.
- As the blood capillaries dilate and bring blood to the skin's surface, the skin colour improves.
- The lymphatic circulation and the venous blood circulation increase. These changes speed up the removal of waste products and toxins, and tend therefore to purify the skin. The removal of excess lymph improves the appearance of a puffy oedematous skin (provided that this does not require medical treatment).
- The increased temperature of the skin relaxes the pores and follicles. This aids the absorption of the massage product, which in turn softens the skin.
- Sensory nerves can be soothed or stimulated, depending on the massage manipulations selected.
- Massage stimulates the sebaceous and sudoriferous glands and increases the production of sebum and sweat. This increase helps to maintain the skin's natural oil and moisture balance.

Reception

ACTIVITY: PLANNING A MASSAGE

How will observations from the skin analysis and the client consultation influence the facial massage treatment?

TIP

The headband can spoil the hair, and the massage medium may enter the hairline. Recommend that the client does not style her hair directly before the facial treatment.

Facial massage is carried out as required, usually once every 4–6 weeks. Before the facial massage is given, the skin is cleansed; afterwards it is usual to apply a cleansing face mask, which absorbs any excess grease from the skin.

When booking a client for this treatment, allow one hour. The facial massage itself should take approximately 20 minutes, but this may vary according to the client's skin type.

Warn the client that the skin may appear slightly red and blotchy after treatment, due to the increase in blood circulation to the area: this reaction will normally subside after 4–6 hours.

Because of the stimulating effect on the skin recommend to the client that she receives this service when she does not have to apply any cosmetic products directly afterwards.

Sometimes the skin develops small blemishes after facial massage; this is due to its cleansing action. If the client is preparing for a special occasion, therefore, such as her wedding, book the appointment at least 5 days in advance.

Massage manipulations

The facial massage is based on a series of classic massage movements, each with different effects. There are four basic groups of massage movements:

- effleurage;
- petrissage;

- percussion (also known as tapotement);
- vibrations.

The therapist can adapt the way each of these movements is applied, according to the needs of the client. Either the *speed of application* or the *depth of pressure* can be altered.

Effleurage

Effleurage is a stroking movement, used to begin the massage, as a link manipulation, and to complete the massage sequence. This manipulation is light, has an even pressure, and is applied in a rhythmical, continuous manner to induce relaxation.

The pressure of application varies according to the underlying structures and the tissue type, but it must *never* be unduly heavy.

Effleurage has these effects:

- desquamation is increased;
- arterial blood circulation is increased, bringing fresh nutrients to the area;
- venous circulation is improved, aiding the removal of congestion from the veins;
- lymphatic circulation is increased, improving the absorption of waste products;
- the underlying muscle fibres are relaxed.

Uses in treatment: to relax tight, contracted muscles.

Petrissage

Petrissage involves a series of movements in which the tissues are lifted away from the underlying structures and compressed. Pressure is intermittent, and should be light yet firm.

Petrissage has these effects:

- improvement of muscle tone, through the compression and relaxation of muscle fibres;
- improvement in blood and lymph circulation, as the application of pressure causes the vessels to empty and fill;
- increased activity of the sebaceous gland, due to the stimulation.

Movements include picking up, kneading, knuckling, pinching, rolling, frictions, and scissoring.

Uses in treatment: to stimulate a sluggish circulation; to increase sebaceous gland and sudoriferous gland activity, when treating a dry skin condition.

TIP

Pressure may be increased when working on larger muscles in the neck or shoulders.

Percussion

Percussion, also known as **tapotement**, is performed in a brisk, stimulating manner. Rhythm is important as the fingers are continually breaking contact with the skin; irritation could occur if the movement were performed incorrectly.

Percussion has these effects:

HEALTH & SAFETY: CONTRA-INDICATIONS

Do not apply percussion over highly sensitive or vascular skin conditions.

TIP

When a stimulating massage is required, incorporate more petrissage and tapotement into the massage sequence.

- a fast vascular reaction because of the skin's nervous response to the stimulus – this reaction, erythema, has a stimulating effect;
- increased blood supply, which nourishes the tissues;
- improvement in muscle and skin tone in the area.

Movements include clapping and tapping. In facial massage, only *light* tapping should be used.

Uses in treatment: to tone areas of loose, crepey skin around the jaw or eyes.

Vibrations

Vibrations are applied on the nerve centre. They are produced by a rapid contraction and relaxation of the muscles of the therapist's arm, resulting in a fine trembling movement.

Vibration has these effects:

- stimulation of the nerves, inducing a feeling of well-being;
- gentle stimulation of the skin.

Movements include *static* vibrations, in which the pads of the fingers are placed on the nerve, and the vibratory effect created by the therapist's arms and hands is applied in one position; and *running* vibrations, in which the vibratory effect is applied along a nerve path.

Uses in treatment: to stimulate a sensitive skin in order to improve the skin's functioning without irritating the surface blood capillaries.

Equipment and materials

The massage is carried out using a **massage medium** which acts as a lubricant. A massage cream or oil may be used; these are slightly penetrating, and soften the skin. Choose a product that contains ingredients to suit the client's skin type and the age of the skin.

Whichever product you choose, it should provide sufficient slip whilst allowing you to control the massage movements.

ACTIVITY: CHOOSING A MASSAGE MEDIUM

Compare two professional skin-care ranges. Look at:

1 the choice of facial-massage preparations;
2 the ingredients used in their formulation, and the effects claimed.

TIP

It is important to maintain good posture during facial massage. If you slouch, you will experience muscle fatigue!

Treatment

There are many different massage sequences, but each uses one or all of the massage manipulations discussed above. What follows is a basic sequence for facial massage.

1 **Effleurage to the neck and shoulders** Slide the hands down the neck, across the pectoral muscles around the deltoid muscle, and across the trapezius muscle. Slide the hands up the back of the neck to the base of the skull.

Repeat step **1** *a further 5 times.*

2 **Thumb kneading to the shoulders** Using the pad of both thumbs, make small circles (frictions) along the trapezius muscle, working towards the spinal vertebrae.

Apply each movement 3 times; then repeat the sequence (step **2***) a further 2 times.*

3 **Finger kneading to the shoulders** Position the fingers of each hand behind the deltoid, and make large rotary movements along the trapezius.

Apply each rotary movement 3 times; repeat the sequence (step **3***) a further 2 times.*

4 **Vibrations** Place the hands, cupped, at the base of the neck: perform running vibrations up the neck to the occipital bone.

Repeat step **4** *a further 6 times.*

5 **Circular massage to the neck** Perform small circular movements over the platysma and the sternomastoid muscle at the neck.

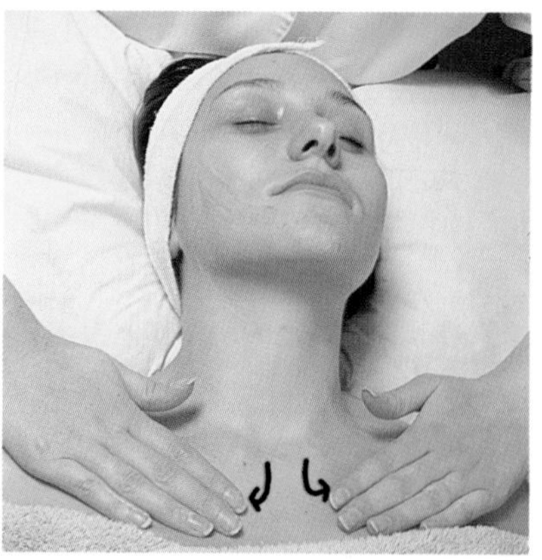

Effleurage to the shoulders

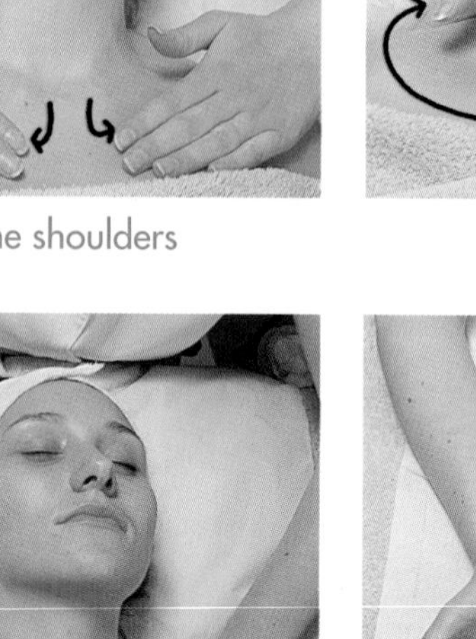

Thumb kneading to the shoulders

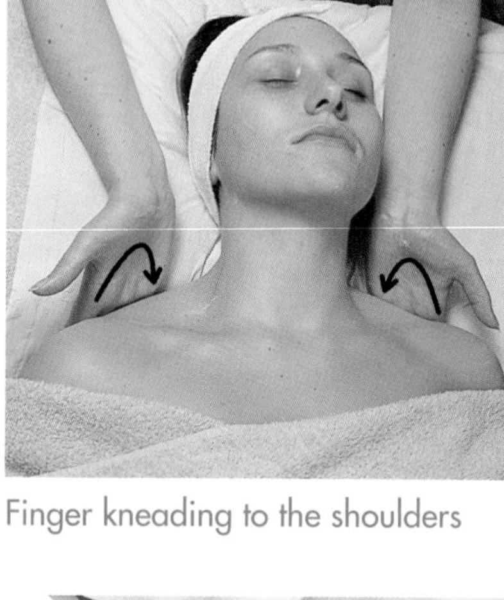

Finger kneading to the shoulders

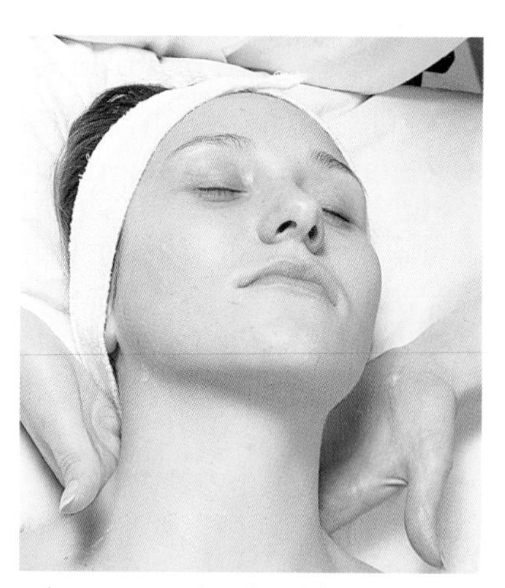

Vibrations to the shoulders

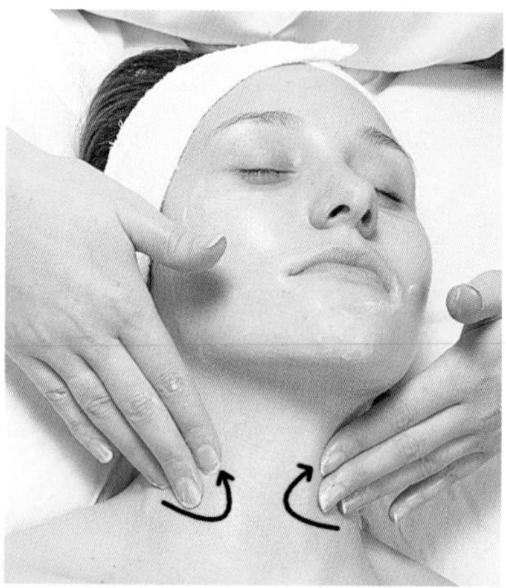

Circular massage to the neck

HEALTH & SAFETY: THE TRACHEA

Never apply pressure when working on the neck over the trachea.

HEALTH & SAFETY: SENSITIVE SKIN

Do not use knuckling on a sensitive skin.

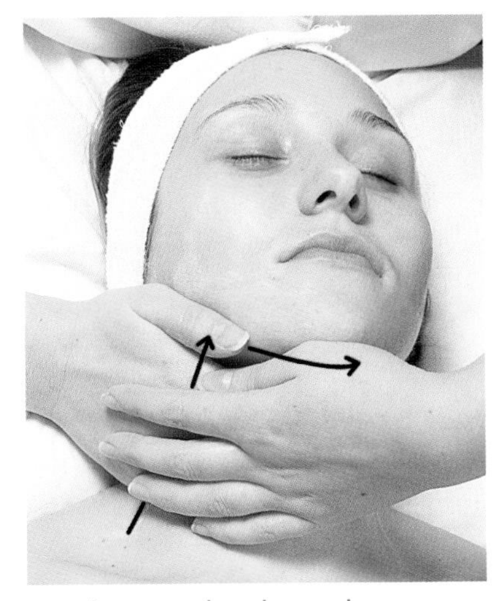

Hands cupped to the neck

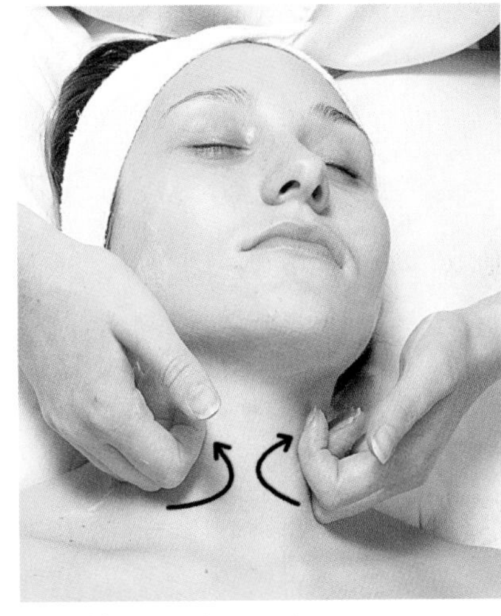

Knuckling to the neck

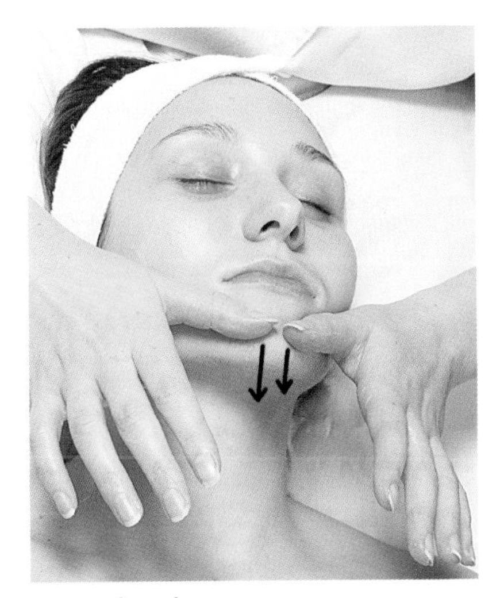

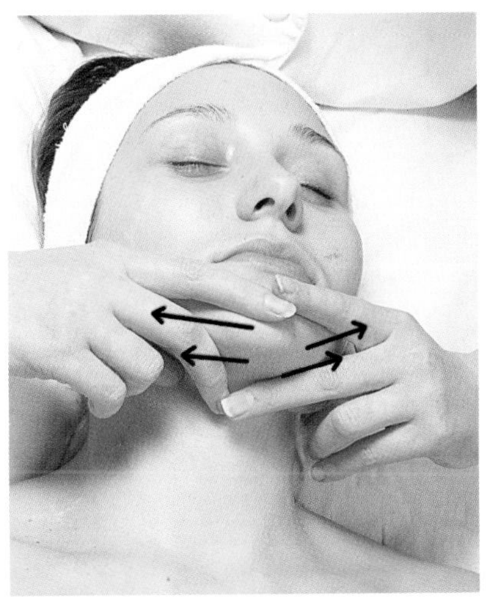

Up and under

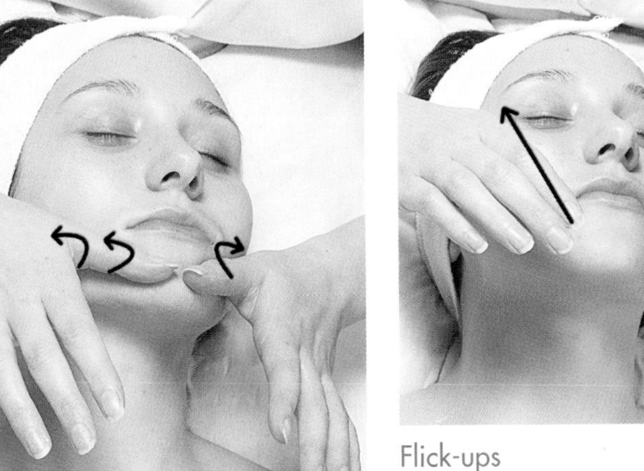

Circling to the mandible

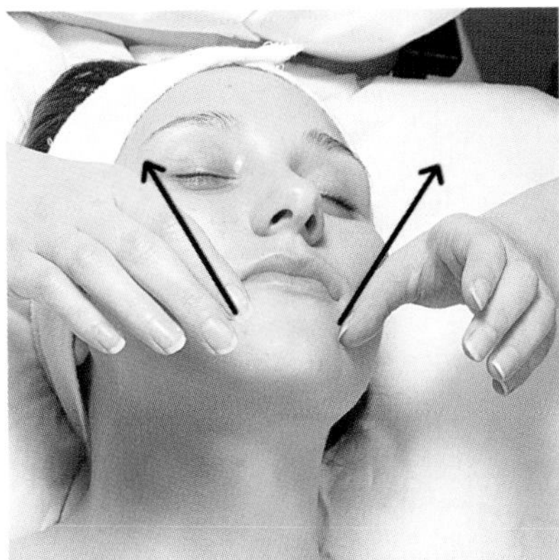

Flick-ups

HEALTH & SAFETY: THE LIPS

Do not flick the lips. Position the thumbs 5mm from the corner of the mouth to avoid this.

6 **Hands cupped to the neck** Cup your hands together. Place the hands at the left side of the neck, above the clavicle. Slide the hands up the side of the neck, across the jawline, and down the right side of the neck; then reverse.

*Repeat step **6** a further 2 times.*

7 **Knuckling to the neck** Make a loose fist: rotate the knuckles up and down the neck area.

*Repeat step **7** to cover, a further 2 times.*

8 **Up and under** Place the thumbs on the centre of the chin, and the index and middle fingers under the mandible. Slide the thumbs firmly over the chin. Bring the index finger onto the chin, and place the middle finger under the mandible forming a V shape. Slide along the jawline to the ear. Replace the index finger with the thumb, and return along the jaw to the chin.

*Repeat step **8** a further 5 times.*

9 **Circling to the mandible** Place the thumbs one above the other on the chin, and proceed with circular kneading along the jawline towards the ear. Reverse and repeat.

*Repeat step **9** a further 2 times.*

10 **Flick-ups** Place the thumbs at the corners of the mouth. Lift the orbicularis oris muscle, with a flicking action of the thumbs.

*Repeat step **10** a further 5 times.*

11 Half face brace Clasp the fingers under the chin; turn the hands so that the fingers point towards the sternum. Unclasp, and slide the hands up the face towards the forehead.

Repeat step **11** *a further 2 times.*

12 Lifting the eyebrows Place the right hand on the forehead at the left temple, and stroke upwards from the eyebrow to the hairline. Repeat the movement with the left hand. Alternate each hand; repeat the movement across the forehead.

Repeat step **12** *a further 2 times.*

13 Inner and outer eye circles Using the ring finger, *gently* draw 3 outer circles and 3 inner circles on each eye, following the fibre direction of the orbicularis oculi muscle.

Repeat step **13** *a further 2 times.*

14 Half face brace Repeat step **11**.

15 Circling to the chin, the nose and the temples Apply circular kneading to the chin, the nose and the temples. Return to the starting position.

Repeat step **15** *a further 2 times.*

16 Thumb kneading under the cheeks Place the thumbs under the zygomatic bones. Carry out a circular kneading over the muscles in the cheek area.

Repeat step **16** *a further 5 times.*

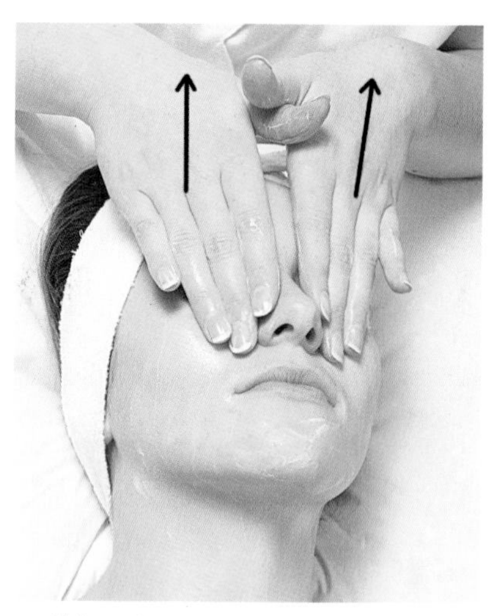

Half face brace

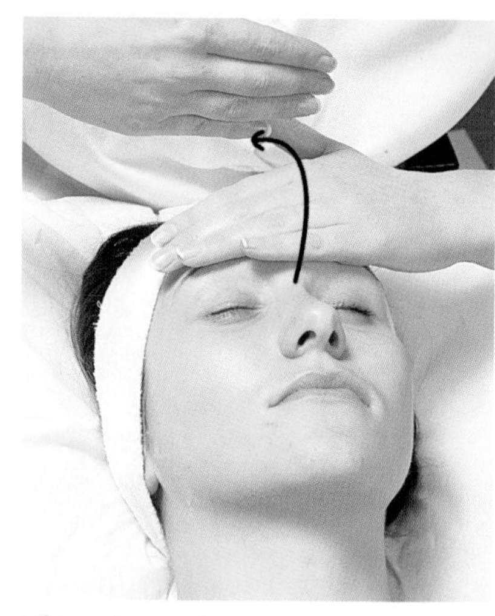

Lifting the eyebrows

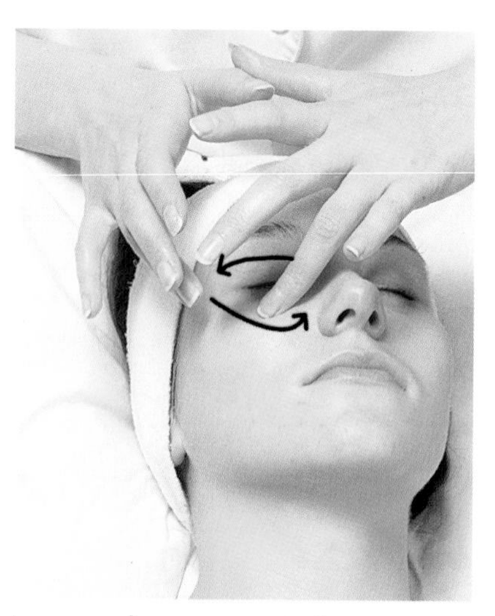

Inner and outer eye circles

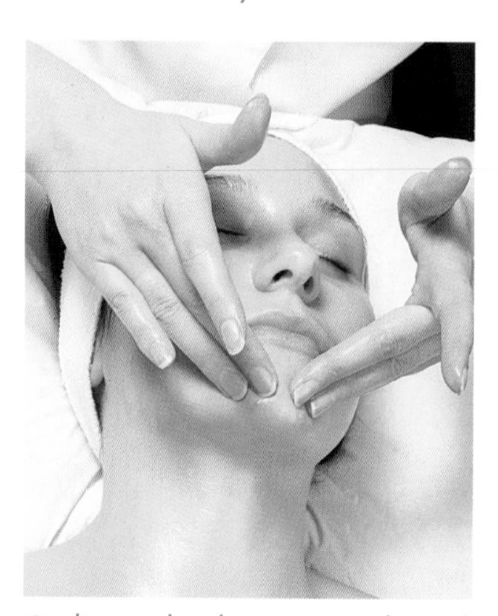

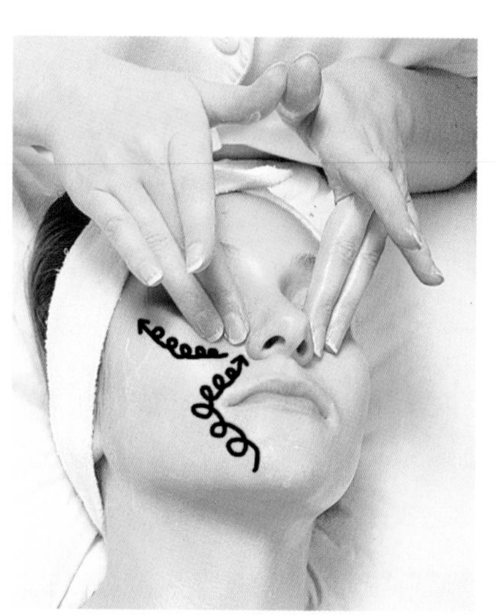

Circling to the chin, nose and temple

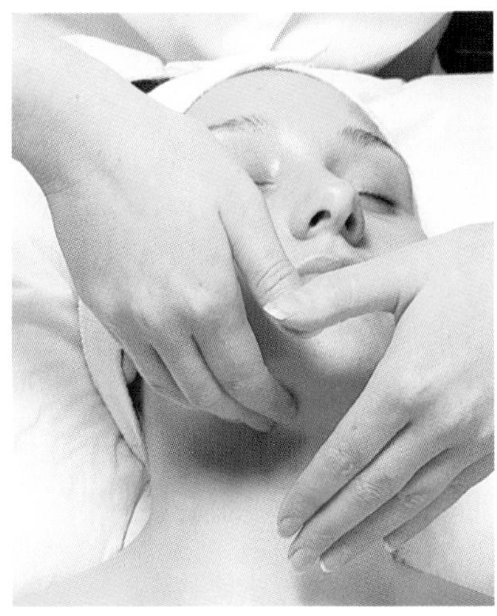

Tapping under the mandible

HEALTH & SAFETY: SENSITIVE SKIN

Avoid the use of tapotement over areas of sensitivity.

TIP

If the skin appears to drag during massage, stop and apply more massage medium. If you keep going you may cause skin irritation or discomfort.

Lifting the masseter

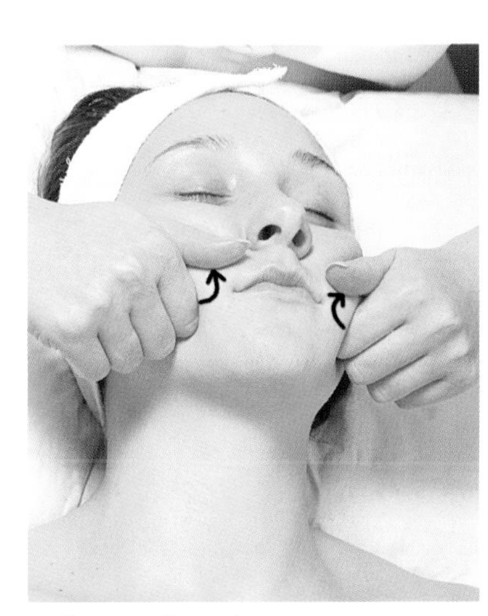

Rolling and pinching

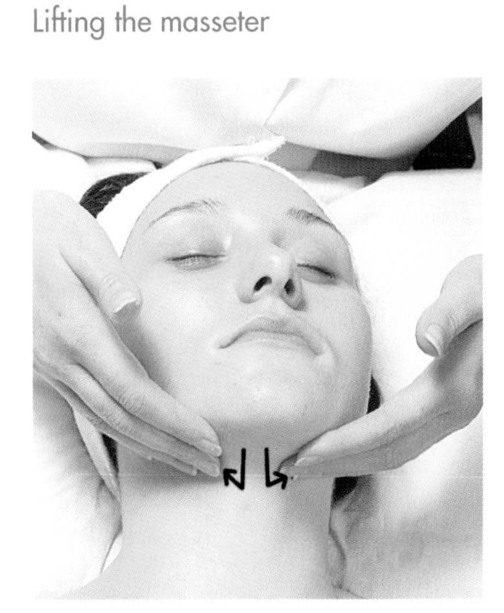

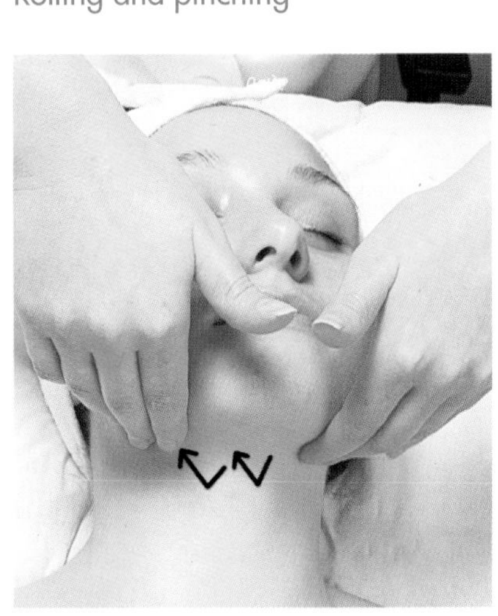

Lifting the mandible

17 **Tapping under the mandible** Tap the tissue under the mandible, using the fingers of both hands. Work from the left side of the jaw to the right; then reverse.

Repeat step **17** *a further 5 times.*

18 **Lifting the masseter** Cup the hands. Using the hands alternately, lift the masseter muscle.

Repeat step **18** *a further 5 times.*

19 **Rolling and pinching** Using a deep rolling movement, draw the muscles of the cheek area towards the thumb in a rolling and pinching movement.

Repeat step **19** *a further 5 times.*

20 **Lifting the mandible** Place the pads of the fingers underneath the mandible and pivot diagonally, lifting the tissues work towards the ear.

Repeat step **20** *a further 2 times.*

TIP

During the facial massage, the client's face should relax. If there are evident signs of tension, such as vertical furrows between the eyebrows, check that the client is warm and comfortable.

21 **Knuckling along the jawline** Knuckle along the jawline and over the cheek area.

Repeat step **21** *a further 2 times.*

22 **Upwards tapping on the face** Using both hands, gently slap along the jawline from ear to ear, lifting the muscles.

Repeat step **22** *a further 5 times.*

23 **Full face brace** Place the fingertips under the chin, and slide both hands up the face to the forehead, lifting all the facial muscles.

Repeat step **23** *a further 2 times.*

24 **Scissor movement to the forehead** Open the index and middle fingers to make a V shape at the outer corner of each eyebrow. Open and close the fingers in a scissor action towards the inner eyebrow.

Repeat step **24** *a further 2 times.*

25 **Tapotement movement around the eyes** Using the pads of the fingers, tap *gently* around the eye area.

Repeat step **25** *a further 2 times.*

26 **Eye circling** Repeat step **13**, 3 times.

27 **Effleurage** Repeat step **1**.

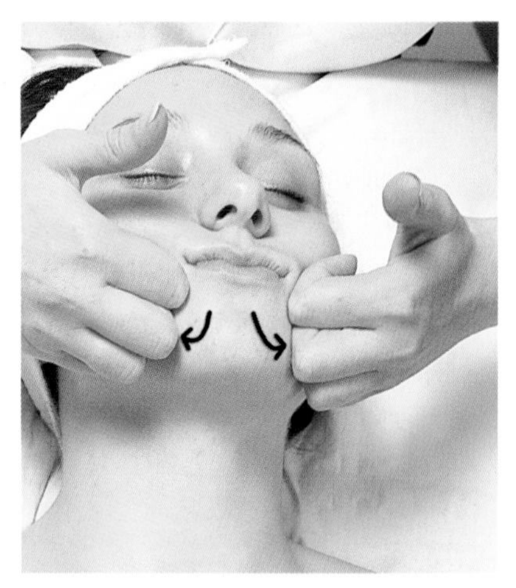

Knuckling along the jawline

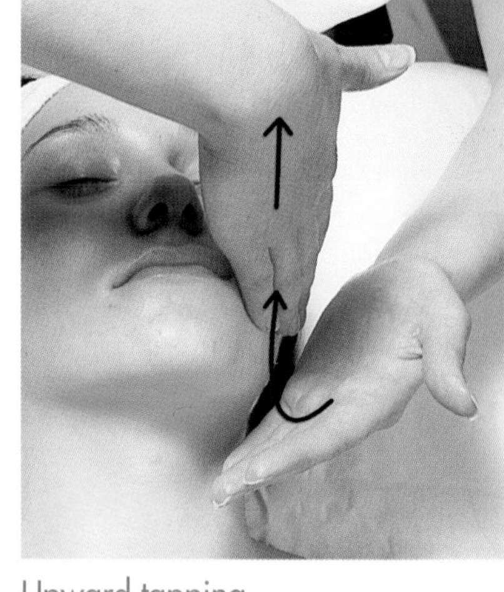

Upward tapping

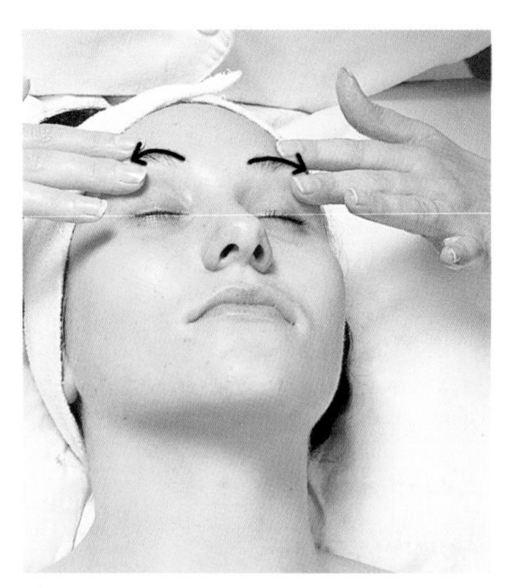

The scissor movement

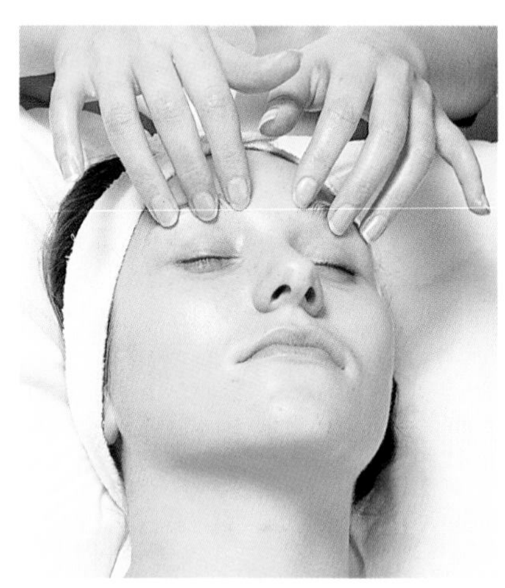

Tapotement around the eyes

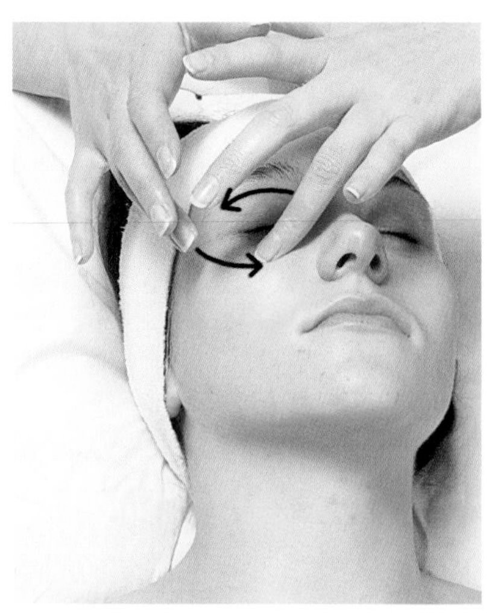

Eye circling

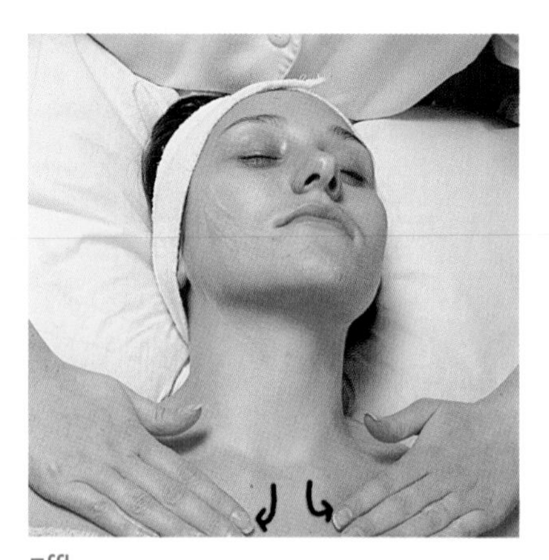

Effleurage

ACTIVITY: DESIGNING MOBILITY EXERCISES

Devise *ten* exercises to increase the strength and mobility of your hands and wrists.

TIP

When learning the facial massage you need to practise. Practise the movements on a styrofoam headblock with facial features, or mannequin head as used in hairdressing, to perfect the manipulations. If these are unavailable, practise each manipulation on your knee – this will increase the agility and strength of your fingers and wrists.

After the massage

After the facial massage, remove the massage medium thoroughly using clean damp cottonwool. Check thoroughly that all product has been removed.

TIP

Massage medium can easily be overlooked in the following areas: the eyebrows; the base of the nostrils; under the chin; in the creases of the neck; behind the ear; and on the shoulders.

Apply toner to remove traces of oil, leaving the skin grease-free. Finally, blot the skin dry.

You may then proceed with further skin treatments, such as a face mask, or simply apply an appropriate moisturiser to conclude the treatment.

Advice on home care

Encourage your client to use massage movements when applying emollient skin-care products. Show her how to perform such movements correctly.

Facial exercises may be given to the client for her to practise at home. These should be carried out at least four times per week.

ACTIVITY: MASSAGE AT HOME

Design a simple massage routine that a client could be taught to use at home.

Which type of manipulation is involved in each movement in your sequence? What effect do you wish to achieve by incorporating it?

ACTIVITY: FACIAL EXERCISES AT HOME

Think of *ten* facial exercises that you could teach a client, to improve the muscle tone of the face and neck.

Teach these to a client after carrying out a facial massage treatment. At a later stage, evaluate the exercises and discuss with her their effectiveness.

KNOWLEDGE REVIEW

Skin care and facial massage treatment

Preparation

1. What details should be recorded on the client's record card?
2. How would you decide the skin type and treatment needs of a client?
3. When would you use a skin-care warming treatment such as facial steaming?

Application

1. What is the purpose of the following skin-care products?
   (a) Cleanser.
   (b) Toning lotion.
   (c) Exfoliant.
   (d) Moisturiser.
   (e) Facemask or facepack.
2. What is the physiological effect on the facial tissues of facial massage?
3. How can you ensure that the client will be relaxed during the facial treatment?

Aftercare

1. What skin-care products would you recommend that a client use as part of her home-care routine?
2. How would you explain to the client the application – and, where relevant, removal – of the following products for home use?
   (a) Cleansing milk.
   (b) Foaming cleanser.
   (c) Eye make-up remover.
   (d) Toning lotion.
   (e) Moisturiser.
   (f) Eye gel.
   (g) Throat cream.
3. What details would be recorded on the client's record card following the facial skin-care treatment?

Health, safety and hygiene

1. How can you ensure that skin-care products and tools are used hygienically?
2. Name *one* infectious and *one* non-infectious skin condition.
3. How would you select and apply skin-care products for a client with an allergic skin type?
4. How would a contra-action to skin-care products be recognised?
5. At what point in the facial treatment would you remove skin blockages, if present?

Case studies

1 A client complains that her skin feels very dry. How can you recognise if the skin is dry or dehydrated?
2 What skin-care products and which treatment mask would you select for a teenage client with a greasy skin prone to blemishes?
3 What would be your treatment aim, and how would you adapt facial massage application, for a client with a mature skin?
4 How would you recognise the following skin types?
   (a) Combination.
   (b) Sensitive.
5 Describe a suitable facial treatment (lasting one hour) to treat each skin type. Justify the choice of the products you select, and explain how you would apply them.

CHAPTER 6

Make-up

MAKE-UP SERVICES

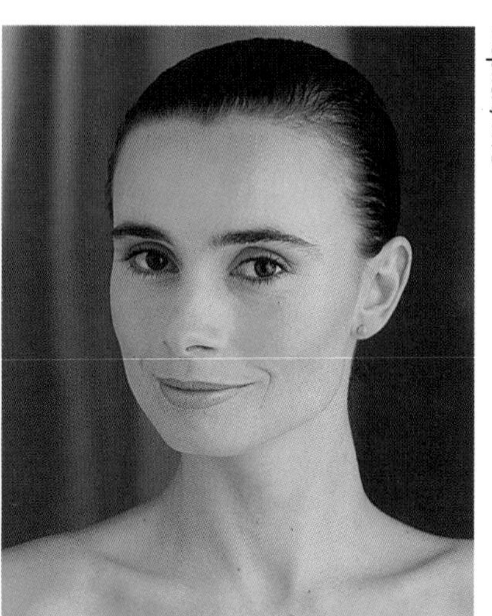

Depilex/RVB

Make-up is used to enhance and accentuate the facial features to make us appear more attractive – which in turn makes us feel more confident. Make-up is used to create balance in the face, by skilful application of different cosmetic products to reduce or to emphasise facial features.

Each client is unique, so each requires an individual approach for her make-up. The overall effect should be attractive, complementing the client's personality, her lifestyle, and the occasion for which the make-up is to be worn.

RECEPTION

Make-up application is offered in the salon for different purposes.

- *A make-up lesson* A chance for the client to learn from a professional how to apply make-up that suits her.
- *Special occasion make-up* Make-up applied by a specialist to suit the occasion for which it is to be worn, such as a wedding. If the make-up is for a bride, advise the client to visit the salon for a consultation and a practice session so that you can decide together on an appropriate make-up. Ask the client if possible to bring a swatch of the dress material with her, so that you can select colours to complement this and to co-ordinate with the accessories.
- *Remedial make-up* Make-up may be applied for remedial purposes, to cover facial disfigurements or birthmarks, and the client can be taught how to do this herself.
- *A professional job* Some clients simply wish to have their make-up professionally applied.

As always, you need to know how long each service will take. Here are some suggested times:

- make-up lesson: 1 hour;
- special occasion make-up: 45 minutes – 1 hour;
- straight make-up: 30–45 minutes.

Advise the client that if she intends to have her hair washed and styled, this should be done before she has the make-up applied.

If a client requests a deep cleansing facial followed by make-up application, suggest that she has the facial at least five days prior to the make-up. The facial will stimulate the skin, increasing its normal physiological functioning. This will affect how long the make-up lasts; it may even cause the colour of the foundation to change.

Do not reshape the eyebrows at the same treatment as make-up application – secondary infection could occur; also the skin in the area will be very pink, altering the colour and thus the effect of eyeshadow.

The consultation

If the client is new, complete a record card noting the client's personal details. Below is an example of a typical record card for make-up.

Depilex/RVB

MAKE-UP CHART	
Client's name	Date
Occasion	
Comments	
Skin type	
Eye colour	Skin colour
Special considerations	Hair colour
Contra-indications	
Recommended products	
Cleanser	Toner
Moisturiser	Foundation
Shading	Highlighting
Cheek colour	
Powder	
Eyeshadow	
Eyeliner	
Mascara	
Lipliner	
Lipstick	
Lipgloss	
Special advice	
Comments/reactions	
Products purchased	
Therapist	

A make-up record card

CONTRA-INDICATIONS

Certain contra-indications preclude make-up application. Check for these at the consultation, and if any of the following are present on inspection of the skin, do not proceed with make-up application:

- *Skin disorders.*
- *Bruising* in the area.
- *Recent haemorrhage.*
- *Swelling and inflammation* in the area.
- *Recent scar tissue.*
- *Sensory nerve disorders.*
- *Cuts or abrasions* in the area.
- *A recent operation* in the area.
- *Eye disorders* such as styes and conjunctivitis.

ACTIVITY: RECOGNISING CONTRA-INDICATIONS
What skin disorders can you think of that would contra-indicate make-up application?

Ask the client whether she has any known allergies to cosmetic preparations: note the answer on the record card.

Discuss your make-up plan with the client to ensure that the make-up will meet her requirements. You may need to ask questions such as those that follow, but of course the questions depend on the purpose of the make-up application.

- 'Do you normally wear make-up?'
- 'For what occasion is the make-up to be worn?'
- 'What colour are the clothing and accessories to be worn on this occasion?'
- 'Are there any colours that you particularly like or dislike?'
- 'What effect would you like the make-up to create?' (This question may be asked in many contexts – the client may wish to achieve a natural or a glamorous effect, or to emphasise or diminish certain facial features.)

ACTIVITY: THE RECORD CARD
Why is it important to complete a record card? What information should be recorded on it?

EQUIPMENT AND MATERIALS

Before beginning the make-up, check that you have to hand the necessary equipment and materials.

- *Couch or beauty chair*, with a reclining back and a head rest and an easy-to-clean surface.
- *Trolley* or other surface on which to place everything.

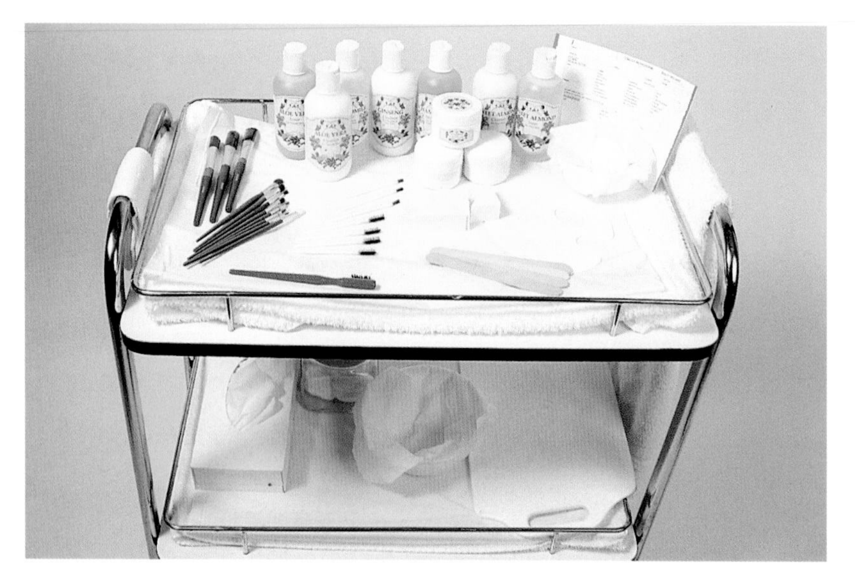

Equipment and materials

TIP

The chair must offer head support, or the client's neck will become strained during the make-up; also the head needs support if it is to be steady during the application of eye and lip make-up.

TIP

As an alternative to a towel, a make-up cape may be used to protect the client's clothing.

TIP

As an alternative to a headband, you may wish simply to clip the hair out of the way with hairclips.

- *Disposable tissue*, such as bedroll – to cover the work surface and the couch or beauty chair.
- *Towels (2)*, freshly laundered for each client – one to be placed over the head of the couch or chair, the other over the client's chest and shoulders to protect her clothing.
- *Headband*, clean – to protect the client's hair whilst cleansing the skin.
- *Hairclips.*
- *Cleansing lotion.*
- *Eye make-up remover.*
- *Toning lotion.*
- *Moisturiser* – lightweight.
- *Magnifying lamp* – to inspect the skin after cleansing.
- *Damp cottonwool* – prepared for each client.
- *Dry cottonwool* – stored in a covered jar.
- *Large white facial tissues* – to blot the skin after facial toning, and to protect the skin during make-up application.
- *Bowls* – to hold the prepared cottonwool.
- *Bowls or lined pedal bin* – for waste materials.
- *Client's record card.*
- *Make-up (a range)* – suitable for clients with known skin allergies to cosmetics, for contact-lens wearers, for different skin types and skin colours, including:
  - concealing and contouring cosmetics (shaders, highlighters and blushers);
  - foundations;
  - translucent powders;
  - eyeshadows;
  - eyeliners;
  - browliners;
  - lipsticks;
  - lipglosses;
  - lipliners.

Depilex/RVB

A make-up stand

TIP

Where possible, use make-up that you also sell in the salon, so that the client can buy the products for home use if she wishes.

Make-up brushes

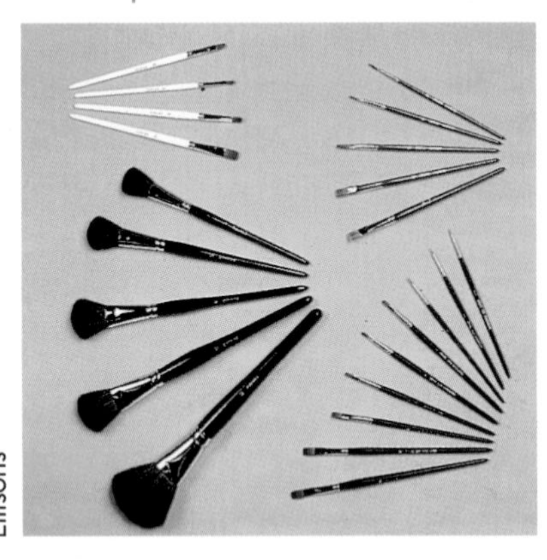

Ellisons

Ellisons

Make-up sponges

- *Bright lighting.*
- *Make-up brushes (assorted)* – at least 3 sets, to allow for sanitisation after use.
- *Disposable applicators and brushes* where possible – for example for mascara and the application of eyeshadow.
- *Cosmetic sponges* – for applying foundation.
- *Make-up palette* – for preparing cosmetic products prior to application.
- *Hand mirror (clean).*
- *Pencil sharpener* – for cosmetic pencils.
- *Eyelash curlers.*
- *Spatulas (several).*
- *Orange sticks (several)* – for removing make-up products from their containers.

Ellisons

Eyelash curlers

TIP

False eyelashes should be available if you are applying an evening make-up – you may wish to apply these to enhance the final effect.

TIP

When you give a make-up lesson, do so in front of a large mirror so that the client can watch the various stages of make-up application.

Record the make-up used and the advice given on a *separate* record, for the client to take away with her.

Depilex/RVB

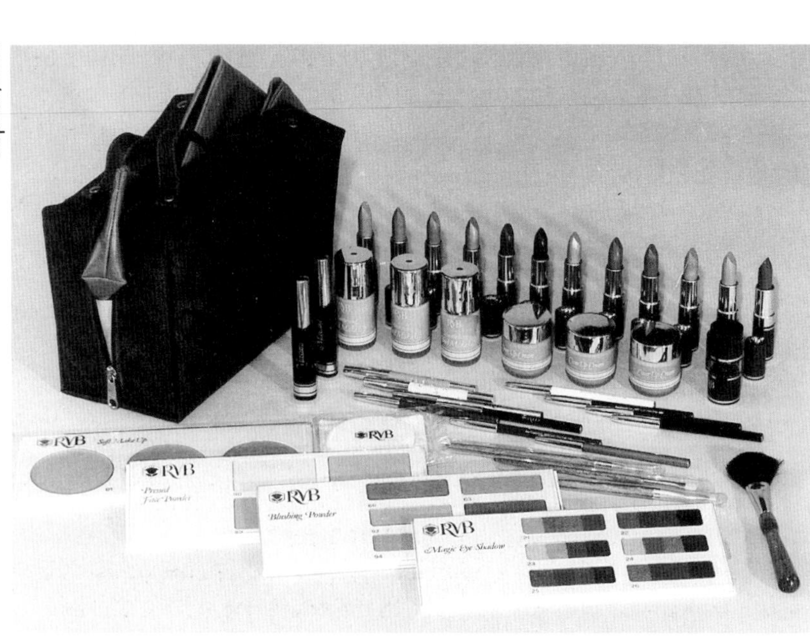

A make-up case

TIP

If you intend to apply make-up at different locations, buy a large make-up case so that you can transport the make-up easily and hygienically.

HEALTH & SAFETY: TRANSPORTING MAKE-UP

If you are transporting your make-up in a box, keep the box clean. Clients won't be impressed if they see soiled, dirty make-up containers.

Sterilisation and sanitisation

Where possible, use disposable applicators during make-up application, costing these into the treatment price. Sanitise make-up brushes after each use: wash them in warm water and detergent, rinse them thoroughly in clean water, and allow them to dry naturally. Once dry, place the brushes in an ultra-violet light cabinet ready for use.

TIP

An *alcohol-based cleanser* may be used to clean make-up brushes. The brushes are first cleaned with a solution of warm water and detergent, then thoroughly rinsed in clean water and allowed to dry. They are then briefly immersed in the alcohol solution and again allowed to dry.

All cosmetic products should be removed from their containers using a clean spatula or orange stick, and placed on the clean plastic make-up palette before application. (This avoids contamination of the make-up with bacteria from unclean make-up applicators.)

The make-up palette should be cleaned with warm water and detergent, then wiped with a disinfectant solution applied using clean cottonwool. It should be stored in the ultra-violet cabinet.

Mascara should be applied using a disposable brush applicator, fresh for each client.

Sharpen cosmetic pencils with a pencil sharpener before each use.

Make-up sponges should be disposed of after use, or washed in warm water and detergent; then placed in a disinfectant solution and rinsed. Allow them to dry, then place them in the ultra-violet cabinet, with each side being exposed for at least 20 minutes.

HEALTH & SAFETY: AVOIDING CROSS-INFECTION

To avoid cross-infection, don't use directly on a client the applicators supplied with cosmetic products.

TIP

Keep a store of spare make-up sponges – they soon crumble with repeated cleaning.

HEALTH & SAFETY: MAINTAINING HYGIENE

If you drop a make-up tool on the floor, discard it – don't use it again until it has been sanitised.

HEALTH & SAFETY: HYGIENE AT HOME

Advise clients to keep their own make-up clean. Dirty brushes spoil effective make-up application, and offer breeding grounds for bacteria.

Preparing the cubicle

The make-up room should be decorated in light, neutral colours to avoid the creation of unnecessary shadows. The area where the make-up is to be applied should be well lit, ideally from the same kind of light as that in which the make-up will be seen.

Place all the equipment and materials required on the trolley or work surface, in front of a make-up mirror. If you are displaying the make-up on a trolley, place the cosmetic products on a lower shelf until required, when they can be moved to the top shelf. This avoids cluttering the working area.

HEALTH & SAFETY: VENTILATION

Good ventilation is important so that the client's skin does not become too warm. It is easier to apply make-up effectively to a cool skin, and the make-up will be more durable.

Lighting

You need to know the *type* of light in which the proposed make-up will be seen: this is important when deciding upon the correct choice of make-up colours, because the appearance of colours may change according to the type of light. Is the make-up to be seen in daylight, in a fluorescent-lit office, or a softly-lit restaurant?

White light (natural daylight) contains all the colours of the rainbow. When white light falls on an object, it absorbs some colours and reflects others: it is the *reflected* colour that we see. Thus, an object that we see as red is an object that absorbs the colours in white light *except* red. A *white* object reflects most of the light that falls upon it; a *black* object absorbs most of the light that falls on it.

If the make-up is to be worn in natural light, choose subtle make-up products as daylight intensifies colours.

If the make-up is to be worn in the office, it will probably be seen under **fluorescent light**. This contains an excess of blue and green, which have a 'cool' effect on the make-up: the red in the face does not show up and the face can look drained of colour. Reds and yellows should be avoided, as these will not show up; blue-toned colours will. Don't apply dark colours, as fluorescent light intensifies these.

TIP

If working under artificial light, use warm fluorescent light for day make-up, as this closely resembles natural light.

Evening make-up is usually seen in **incandescent light** – light produced by a filament lamp. This produces an excess of red and yellow light, which creates a warm, flattering effect. Almost all colours can be used in this light, except that browns and purples will appear darker. Choose a lighter foundation than normal to reflect the light, and use frosted highlighting products where possible for the same reason.

Because it is necessary to choose brighter colours and to emphasise facial features using contouring cosmetics, evening make-up will appear very obvious and dramatic in daylight.

Explain to the client the reasons for the effect created, so that when she leaves the salon in daylight she won't feel that the make-up is inappropriate.

PREPARING THE CLIENT

Take the client through to the treatment area. Make-up application may take place either at the make-up chair, in front of a mirror, or at the treatment couch. Before you start the make-up, discuss and plan the make-up with the client, recording significant details on her record card.

Before preparing the client for the treatment, wash your hands. The client need remove only her upper outer clothing, to her underwear. Offer the client a gown, or drape a clean towel or make-up cape across her chest and shoulders. Place a headband or hairclip around the hairline, to protect the hair and keep it away from the face. Any jewellery in the treatment area should be removed. Refer to the record card to check for any known allergies to cosmetic products.

TIP

If a headband is used, remove it directly after the facial cleanse so that it doesn't flatten and spoil the hair.

HEALTH & SAFETY: CONTACT LENSES

If the client wears contact lenses, ask her whether she wishes to remove them before the skin is cleansed. (The need for this will depend on the sensitivity of the eyes.)

After preparing the client, and before touching the skin, wash your hands again. Now cleanse and tone the skin, using products appropriate to her skin type. Just as skin-care products vary in their formulation to suit the various skin types, so make-up products are designed for different skins. Record all relevant details on the record card.

Inspect the skin using a magnifying lamp. Identify any areas that require specific attention.

Apply a light-textured moisturiser before make-up application. This has the following benefits:

- it prevents the natural secretions of the skin changing the colour of the foundation;
- it seals the surface of the skin, and prevents absorption of the foundation into the skin;
- it facilitates make-up application by providing a smooth base.

Remove excess moisturiser by blotting the skin with a facial tissue.

TIP

Existing cosmetic eyeliner is sometimes difficult to remove. Use a cotton bud soaked in eye make-up remover: gently stroke this along the base of the lower eyelashes, in towards the nose.

APPLYING THE MAKE-UP

The make-up sequence

- Conceal any blemishes.
- Apply foundation.
- Contour the face.

TIP

Concealing may be completed before or after foundation application.

- Apply powder.
- Apply blusher.
- Apply eyeshadow.
- Make up the eyebrows.
- Apply mascara.
- Make up the lips.

Concealing blemishes

Before you begin to apply make-up to the face, inspect the skin and identify any areas that require concealing, such as blemishes, uneven skin colour or shadows.

Foundation may be used to disguise minor skin imperfections, but where extra coverage is required it is necessary to apply a special **concealer**, a cosmetic designed to provide maximum skin coverage. The concealer may be applied directly to the skin after skin moisturising, or following application of the foundation.

Choose a concealer that matches the client's skin tone as closely as possible. If the concealer is to be applied *after* the foundation, it should match the colour of the *foundation*.

Concealer can contain pigment to help correct skin tone.

- *Green* helps to counteract high colouring, and to conceal dilated capillaries.
- *Lilac* counteracts a sallow skin colour.
- *White or cream* helps to correct unevenness in the skin pigmentation.

Concealers come in a range of colours, to suit all skins. Mix different colours together to obtain the required colour.

TIP

Asian clients very often have darker skin underneath the eye area, and this may require concealing.

Applying concealer

Remove a small quantity of the concealer from its container, using a clean disposable spatula.

Apply the concealer to the area to be disguised, using either a clean make-up sponge or a soft make-up brush. Blend the concealer to achieve a realistic effect.

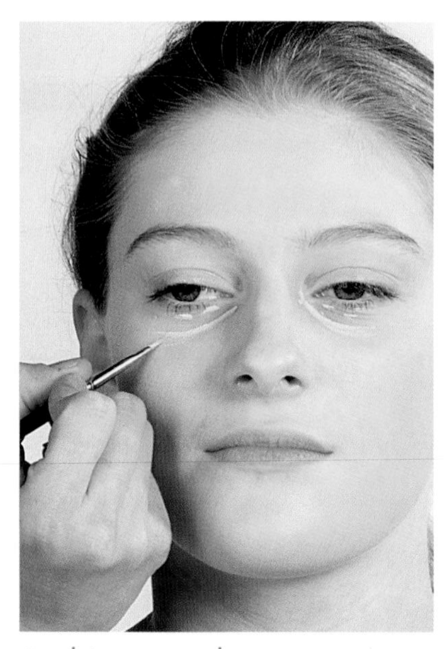

Applying concealer

Remedial camouflage

Scars, birthmarks and tattoos can be covered using a densely pigmented concealer. This technique is called **remedial camouflage**.

HEALTH & SAFETY: MEDICATED CONCEALERS

Advise clients to avoid medicated 'lipstick'-type concealers – these are too thick for general concealing. They are also unhygienic, as they are designed to be directly applied to a blemish: this means that the product becomes a breeding ground for bacteria.

TIP

Avoid rubbing the product whilst blending it, or it will wipe off.

The foundation

Foundation is applied to produce an even skin tone, to disguise minor skin blemishes, and as a contour cosmetic. Black skin in particular often has an uneven skin tone, requiring certain parts of the face to be lightened and others darkened to produce an even skin tone.

Foundation is available as a cream, a liquid, a gel or a cake. It is composed of water, powder, oil, a humectant (such as glycerol), inorganic pigments, and additives which protect the skin from the environment (such as sunscreens and moisturisers).

Some foundations contain **'anti-ageing' ingredients** such as vitamin E. These are to neutralise **free radicals**, natural chemicals thought to be responsible for damaging the skin and producing the signs of ageing – the lines and wrinkles!

ACTIVITY: COMPARING FOUNDATIONS

Compare the proportions of ingredients contained in foundations for each skin type. How, and why, are they different?

TIP

To achieve a light, healthy, natural appearance, the client may apply a tinted moisturiser.

Kinds of foundation

Each foundation differs in its formulation to suit a particular skin type. The correct choice will guarantee that the foundation lasts throughout the day.

Cream foundations

Cream foundations are oil-based, and blend easily on application. They provide a heavy coverage, and have these specific treatment uses:

- dry skin;
- normal skin;
- mature skin.

TIP

Cream and cake foundation can settle in creases and accentuate wrinkles. Apply only a very fine film of such foundations over these areas.

Liquid foundations

Liquid foundations are oil- or water-based, providing a light to medium coverage. Oil-based liquid foundations have these specific treatment uses:

- dry skin;
- normal skin;
- mature skin;
- combination skin (apply the foundation to the *dry* areas).

Water-based liquid foundations have the following uses:

- normal skin,
- oily skin,
- combination skin (apply the foundation to the *greasy* areas).

Water-based foundations do not spread very easily because the water content rapidly evaporates, so these foundations must be applied quickly.

TIP

Special ingredients may be included in a treatment foundation for a greasy skin, selected to avoid clogging the pores and thereby causing congestion.

Gel foundations

Gel foundations provide a sheer, non-greasy coverage. They have these specific treatment uses:

- black, unblemished skin;
- tanned skin;
- skin on which a natural effect is required.

TIP

On an ebony black skin, avoid powder-based foundations as these will make the skin look grey and dull. Use a transparent gel foundation instead.

Cake foundations

Cake foundations may have an oil, a wax or a powder base. They give a heavy coverage, and have these specific treatment uses:

- dry skin;
- normal skin;
- badly blemished or scarred skin.

HEALTH & SAFETY: CHOOSING A FOUNDATION

- If the skin is greasy and blemished, a *medicated* foundation may be used.
- If the client suffers from acne vulgaris, it is preferable not to apply foundation at all – bacterial infections might be aggravated.
- If the skin is sensitive, select a *hypoallergenic* foundation.

The colour of the foundation

The colour of the foundation should match the client's natural skin colour. Test the foundation for compatibility on the client's jawline or forehead. If an incorrect colour is selected, or if the foundation is insufficiently blended on application, there will be a noticeable **demarcation line**.

ACTIVITY: SELECTING A FOUNDATION

Name a suitable foundation for each of the following skin types:

- normal;
- dry;
- greasy;
- combination;
- sensitive;
- blemished.

Skin colour	*Foundation colour*
Fair	Ivory or light beige, with warm tones of pink or peach
Olive	Dark beige or bronze
Suntanned	Bronze
Florid	Matt beige with a green tint
Sallow	Beige with a pink tint
Light brown	Light brown foundation with a warm tone
Medium brown	Light brown with an orange tone
Dark brown	Deep bronze foundation with an orange tone
Black	Dark golden bronze (usually a gel)

TIP

The make-up palette is useful when mixing foundations to match the colour of your client's skin.

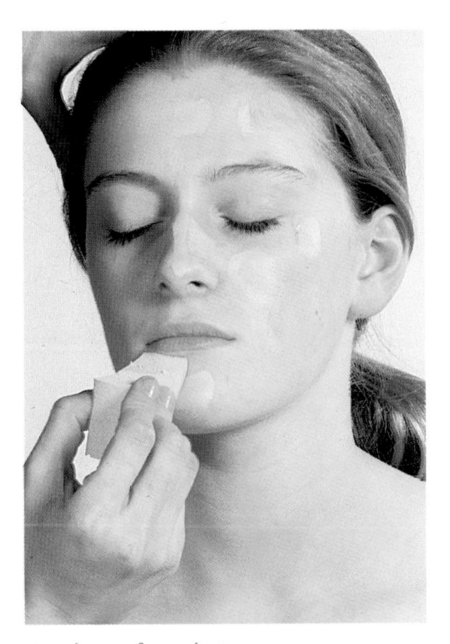
Applying foundation

TIP

Avoid applying foundation with the fingers. Apart from being less hygienic, with this method the warmth of your hands may cause streaking.

Applying the foundation

If the foundation is in a jar, remove some from its container using a clean disposable spatula. Put it on a clean make-up palette.

Foundation may be applied using either a large soft brush, which is stroked over the surface of the skin, or a cosmetic sponge. It should be applied to one area of the face at a time, with an outward stroking movement.

Use a cosmetic sponge to blend the foundation. Take care that you blend it at the hairline and at the jawline. Avoid clogging the eyebrows with foundation. The **cosmetic make-up wedge** is designed to apply varying amounts of pressure to the different areas of the face, and to ensure even coverage of the foundation.

When applying foundation around the eye area, use a small soft brush or the angular edge of a cosmetic sponge. This will help you achieve accuracy in application.

The extent of coverage can be controlled by the method of application. If the cosmetic sponge is damp, coverage is light and sheer. To achieve a heavier coverage, use a dry latex sponge.

Apply foundation to cover the entire face, including the lips and the eyelids. Do not extend the make-up past the jawline unless the occasion requires this – for example, if a bride's dress exposes part of the upper chest – as the foundation will mark clothes at the neckline.

Contouring

Contour cosmetics

Changing the shape of the face and the facial features can be achieved with the careful application of **contour cosmetics**. These products draw attention either towards or away from facial features, and can create the optical illusion of perfection.

Contour cosmetics include **highlighters**, **shaders** and **blushers**. They are available in powder, liquid and cream forms.

- Highlighters draw attention towards – they emphasise.
- Shaders draw attention away – they minimise.
- Blushers add warmth to the face, and emphasise the facial contours.

Each face differs in shape and size, so each requires a different application technique.

Applying blusher

Some blushers appear very vibrant in the container, yet when they are applied to the skin they appear very subtle.

Powder blushers

1. Stroke the contour brush over the powder blusher. Tap the brush gently, to dislodge excess blusher.
2. Apply the blusher to the cheek area, carefully placing the product according to the effect you wish to create. The direction of brush strokes should be upwards and outwards, towards the hairline. Keep the blusher away from the nose, and avoid applying blusher too near the outer eye.
3. Apply more blusher if necessary. The key to successful blusher application is to build up colour slowly until you have achieved the optimum effect.

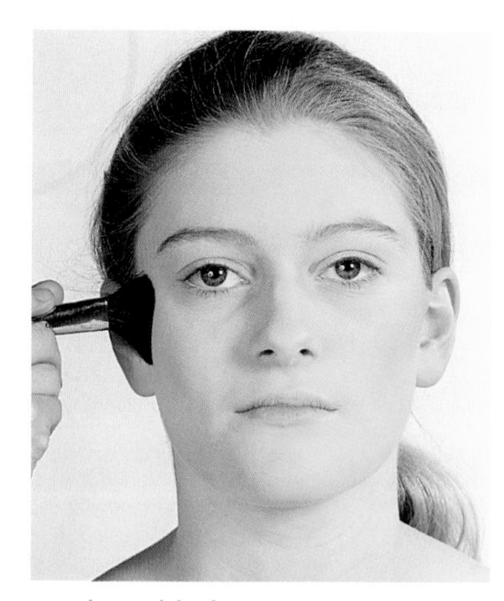

Applying blusher

Cream blushers

Apply the cream blusher after foundation application, then place a loose translucent powder over the cream blusher.

If liquid or cream cosmetics are used, these must be applied on top of a liquid or cream foundation, *before* powder application. (If powder contour products are used, these should be applied *after* the application of the loose face powder. The rule of contour cosmetic application is: powder on powder; cream on cream.)

Before applying these products, decide on the effect you wish to achieve. Study the client's face from the front and side profiles, and determine what facial corrective work is necessary.

TIP

Remember – it is easier to apply more blusher than to remove excess, which would disturb the foundation!

TIP

Foundation may be suitable as a contour cosmetic. Choose a foundation either two shades lighter (as a highlighter) or two shades darker (as a shader) then the base foundation.

Face shapes

To assess the client's face shape, take the hair away from the face – hairstyles often disguise the face shape. Study the size and shape of the facial bone structure. Consider the amount of excess fat, and the muscle tone.

ACTIVITY: FACIAL BONE STRUCTURE

Draw and label the main facial bones. It is the differing sizes and proportions of these bones that give us our individual features.

To discover the size of each facial feature, feel the bony prominences of your own face with your fingers.

Oval

This is regarded as the perfect face shape. Corrective make-up application usually attempts to create the *appearance* of an oval face shape.

Oval

TIP

Keep blusher away from the centre of the face, to avoid accentuating the breadth of the face.

Round

Square

Draw attention to the cheekbones by applying shader beneath the cheekbone, and highlighter above. Blusher should be drawn along the cheekbone and blended up towards the temples.

Round

Bone structure Broad and short.
Corrective make-up Apply highlighter in a thin band down the central portion of the face, to create the illusion of length. Shader may be applied over the angle of the jaw to the temples. Apply blusher in a triangular shape, with the base of the triangle running parallel to the ear.

Square

Bone structure A broad forehead and a broad, angular jawline.
Corrective make-up Shade the angles of the jawbone, up and towards the cheekbone. Apply blusher in a circular pattern on the cheekbones, taking it towards the temples.

Heart

Bone structure A wide forehead, with the face tapering to a narrow, pointed chin, like an inverted triangle.
Corrective make-up Highlight the angles of the jawbone and shade the point of the chin, the temples and the sides of the forehead. Apply blusher under the cheekbones, in an upward and outward direction towards the temples.

Diamond

Bone structure A narrow forehead, with wide cheekbones tapering to a narrow chin.
Corrective make-up Apply shader to the tip of the chin and the height of the forehead, to reduce length. Highlight the narrow sides of the temples and the lower jaw. Apply blusher to the fullness of the cheekbones, to draw attention to the centre of the face.

Heart

Diamond

Oblong

Bone structure Long and narrow, tapering to a pointed chin.
Corrective make-up Apply shader to the hairline and the point of the chin, to reduce the length of the face. Highlight the angle of the jawbone and the temples, to create width. Blend blusher along the cheekbones, outwards towards the ears.

Oblong

Pear

Bone structure A wide jawline, tapering to a narrow forehead.
Corrective make-up Highlight the forehead, and shade the sides of the chin and the angle of the jaw. Apply blusher to the fullness of the cheeks; or blend it along the cheekbones, up towards the temples.

Pear

ACTIVITY: CONTOURING
Study three different clients or colleagues. Identify their face shapes. Where would you apply the contouring cosmetics for each face shape, and why?

Features

Noses

- *If the nose is too broad* Apply shader to the sides of the nose.
- *If the nose is too short* Apply highlighter down the length of the nose, from the bridge to the tip.
- *If the nose is too long* Apply shader to the tip of the nose.
- *If there is a bump on the nose* Apply shader over the area.
- *If there is a hollow along the bridge of the nose* Apply highlighter over the hollow area.
- *If the nose is crooked* Apply shader over the crooked side.

ACTIVITY: CORRECTING NOSE SHAPES
Think of the different nose shapes you may encounter, such as Roman, turned up, bulbous, or with a long tip. Which contour cosmetics would you select to correct each? Where would you apply them?

TIP
An Asian face may appear as a rather flat plane: the skilful application of shading and highlighting products can create highs and lows.

Foreheads

- *If the forehead is prominent* Apply shader centrally over the prominent area, blending it outwards toward the temples.
- *If the forehead is shallow* Apply highlighter in a narrow band below the hairline.
- *If the forehead is deep* Apply shader in a narrow band below the hairline.

TIP
Foreheads can be improved by a flattering hairstyle.

Chins

- *If the jaw is too wide* Apply shader from beneath the cheekbones and along the jawline, blending it at the neck.
- *If the chin is double* Apply shader to the centre of the chin,

blending it outwards along the jawbone and under the chin.

- *If the chin is prominent* Apply foundation to the tip of the chin.
- *If the chin is long* Apply shader over the prominent area.
- *If the chin recedes* Apply highlighter along the jawline and at the centre of the chin.

Necks

- *If the neck is thin* Apply highlighter down each side of the neck.
- *If the neck is thick* Apply shader to both sides of the neck.

Powdering

Face powder is applied to set the foundation, disguising minor blemishes and making the skin appear smooth and oil-free. It also protects the skin from the environment by acting as a barrier.

Most powders are based on **talc** as the main ingredient, but talc particles are of uneven size, and substitutes such as **mica** are now becoming popular. These give a more natural, flattering effect to the skin's appearance.

Powder adheres to the foundation through the addition of **zinc**, **magnesium stearate** or **fatty esters**. These chemicals set the make-up and remove tackiness. Further powder products can then be applied to the skin.

Face powder contains absorbent materials such as **precipitated chalk** or **nylon derivatives**. These absorb sweat and sebum throughout the day, reducing shine and giving the foundation greater durability.

During manufacture, the insoluble substances in face powder go through a process called **micronisation** in which the particles are finely ground to make a powder, which is then thoroughly blended. Colour pigment may be added to produce different shades and effects.

Face powders

There are two basic products: loose powders and pressed powders.

Loose powders

Loose powders do not contain any oils or gums to bind the powder together. They are available in a range of shades, with different pastel pigments being added to counteract skin imperfections. Colours include pink and lilac, which are flattering when viewed under artificial lights; yellow, which enhances a

HEALTH & SAFETY: AVOIDING CONTAMINATION
Before application, always remove sufficient loose powder from the container. This minimises the chances of bacteria entering the powder.

tanned skin; and green, which counteracts a red skin. Iridescent ingredients may be included to produce shimmering and highlighting effects.

Many cosmetic products contain **titanium dioxide**, an opaque white pigment, to provide coverage. When applied to a black skin, this can give the skin a chalky appearance. When selecting products for a black skin, bear in mind not just the shade but also the ingredients.

TIP

Some powders reflect light whilst appearing subtle and non-shiny. These are most flattering for a mature skin, as wrinkles appear less obvious.

Pressed powders

Pressed powders contain a gum, mixed with the ingredients to bind them together. These powders provide a greater coverage, especially if they contain titanium dioxide. Pressed powders should be recommended only for a client's personal use, and then to only remove shine from the skin during the day, as required.

HEALTH & SAFETY: HOME USE

If the client uses a pressed powder, advise her to wash the powder applicator regularly to minimise the reproduction of bacteria.

TIPS

Don't apply face powder to an excessively dry skin, as it would aggravate and emphasise the dry skin condition.

Beware of applying powder if a client has superfluous facial hair, as it may emphasise this.

Applying powder

Face powder is applied *after* the foundation, unless a water-based foundation or a combination powder foundation has been selected. Select a matt powder for a daytime make-up, and an iridescent powder for an evening make-up.

1. Remove the loose powder from its container, using a clean spatula or – if the powder is in a shaker – by sprinkling it out. Place the powder on a clean facial tissue.
2. Ask the client to keep her eyes closed. Using a clean piece of cottonwool, press the cottonwool into the powder and then press the powder all over the face.
3. Remove excess powder using a large, sterilised facial powder brush. Direct the brush strokes first up the face, which dislodges the powder, then down the face, which flattens the facial hair and removes the final residue of excess powder.

Facial contouring using powder products may now be carried out.

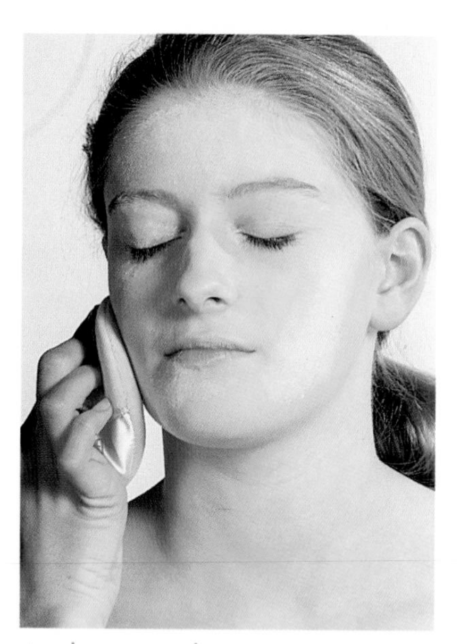

Applying powder

The eyes

Make-up is applied to the eye area to complement the natural eye colour, to give definition to the eye area, and to enhance the natural shape of the eye.

HEALTH & SAFETY: EYE COSMETICS

The eye tissue is particularly sensitive. Eye cosmetic products should be of the highest quality, and be permitted for use according to the EC Cosmetics Directive 1976.

Eyeshadow

Eyeshadow adds colour and definition to the eye area. The different types include matt, pearlised, metallic and pastel. They are available in cream, crayon or powder form. Eyeshadows are composed of either oil-and-water emulsions or waxes containing inorganic pigments to give colour.

TIP

Cream eyeshadows are not very popular – they are difficult to blend, and quickly settle into creases. They are usually used by clients who have a dry, mature skin.

- **Cream eyeshadows** These contain wax and oil.
- **Crayon eyeshadows** These are composed of wax and oil, and are similar in appearance and application to an eyepencil.
- **Powder eyeshadows** These have a talc base, mixed with oils to facilitate application. Lighter shades are produced by the addition of **titanium dioxide** – avoid these on a dark skin as they contrast too harshly with the natural skin colour.

Pearlised eyeshadows are created by the addition of **bismuth oxychloride** or **mica**; a *metallic* effect is created by the addition of fine particles of **gold leaf**, **aluminium** or **bronze**.

Applying eyeshadow

Applying eyeshadow

Eyeshadow application will differ according to the eye shape of the client.

1. Protect the skin beneath the eye with a clean tissue – this is to collect small particles of eyeshadow that may fall during application.
2. Lift the skin at the brow slightly to keep the eye tissue taut, enabling you to reach the skin near to the base of the eyelashes.
3. Apply the selected eyeshadow to the eyelid, using a sponge or a brush applicator.
4. Highlight beneath the browbone.
5. Using a brush, apply a darker eyeshadow to the socket area, beginning at the outer corner of the eye. Blend the colour evenly, to avoid harsh lines.

TIP

When applying powder colours, always tap the brush before application, to remove excess eyeshadow. If too much colour is deposited on the applicator, stroke it over a clean tissue to remove the excess.

HEALTH & SAFETY: EYEPENCIL

A good-quality eyepencil will be quite soft when applied to the skin, to avoid dragging the delicate eye tissue.

Eyeliner

Eyeliner defines and emphasises the eye area. It is available in pencil, liquid or powder form.

- **Eyepencil** This is made of wax and oil, and contains different pigments which give it its colour.

- **Liquid eyeliner** This is a gum solution, in which the pigment is suspended.
- **Powder eyeliner** This has a powder base with the addition of mineral oil.

Powder eyeliner is the most suitable choice for a client who lives in a hot country, as it will not smudge.

TIP

Have a clean cotton bud available so that you can remove the powder from any minor mistakes during application.

Applying eyeliner

If you are using a powder or liquid eyeliner, apply it with a clean eyeliner brush.

1. Lift the skin gently upwards at the eyebrows, to keep the eyelid firm and make application easier.
2. Draw a fine line along the base of the eyelashes, as required.
3. Lightly smudge the eyeliner to soften the effect of the line.

Applying eyeliner to the upper lashline

Applying eyeliner to the base lashline

Eyebrow colour

Eyebrow colour emphasises the eyebrow, and can make sparse eyebrows look thicker. It is available in pencil or powder form.

- **Eyebrow pencil** This is firmer than an eyepencil, and is composed of waxes which hold the inorganic pigments.
- **Powder brow colour** This is composed of a talc base, mixed with mineral oil and pigments.

Applying eyebrow colour

1. Select an appropriate colour of powder or eyebrow pencil.

Applying eyebrow colour

Brush the eyebrows with a clean brow brush, to remove excess face powder and eyeshadow.

2. Simulate the appearance of brow hair by using fine strokes of colour, or disguise bald patches with a denser application.
3. Brush the eyebrows into shape.

HEALTH & SAFETY: MASCARA
Mascara when purchased is usually provided with a brush applicator. This applicator cannot be effectively cleaned and sanitised, however, so it should not be used. Instead use a disposable mascara brush for each client.

TIP
Clear mascara makes the lashes appear thicker, whilst appearing very natural.

Mascara

Mascara enhances the natural eyelashes, making them appear longer, darker and thicker. It is available in liquid, cream and block-cake forms. It is composed of waxes or an oil-and-water emulsion, and contains pigments which give it its colour.

- **Liquid mascara** This is a mixture of gum in water or alcohol; the pigment is suspended in this. It may also contain short textile filaments which adhere to the lashes, having a thickening, lengthening effect. Water-resistant mascara contains resin instead of gum, so that it will not run or smudge.
- **Cream mascara** This is an emulsion of oil and water, with the pigment suspended in this.
- **Block mascara** This is composed of mineral oil, lanolin and waxes, which are melted together to form a block on setting. It must be dampened with water before application.

HEALTH & SAFETY: ALLERGIES
If the client has hypersensitive eyes or skin, use hypoallergenic cosmetics that contain no known sensitisers.

HEALTH & SAFETY: CONTACT LENSES
If the client wears contact lenses, don't use either lash-building filament mascaras or loose-particled eyeshadows, which have a tendency to flake and may enter the eye.

Applying mascara

Applying mascara

Using a disposable mascara brush, apply mascara to the eyelashes:

1. Hold the brush horizontally to apply colour to the length of the lashes. Where the lashes are short and curly, or difficult to reach, hold the brush vertically and use the *point* of the brush.
2. Place a clean tissue underneath the base of the lower eyelashes, and stroke the brush down the length of the lashes from the base the tips.
3. Lift the eyelid at the brow bone. Ask the client to look down slightly whilst keeping her eyes open. From above, stroke down the length of the lashes from the base to the tips.

4. Using a zigzag motion, draw the brush upwards through the upper and lower surfaces of the lashes, from the base to the tips.
5. Finally, separate the eyelashes with a clean brush or lash comb.

TIP

Curly lashes will require brushing, using a clean brush, both before mascara application and after each coat, to separate the lashes.

ACTIVITY: CHOOSING MASCARA

What colour mascara should be applied if the client had the following hair colouring: brown, auburn, black or grey?

Eye make-up for the client who wears glasses

If the client wears glasses, check the function of the lens, as this can alter the appearance and effect of the eye make-up.

- *If the client is short-sighted, the lens makes the eye appear smaller.* Draw attention to the eyes by selecting brighter, lighter colours. When applying eyeshadow and eyeliner, use the corrective techniques for small eyes. Apply mascara to emphasise the eyelashes.
- *If the client is long-sighted, the lens will magnify the eye.* Make-up should therefore be subtle, avoiding frosted colours and lash-building mascaras. Careful blending is important, as any mistakes will be magnified!

Corrective eye make-up

Dark circles

1. Minimise the circles by applying a concealing product.

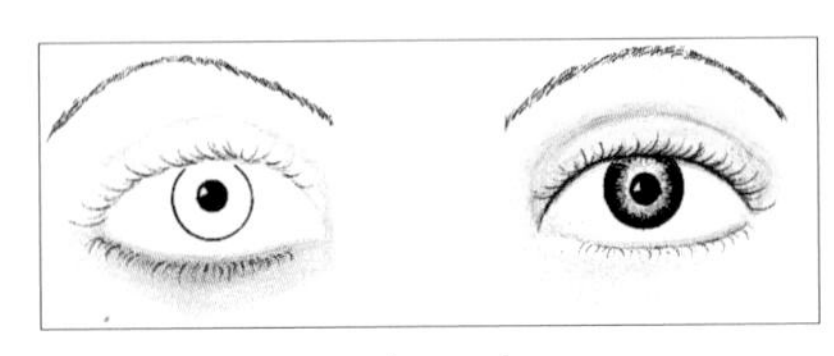

Dark circles

Wide-set eyes

1. Apply a darker eye colour to the inner portion of the upper eyelid.
2. Apply lighter eyeshadow to the outer portion of the eyelid.
3. Apply eyeliner in a darker colour to the inner half of the upper eyelid.
4. Eyebrow pencil may be applied to extend the inner browline.

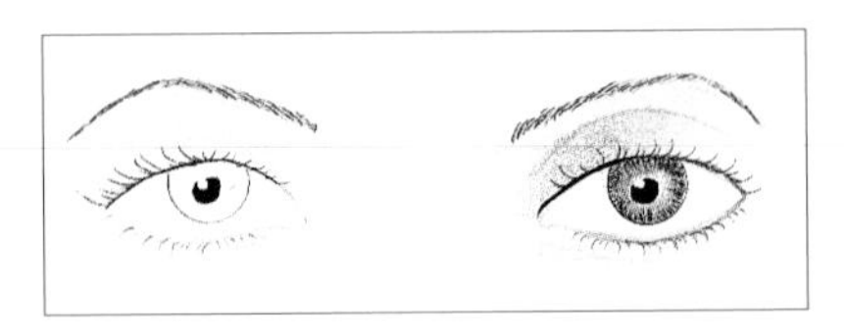

Wide-set eyes

Close-set eyes

1. Lighten the inner portion of the upper eyelid.
2. Use a darker colour at the outer eye.
3. Apply eyeliner to the outer corner of the upper eyelid.
4. Pluck brow hairs at the inner eyebrow – this helps to create the illusion of the eyes being further apart.

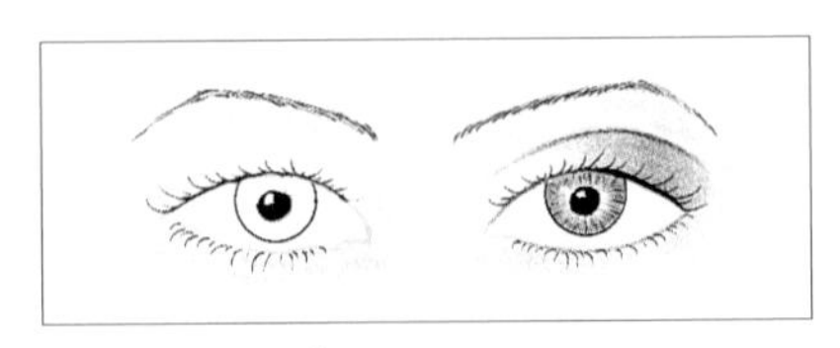

Close-set eyes

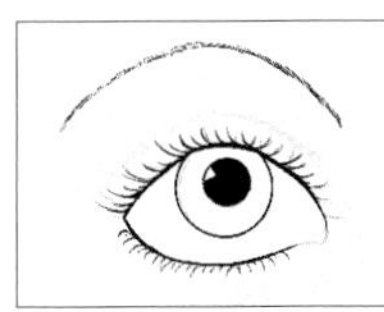 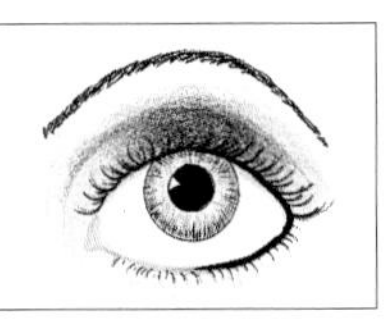

Round eyes

Round eyes

1. Apply a darker colour over the prominent central upper-lid area.
2. Elongate the eyes by applying eyeliner to the outer corners of the upper and lower eyelids.

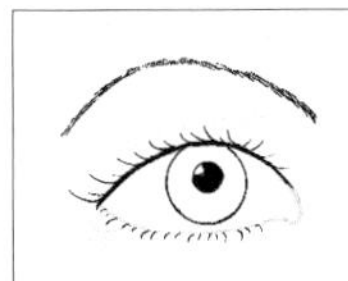 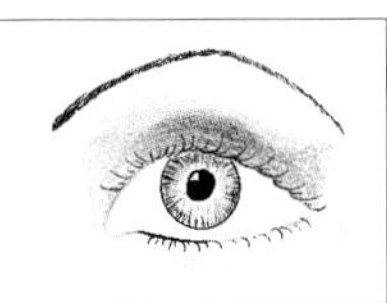

Prominent eyes

Prominent eyes

1. Apply dark matt eyeshadow over the prominent upper eyelid.
2. Apply a darker shade to the outer portion of the eyelid, and blend it upwards and outwards.
3. Highlight the browbone, drawing attention to this area.
4. Eyeliner may be applied to the inner lower eyelid.

TIP

False eyelashes may be effective in enhancing the eye's natural shape.

TIP

To make the eyes appear less prominent, select matt eyeshadows – pearlised and frosted eyeshadows will highlight and emphasise the eye.

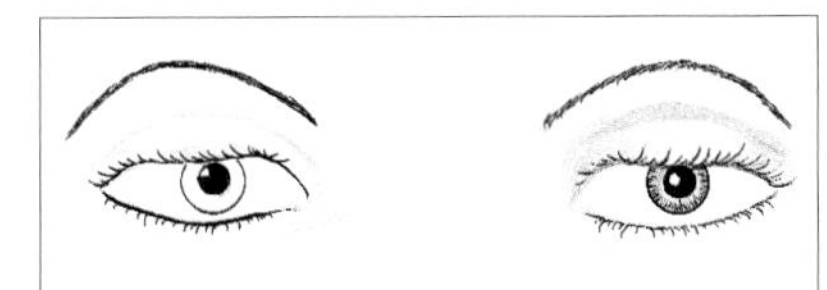

Overhanging lids

Overhanging lids

1. Apply a pale highlighter to the middle of the eyelid.
2. Apply a darker eyeshadow to contour the socket area, creating a higher crease (which disguises the hooded appearance).

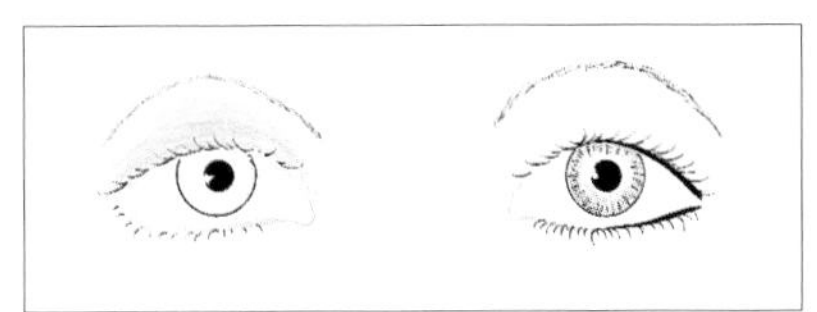

Deep-set eyes

Deep-set eyes

1. Use light-coloured eyeshadows.
2. Eyeshadow may also be applied in a fine line to the inner half of the lower eyelid, beneath the lashes.
3. Apply eyeliner to the outer halves of the upper and lower eyelids, broadening the line as you extend outwards.

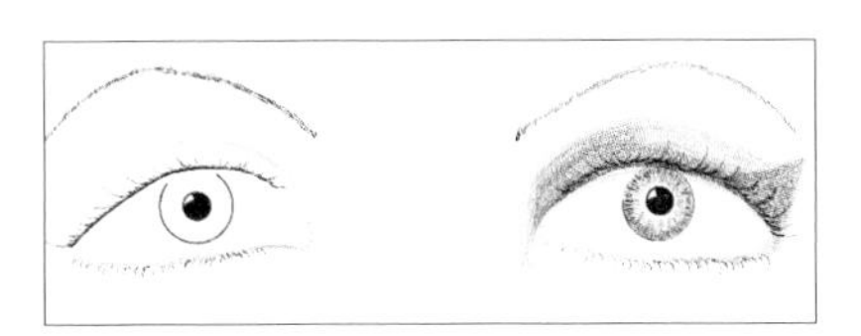

Downward-slanting eyes

Downward-slanting eyes

1. Create lift by applying the eyeshadow upwards and outwards at the outer corners of the upper eyelid.
2. Apply eyeliner to the upper eyelid, applying it upwards at the outer corner.
3. Confine mascara to the outer lashes.

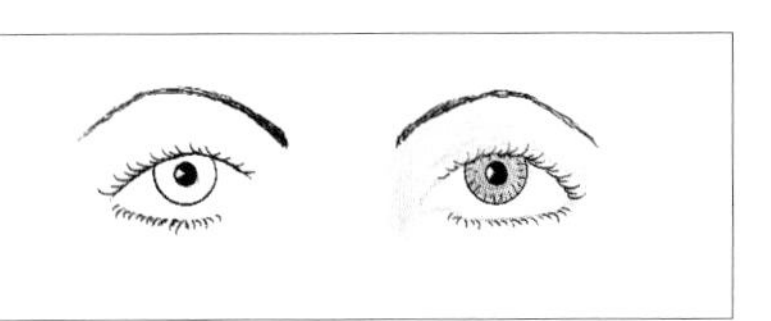

Small eyes

Small eyes

1. Choose a light colour for the upper eyelid.
2. Highlight under the brow, to open up the eye area.
3. Curl the lashes before applying mascara.

4 Apply a light-coloured eyeliner to the outer third of the lower eyelid.
5 A white eyeliner may be applied to the inner lid, to make the eye appear larger.

Narrow eyes

1 Apply a lighter colour in the centre of the eyelid, to open up the eye.
2 Apply a shader to the inner and outer portions of the eyelid.

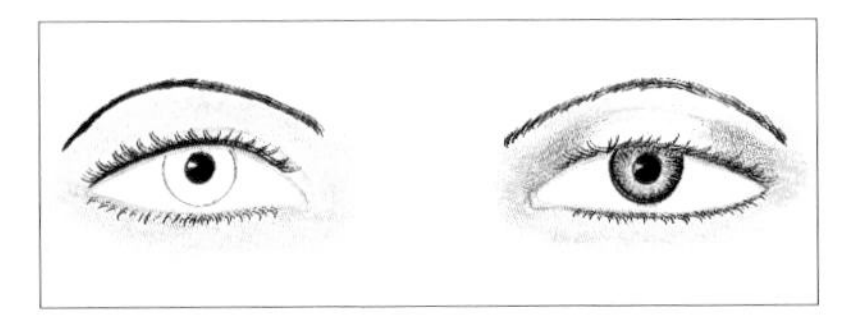

Narrow eyes

Oriental eyes

1 Divide the upper eyelid in two vertically. Place a lighter colour over the inner half of the eyelid and a darker colour at the outer half.
2 Apply a highlighter under the eyebrow.
3 White eyeliner may be applied at the base of the lashline, on the lower inner eyelid.

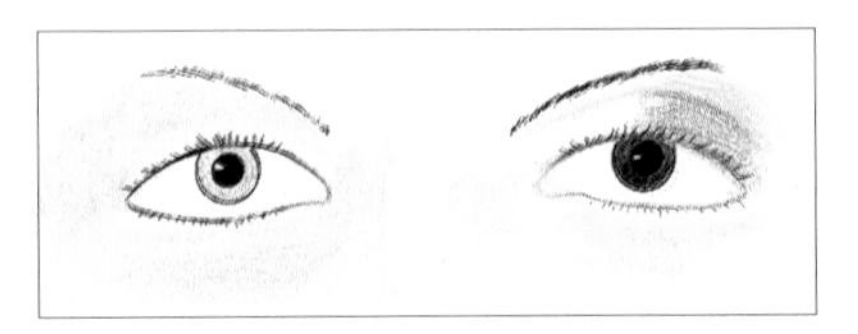

Oriental eyes

The eyelashes

To emphasise the eyelashes, making them appear longer temporarily, curl them using **eyelash curlers**.

> **TIP**
> Eyelash curling is beneficial for oriental clients who have short lashes that grow downwards.

Curling the eyelashes

1 Rest the upper lashes between the upper and lower portions of the eyelash curlers.
2 Bring the two portions gently together with a squeezing action.
3 Hold the lashes in the curlers for approximately 10 seconds, then release them.
4 If the lashes are not sufficiently curled, repeat the action.

> **HEALTH & SAFETY: EYELASH CURLING**
> Repeated eyelash curling can lead to breakage. The technique should therefore be used only for special occasions.

The lips

Lip cosmetics add colour and draw attention to the lips. As the lips have no protective sebum, the use of lip cosmetics also helps to prevent them from drying and becoming chapped.

It is not uncommon for the lips to be out of proportion in some way. Using lip cosmetics and corrective techniques, symmetrical lips can be created. A careful choice of product and accurate application are required to achieve a professional effect.

The main lip cosmetics are lip pencils, lipsticks and lipglosses.

Sometimes the lips may be unevenly pigmented. The application of a lip toner or foundation over the lips corrects this.

HEALTH & SAFETY: LIP PENCILS
Always sharpen the lip pencil before use on each client, to provide a clean, uncontaminated cosmetic surface.

Lip pencils

Lip pencil is used to define the lips, creating a perfectly symmetrical outline. This is coloured in with another lip cosmetic, either a lipstick or a lipgloss. The lipliner also helps to prevent the lipstick from 'bleeding' into lines around the lips.

Lipliner has a wax base which does not melt and can be applied easily. It contains pigments which give the pencil its colour.

When choosing a lipliner, select one that is the same colour as, or slightly darker than, the lipstick to be used with it.

HEALTH & SAFETY: LIPSTICK
For reasons of hygiene, remove a small quantity of lipstick by scraping the stick with a clean spatula – don't apply the lipstick directly.

Lipsticks

Lipstick contains a blend of oils and waxes, which give it its firmness, and silicone, essential for easy application. It also contains pigment, to add colour; an emollient moisturiser, to keep the lips soft and supple; and perfume, to improve its appeal. In addition it may include vitamins, to condition the lips, or sunscreens, to protect the lips from ultra-violet rays. Some lipsticks contain a relatively large proportion of water – these moisturise the lips and provide a natural look. The coverage provided by a lipstick depends on its formulation.

Lipsticks are available in the following forms: cream, matt, frosted and translucent. Frosted lipstick has good durability, as it is very dry. Some other lipsticks also offer extended durability, and are suitable for clients who are unable to renew their lipstick regularly.

When choosing the colour of lipstick, take into account the natural colour of the client's lips, the skin and hair colours, and the colours selected for the rest of the make-up.

Lipglosses

Lipgloss provides a moist, shiny look to the lips. It may be worn alone, or applied on top of a lipstick. Its effect is short-lived. Lipgloss is made of mineral oils, with pigment suspended in the oil.

Note that black women often have creases on the lips that extend to the surrounding skin. If lipgloss is used it will often bleed into these lines.

HEALTH & SAFETY: ALLERGIES
Lipsticks often contain ingredients that can cause allergic reactions, such as lanolin and certain pigment dyes. If the client has known allergies, use a hypoallergenic product instead.

Dry lips

Sometimes the lips become dry and chapped. Recommend that the client keeps them moisturised at all times, especially in extremes of heat, cold or wind. Some facial exfoliants can be professionally applied over the lips to remove dead skin.

If the client does not like to wear make-up during the day, or if the client is male, recommend that the lips be protected with a lip-care product.

Corrective lip make-up

Thick lips

Select natural colours and darker shades, avoiding bright, glossy colours.

1. Blend foundation over the lips to disguise the natural lip line.
2. Apply a darker lipliner inside the natural lip line to create a new line.

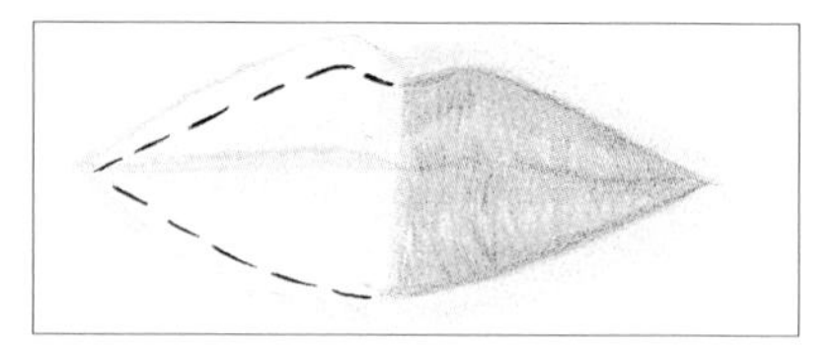

Thick lips

Thicker upper or lower lip

1. Use the technique described above to make the larger lip appear smaller.
2. Apply a slightly darker lipstick to the larger lip.
3. If the lips droop at the corners, raise the corners by applying lipliner to the corners of the upper lip, to turn them upwards.

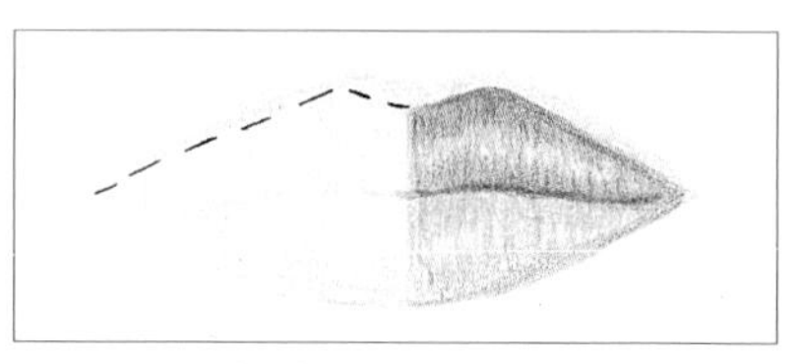

Thicker upper lip

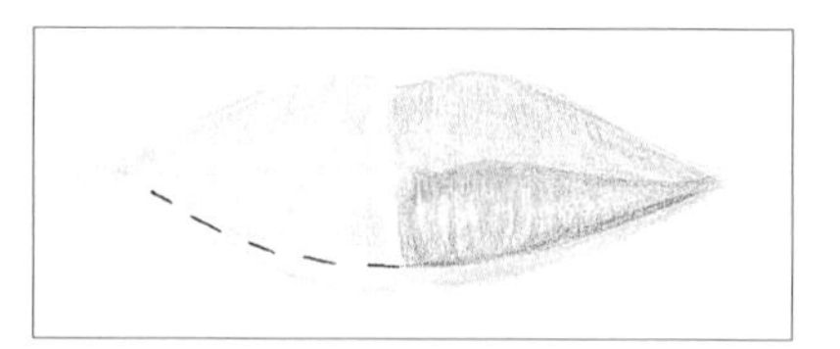

Thicker lower lip

Thin lips

Select brighter, pearlised colours. Avoid darker lipsticks, which will make the mouth appear smaller.

1. Apply a neutral lipliner just outside the natural lip line.

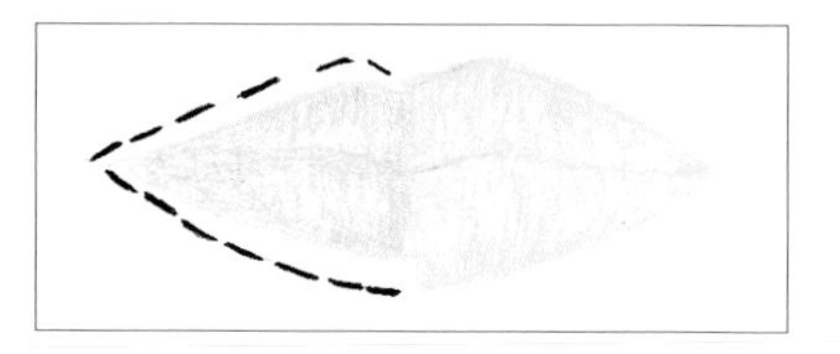

Thin lips

Small mouth

1. Extend the line slightly at the corners of the mouth, with both the upper and the lower lips.

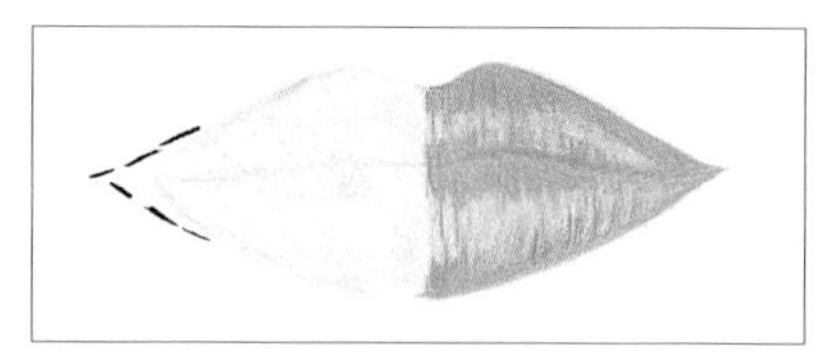

Small mouth

Uneven lips

1. Use a lipliner to draw in a new line.
2. Apply lipstick to the area.

Lines around the mouth

1. Apply lipliner around the natural lip line.
2. Apply a matt cream lipstick to the lips. (Don't use gloss, which might bleed into the lines around the mouth.)

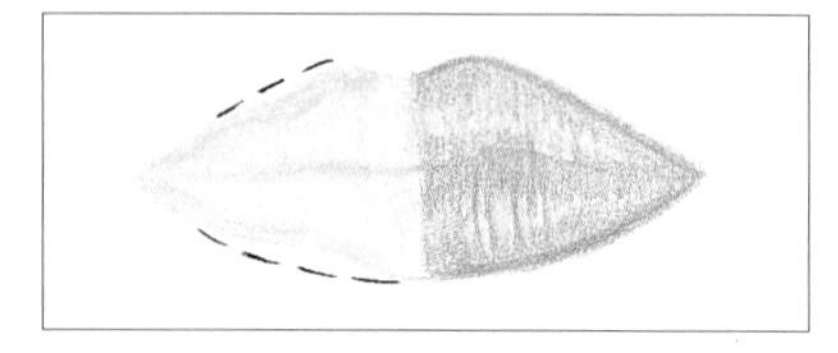

Uneven lips

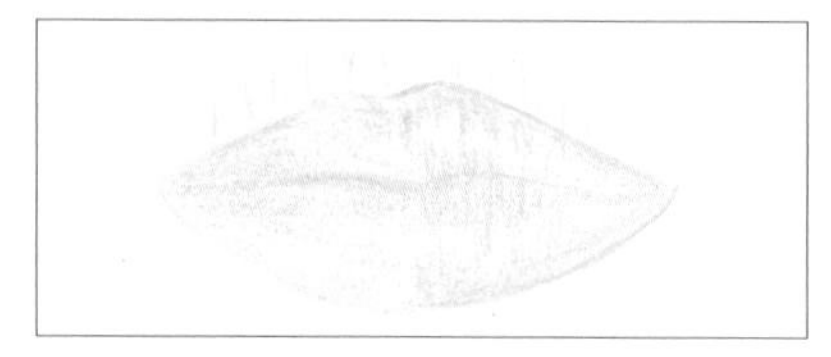

Lines around the mouth

TIP

To obtain the correct colour of lipstick, you may need to mix different lipstick shades together.

Applying lipliner

Applying lipstick

Applying lipstick

1. Select a lip pencil and lipstick to complement the client's colouring and the colour theme of the make-up.
2. Using a pencil sharpener, sharpen the lip pencil to expose a clean surface.
3. Ask the client to open her mouth slightly.
4. Outline the lips, carrying out lip correction as necessary. Begin the lip line at the outer corner of the mouth, and continue it to the centre of the lips. Repeat the process on the other side of the lip, commencing at the outer corner of the mouth.
5. Remove sufficient lipstick using a clean spatula.
6. Using a sterile lipbrush, apply the lipstick to the lips.
7. Apply a clean facial tissue over the lip area, and *gently* press it onto the lips. This process, known as **blotting**, removes excess lipstick and fixes the colour on the lips.
8. A second light application of lipstick may then be applied.
9. If desired, lipgloss may be applied over the lipstick to add sheen, again using a sterile lipbrush.

HEALTH & SAFETY: LIPBRUSHES

Disposable lipbrushes are available to enable lip cosmetics to be applied hygienically.

When you have finished . . .

After applying the make-up, fix the client's hair and then discuss the finished result in front of the make-up mirror.

Wash your hands. Record details of the treatment on the client's record card.

The completed make-up

CONTRA-ACTIONS

Certain cosmetic ingredients are known to provoke allergic reactions in some people. These allergens may cause irritation, inflammation and swelling.

Known cosmetic allergens include the following:

- *Lanolin* This is similar to sebum, and is obtained from sheep's wool. It is added to many cosmetics as an emollient.
- *Mineral oils* Examples are oleic acid and butyl stearate.
- *Eosin (bromo-acid dye)* A staining pigment, used in some lip cosmetics and perfumes.
- *Paraben* An antiseptic ingredient, used as a preservative in facial cosmetics.
- *Certain colourants* One example is carmine.
- *Perfume* This is added to most cosmetics, and is a common sensitiser.

External contact with an allergen may cause urticaria (hives or nettle rash), eczema or dermatitis. If an allergy occurs, the product should be removed from the skin and a soothing substance applied. The client should be advised not to use the product again.

HEALTH & SAFETY: ALLERGIES

Where possible, before using a new cosmetic, a client with known allergies should first receive a patch test, or be given a small sample of a product to try on a less sensitive area of the skin, to assess the skin's tolerance. If a severe allergic reaction occurs, medical attention must be sought immediately.

It is possible that clients may grow out of an allergy to a given product. It is also possible, however, suddenly to become allergic to a product that has not given problems before.

AFTERCARE AND ADVICE

Explain to the client that you can offer a make-up lesson in which you would discuss the reason for the selection of make-up products and colours, and the techniques for applying them. Recommend the correct skin-care products to remove the products you have applied.

Following make-up application it is a good idea to record on a personal make-up chart the products and colours you have applied. This chart can be given to the client, to reinforce the discussion that has taken place during any make-up lesson; it will also be invaluable if the client returns wishing to purchase any of the make-up cosmetics you have used.

TIP

Advise the client if necessary to remove facial shine from the face with pressed powder.

A make-up palette available for retail

Depilex/RVB

THE CLIENT'S NEEDS

Day make-up

The effect should be natural. Any corrective work carried out should be very subtle and kept to the minimum, as natural light makes any imperfections appear obvious.

Select a foundation the same colour as the skin – aim to even out the skin tone. Set the foundation with a translucent face powder.

Apply a subtle, warm blusher to add colour to the face. Avoid strong colours of cosmetics, especially on the eyes. The mascara colour should be chosen to complement the client's natural lash and skin colours. Mascara should be used to emphasise, but not exaggerate, the length and thickness of the eyelashes. Eyeliner may be used, but it should be carefully placed and blended.

Line the lips in a colour that will coordinate with the lipstick to be applied, which again should be quite natural.

TIP

In poor lighting, face shading should be subtle as it makes shadows appear darker.

Derrick Mullings/Media Image

Evening make-up

This should be applied bearing in mind the type of lighting in which the client will be seen. Artificial light dulls the effect, and changes the colour of the make-up: dark shades lose their brilliance, appearing 'muddy', so you need to use brighter colours. Emphasise the facial features with the careful placement of contouring cosmetics.

Areas where shadows may be created, such as the eyes, should be emphasised using light, bright, and highlighting cosmetic products. Add warmth to the face with an intense colour of blusher placed on the cheekbones. A highlighting powder in a pearlised or metallic shade may be applied directly on top of or over the blusher. The client may like to try adventurous cosmetics such as metallics and frosted eye products.

Curl the eyelashes with eyelash curlers, or apply false eyelashes to emphasise the eyes. Fashion shades of mascara may be selected, in purples, greens and blues, to complement the make-up and produce the effect required. Light shades of eyeliner may be used to frame the eyes and to 'open' them up.

Add a lipgloss to the lips, or apply a frosted lipstick to emphasise the mouth.

Colour the eyebrows, and carefully groom them to frame the eye area.

Make-up to suit skin and hair colouring

Fair skin and blonde hair

If the client has fair hair and fair skin, keep the skin colour natural. Apply a blusher in rose pink or beige.

Define the eyes with soft tones of browns and pinks. Apply a brown-black mascara.

Colour the lips with a rose-pink or peach lip colour. Avoid lip colours lighter than the natural skin tone.

Oriental skin and black hair

For creamy, sallow skin with dark hair, use blusher to add warmth and to brighten the skin, in either pink or brown.

The eyes are dark, with a prominent browbone. Emphasise the socket of the eye with careful shading; extend this upwards and outwards. Place highlighter along the browbone.

Pastel colours complement the eye colour. Select black mascara to emphasise the eyes. Deep pinks and orangey-reds suit the lips.

Philippine client (oriental skin and black hair)

A yellow-toned make-up base. Pinky shades are chosen to contrast with the natural skin tone. Pinks and brown-green are selected for the eyeshadow. Deep pink lipstick is applied

Fair skin and red hair

Redheads usually have fair skin with freckles. The skin will flush and colour easily, probably requiring the application of a green-tinted moisturiser, concealer or face powder. Apply blusher in a warm rose or peach colour. Browns, rusts, greens and peach eyeshadow colours suit this skin and complement the eyes. Brown mascara is preferable, to avoid making the eyes appear hard.

For the lips, select a lipstick in peach, golden rust or pink.

Black skin and black hair

A yellow-toned foundation is required: it may be necessary to blend foundations to obtain the correct colour. Avoid pink-toned foundations, which make the skin appear chalky.

Women with dark skin tend to have dark brown to brown-black eyes, and can use a wide range of heavily pigmented colours, especially browns and bronzes. Dark shades of blusher in red and plum may be chosen; eyeliner and mascara can be black, or any other dark shade.

Avoid lip colours lighter than the skin tone. A lipliner darker than the lip colour may be used.

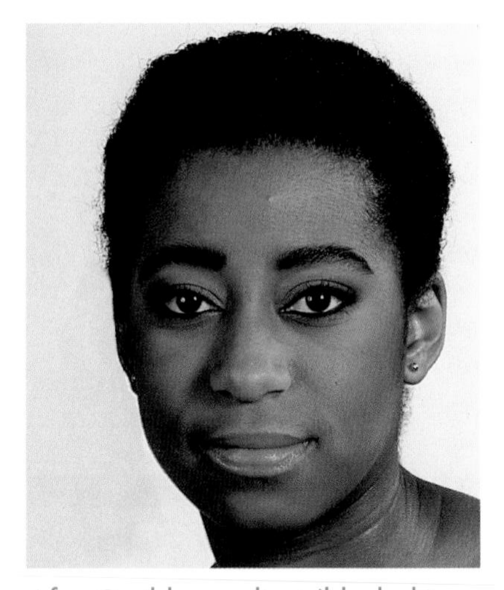

Afro-Caribbean client (black skin and black hair)

The skin tone of the face is very uneven. Mixtures of base shades are chosen to lighten some areas (especially the forehead) and to darken other areas, to achieve an even skin tone. The face is slimmed slightly with a dark brown shader at the sides of the face. Brown shades are selected for the eyeshadow, and a brown/tan blusher. Reddish-brown lipstick is applied

Mexican client (olive skin and dark hair)

Corrective work is required to conceal shadows under the eyes. A light foundation is used. The natural colouring allows the use of strong, bright colours. Orange-red lipstick is applied

Indian client (dark skin and black hair)

Corrective work is required around the eyes to even the skin tone. Browns and purples are selected for the eyeshadow. Purple-toned lipstick is applied

Caucasian client (fair skin and blonde hair)

Blemishes are covered with a skin-tone concealer. An ivory base foundation is used. Ivory, peaches and brown colours are selected for the eyeshadow. Brown shadow is used to define the eyebrows. A dark grey pencil is used to define the upper and lower outer eyelids. Dark brown mascara is applied. Peachy-brown blusher is applied. Translucent fixing facepowder is applied. Natural brown lipstick is applied

Olive or fair skin and dark hair

Select a foundation to suit the basic skin tone. If the skin is fair, choose an ivory base; if it is sallow, select a foundation with a rusty, yellow tone. (With a sallow skin, avoid the use of pinks on the eyes – they make the eyes appear sore.)

A beige blusher suits this skin colour, and is complemented by the selection of brown or green shades for the eyes. Black mascara should be used for the eyelashes.

For the lips, choose warm reds or beige.

Make-up for the mature skin

As the skin ages it becomes sallow in colour and appears thinner. Small capillaries can be seen, commonly on the cheek area, and small veins may appear around the eyes. Pigment changes in the skin become obvious, and remain permanently.

At the make-up consultation, discuss your ideas with your client. Very often a mature client will have been using the same colours and the same cosmetics for many years, and they may not even be complimentary. You will need to advise her tactfully on a fresh approach.

Select a foundation that matches the skin colour yet enhances the skin's appearance. An oil-based foundation is appropriate for use on a mature skin: it keeps the skin supple and prevents the foundation from settling into the creases and emphasising the lines and wrinkles.

A concealer may be applied to cover obvious capillaries and small veins, or a foundation may be selected which provides adequate coverage.

A lighter foundation may be applied over wrinkled areas, to make them less obvious. These areas include:

- around the eyes (crow's feet);
- between the brows;
- across the forehead;
- between the nose and the mouth (naso-labial folds);
- around the mouth (the lipline).

With age, the contours of the face lose their firmness as the fat cells that plump the face reduce, and the facial muscles lose tone and sag. Poor muscle tone can be seen in the following areas:

- the cheek area;
- loose skin along the jawline;
- loose skin under the brow and overhanging the lid;
- loose skin on the neck.

The application of a shader, subtly blended, can improve the appearance of such areas.

TIP

Dark circles under the eyes can be minimised with concealer. Select a concealer lighter than the foundation to be applied.

ACTIVITY: SHADING

Where would you place the shading product in order to correct poor muscle tone in the areas discussed above?

Apply translucent powder. It may be preferable to avoid doing so in the eye area as it can emphasise lines around the eyes. To reduce the powdery effect – which may make the skin appear dry – you may direct a fine water spray from a suitable distance, to set the make-up.

Apply a blusher with a warm tone – avoid harsh, bright shades. A cream blusher may be applied after the foundation. Place it high on the cheekbone and blend it upwards at the temples, drawing attention upwards rather than downwards.

The lip line becomes less obvious as one grows older, and lines often appear along it. Lipliner should be applied, to redefine the lips and to prevent the lipstick from 'bleeding' into the lines. Select a lipliner that is the same colour as, or slightly lighter than, the lip colour. (A darker lip colour would create an unwanted shadow.) A special lip fixative may be recommended, and a durable cream lipstick applied. Avoid the use of a gloss lipstick which would emphasise lines around the mouth.

The angle of the mouth may droop. Corrective techniques may be used to disguise this.

Powder matt eyeshadows should be selected for use on the

TIP

Cream eyeshadow settles into creases, which emphasise a crepey eyelid. Frosted or pearlised eyeshadows also emphasise a crepey eyelid.

eyelid: these soften the appearance of any lines in the area. Choose natural, light shades. Dark colours can make the eyes appear small and tired.

Eyeliner should be used, in neutral shades of brown and grey – avoid harsh, bright colours, which will give a hard appearance.

Eyebrows should be perfectly groomed, and arched to give lift to the eye. Bushy eyebrows give the eyes a hooded appearance. If eyebrow colour is required, select a colour that is slightly lighter than the brow colour, or a blend of two colours. If the client has grey hair, use grey and charcoal to provide a natural effect.

Eyelashes should be emphasised with a natural-looking shade of mascara, lightly applied. (If the eyelashes lose their colour, you can recommend that the client has her lashes professionally – and permanently – tinted. Individual false eyelashes may be recommended for corrective purposes.)

TIP

If the client is visiting the salon for a make-up, you could recommend that she have a professional make-up lesson – at which you could discuss a fresh approach to her make-up.

HEALTH & SAFETY: ALLERGIES

Sometimes the eyes become quite sensitive with age. Apply hypoallergenic mascara in such cases.

KNOWLEDGE REVIEW

Make-up

Preparation

1. Why is it important to have a variety of make-up products and colours available?
2. What should be considered when planning a make-up?
3. Why is it important to discuss your make-up plan with the client?
4. How should the skin be prepared before the application of make-up?

Application

1. For what purposes would you apply concealer?
2. What corrective make-up techniques would be applied for each of the following?
   - (a) Square face.
   - (b) High forehead.
   - (c) Narrow eyes.
   - (d) Crepey eyelid.
   - (e) Broad, short nose.
   - (f) Lined mouth.
   - (g) Protruding chin.
   - (h) Loose muscle tone at the jawline (jowls).
3. Name *three* facial bones to which you may apply contouring products during facial corrective work.

Aftercare

1. How can you promote the sale of skin-care and make-up products during a make-up application?

Health, safety and hygiene

1. What skin and eye disorders would contra-indicate make-up application?
2. How can you ensure hygienic practice when applying the following products?
   (a) Foundation.
   (b) Face powder.
   (c) Mascara.
   (d) Lipstick.
3. What product ingredients are known to cause allergic reactions, and should therefore not be used on a client with sensitive skin?

Case studies

1. What should you consider in your choice of make-up products and in your application technique when applying make-up to the following clients?
   (a) A teenage Caucasian bride, who will be getting married in white, and who does not normally wear make-up.
   (b) The Caucasian mother of the bride, who will be wearing a coral suit: she normally does wear make-up, and emphasises her eyes with bright pearlised eyeshadow colours.
   (c) A mature Asian wedding guest, who normally wears make-up.
   (d) A young Afro-Caribbean colleague of the bride, who is attending the evening function.
2. After returning from her honeymoon, the bride is to attend a job interview. She has booked an appointment for a make-up application prior to the interview; this will also be a make-up lesson. Consider an appropriate make-up for the client, and explain what should be discussed at the make-up lesson.

CHAPTER 7

Eye treatments

EYEBROW SHAPING

EYEBROW HAIR REMOVAL

The eyebrows, situated above the bony eye orbits of the face, help to protect the eyes from moisture and dust, and to cushion the skin from physical injury. Misshapen bushy brows give an untidy appearance to the face; but when correctly shaped, the brows give balance to the facial features and enhance the eyes – the most expressive feature of the face.

Hair removal is a popular treatment in the beauty salon, where both temporary and permanent methods of removal are usually available. Temporary removal must be repeated regularly, as the removed hair will regrow. With permanent methods, the client needs to visit the salon regularly to have the hair removed; thereafter, the part of the hair responsible for its growth has been destroyed and the hair will not grow again.

Temporary measures

Depilatory waxing

Depilatory waxing, using a warm, hot or cold wax, involves applying wax to the treatment area, embedding the hairs in it. When the wax is removed from the area, the hairs are removed also, at their roots. They grow again in approximately four weeks. (See Chapter 9.)

Plucking

Plucking or **tweezing** uses a pair of tweezers to remove the hair. These grasp the hair near the surface of the skin, and the hair is then plucked in the direction of growth, again removing it at its root. The hair grows again in approximately four weeks.

Threading

Threading involves the use of a thread of twisted cotton, which is rolled over the area from which the hair is to be removed: the hairs catch in the cotton, and are pulled out. This skill is frequently practised by people of Asian or Mediterranean origin.

Due to the sensitive nature of the eye tissue, tweezing is often considered the most suitable choice for temporary hair removal from eyebrows.

Permanent measures

Electrical methods

Galvanic electrolysis, **electrical epilation** and the **blend epilation technique** are all techniques that use an electrical current. The current is passed to the hair root via a fine needle inserted into the hair follicle. The current destroys the hair root, preventing hair regrowth.

The client needs to understand that the hair will never grow back if effectively treated with any of the above permanent methods. This is an important consideration when treating the brow hair, as the desired shape and thickness of the brows change frequently under the influence of fashion.

Methods to be avoided

There are other methods of hair removal, but these are unsuitable for use in the eye area and for the type of hair found there:

- *Cutting the hairs with scissors* Scissors are used to trim the hair close to the skin's surface.
- *Shaving* A razor blade is stroked over the skin, against the natural hair growth. This removes the hair at the skin's surface.
- *Depilatory cream* A strong alkaline chemical cream is applied to the hair, and removed after 5–10 minutes: the hair will have been dissolved at the skin's surface.

ACTIVITY: HAIR REMOVAL

Make short notes on the suitability and effectiveness of the different methods of hair removal for shaping eyebrow hair.

Can you think of any other methods of removing hair that might be available to the client?

RECEPTION

Eyebrow shaping is offered in the salon as either an **eyebrow reshape** or an **eyebrow trim** – the former involves removing eyebrow hair to create a new shape, the latter involves removing only a few stray hairs in order to maintain the existing shape. When making an appointment it is usual to allow 15 minutes for each service.

It is wise to have a designated time between treatments so that there is no confusion between the two services. For example, under two weeks could be regarded as a trim, and over two weeks as a reshape.

TIP

Whilst the client is having her eyebrows shaped, you have an ideal opportunity to discuss further possible treatments, such as an eyebrow tint.

> **TIP**
> At the consultation, identify any peculiarities such as bald patches or scarring in the brow area to avoid any argument later.

If the client has thick, heavy brows, or if she does not have her brows shaped regularly, she should be encouraged to have them shaped gradually over a period of weeks, until the desired shape is achieved. This will allow the client to become accustomed to the new shape and will minimise any discomfort.

Eyebrow shaping may be carried out as an independent treatment, or combined with other treatments such as permanent tinting of the brows. In the latter case, the brows should be tinted before shaping, to avoid the tint coming into contact with the open follicle and perhaps causing an allergic reaction.

Before brow shaping treatment commences, carry out a consultation. Discuss the shape and the effect that might be achieved. Consider such factors as age, the natural shape of the brow, and fashion.

If a client wears contact lenses, these must be removed before treatment commences.

CONTRA-INDICATIONS

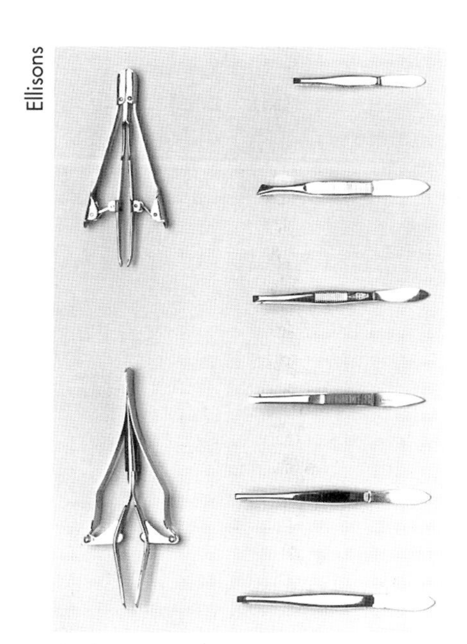
Ellisons

Eyebrow tweezers

If whilst completing the record card or on visual inspection of the skin the client is found to have any of the following, eyebrow shaping treatment must not be carried out:

- *Hypersensitive skin.*
- *Any eye disorder*, such as conjunctivitis, styes or hordeola, watery eye, blepharitis or cysts.
- *Inflammation or swelling* of the surrounding eye tissue.
- *Skin disease* in the area.
- *Skin disorder* in the area, such as psoriasis or eczema.
- *Bruising, cuts or abrasions* in the eye area.

If you are unsure about the wisdom of proceeding with treatment – for example, if there is an undiagnosed lump in the area – ask your client to seek medical approval first.

EQUIPMENT AND MATERIALS

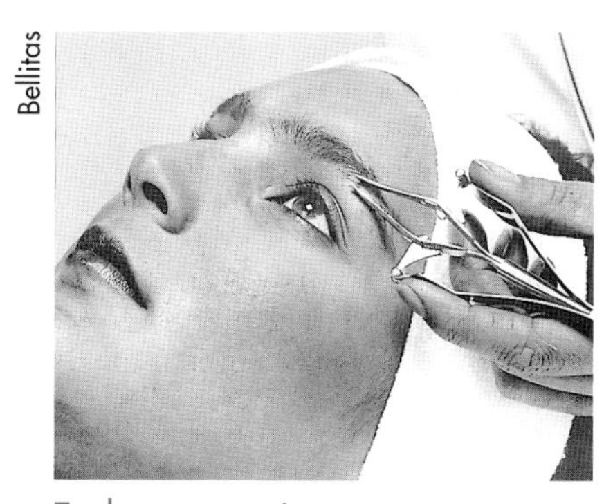
Bellitas

Eyebrow tweezing

There are two sorts of tweezers. **Automatic tweezers** are designed to remove the bulk of excess hair; they have a spring-loaded action. **Manual tweezers** are used to remove stray hairs, and to accentuate the brow shape where more accurate care is required. They are available with various ends, as illustrated – which you use is a matter of personal preference, but slanted ends are generally considered to be the best for eyebrow shaping.

Although many therapists may complete an eyebrow shaping using only one of these – automatic or manual tweezers – it is important to be competent in the use of both these tools.

Disposable mascara wands are ideal for brushing the hairs during brow shaping.

To carry out eyebrow shaping, you will need the following equipment and materials:

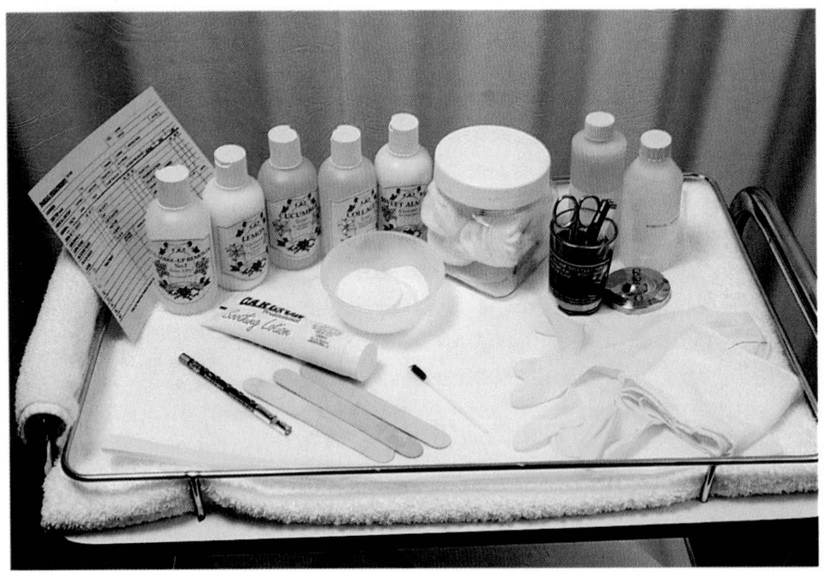

Equipment and materials

HEALTH & SAFETY: MAINTAINING HYGIENE

Several pairs of tweezers must be purchased (perhaps five), due to the length of time required for sterilisation. Buy good-quality stainless-steel tweezers: cheaper metals rust after repeated sterilisation.

TIP

When purchasing tweezers, make sure that the ends meet accurately so they will grasp the hair effectively.

- *Couch or beauty chair* – with sit-up and lie-down positions and an easy-to-clean surface.
- *Disposable tissue* – such as bed roll.
- *Towels (2)* – freshly laundered for each client.
- *Tweezers (sterilised)* – both automatic and manual.
- *Eyebrow pencil* – used to mark the skin when measuring brow length.
- *Pencil sharpener (stainless steel)* – suitable for use in the autoclave.
- *Disposable spatula.*
- *Antiseptic cleansing solution* – for use on the skin.
- *Soothing lotion or cream* – with healing and antiseptic properties, for the skin of the face.
- *Dry cottonwool* – stored in a covered jar.
- *Damp cottonwool* – prepared for each client.
- *Hand mirror (clean)* – used when discussing the brow shaping requirements and the finished result.
- *Record card.*
- *Scissors* – for trimming hairs.
- *Trolley* – on which to place everything.
- *Light magnifier (cold).*
- *Facial tissues (white)* – for blotting the skin dry and for stretching the skin during hair removal.
- *Surgical spirit* – for cleansing the tweezers before sterilisation.
- *Disposable surgical gloves.*

Sterilisation and sanitisation

Sterilise tweezers at an appropriate time during the working day. Ensure that you always have sterile tweezers ready for use with each client. After they have been sterilised in the autoclave, the tweezers should be stored in the ultra-violet cabinet.

A fresh disinfectant solution may be used to store a spare pair of tweezers while carrying out an eyebrow treatment. This solution is usually dispensed into a small container stored on the trolley. (Spare tweezers are necessary in case you should accidentally drop the others during the treatment.)

Disposable rubber gloves may be worn for protection during the eyebrow shaping treatment – the therapist may come into contact with tissue fluids from the client's skin.

As the waste from the treatment may contain body fluids and pose a health threat, it must be collected and disposed of carefully, in accordance with the local health officer's rules and regulations.

Preparing the cubicle

Before the client is shown through to the cubicle, it should be checked to ensure that the required equipment and materials are available and the area is clean and tidy.

The plastic-covered couch should be clean, having been thoroughly washed with hot soapy water, or wiped thoroughly with surgical spirit or a professional, alcohol-based cleaner. The couch or chair should be protected with a long strip of disposable paper bedroll or a freshly laundered sheet and bath towel. A small towel should be placed neatly at the head of the couch – for hygiene and protection during treatment, this will be draped across the client's chest. The tissue will need changing and the towels should be freshly laundered for each client.

The couch or beauty chair should be flat or slightly elevated.

PLANNING THE TREATMENT

Factors to be considered

Before shaping the brows you must consider the following factors.

The natural shape of the brow

The natural brow follows the line of the eye socket. This varies greatly between clients, and affects what is achievable.

If the client has been shaping her own brows, it may be necessary to let them grow for a short period before shaping them professionally. If the brows are very thin or very thick, it may take several sessions before the desired shape is achieved.

If the brows have been plucked over a long period of time, they may not grow back successfully; this should be discussed with the client. In such instances, temporary eyebrow pencil or matt

TIP

A sharpened eyebrow pencil should be used to simulate brow hairs – apply feathery strokes using the pencil point. To create a natural effect, two different pencil colours may be used, for example brown and grey.

powder eyeshadow may be used to achieve the desired effect (see the photograph on page 164).

TIP

By permanently tinting brow hair before shaping you will colour any finer lighter hairs, which will then form part of the brow.

Fashion

Each season sees new fashion trends, which also affect eye make-up and eyebrow shapes. This should be considered before using any form of permanent hair removal to shape the brows.

The age of the client

The brow hair of older clients may include a few coarse, long, discoloured white or grey hairs. These may be removed provided that this does not alter the brow line or leave bald patches.

In general, thick eyebrows make the client look older, by creating a hooded appearance; and thin eyebrows will make the client look severe. Ideally the brows should therefore be shaped to a medium thickness.

The natural growth pattern

The shape and effect that the client requests may be made impossible by the pattern of the natural growth of the hair, which is genetically determined. When shaping the brows of the oriental client, for example, you will notice that the eyebrow hair grows in a downward direction. To create an arch it may be necessary to trim the hairs, using a small, sterile pair of sharp nail scissors. Alternatively, you may remove the outer eyebrow length and use a cosmetic pencil to create a new brow line. Use your professional judgement to advise the client.

Choosing an eyebrow shape

The eyebrows should be in balance with the rest of the facial features: the right brow shape for each client will depend on her facial proportions and her natural brow shape.

There is no single ideal brow shape; different brow shapes are shown on the right.

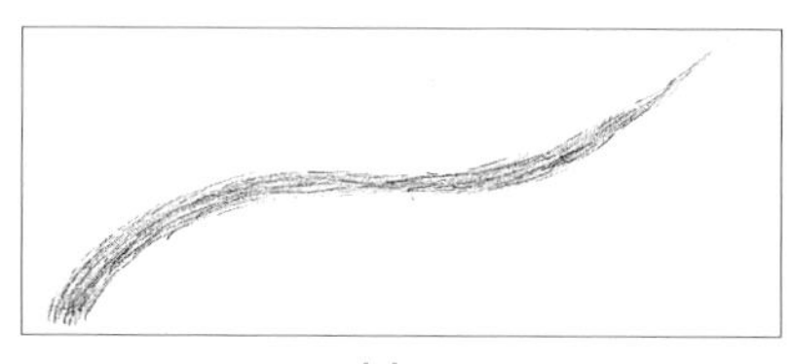

Oblique

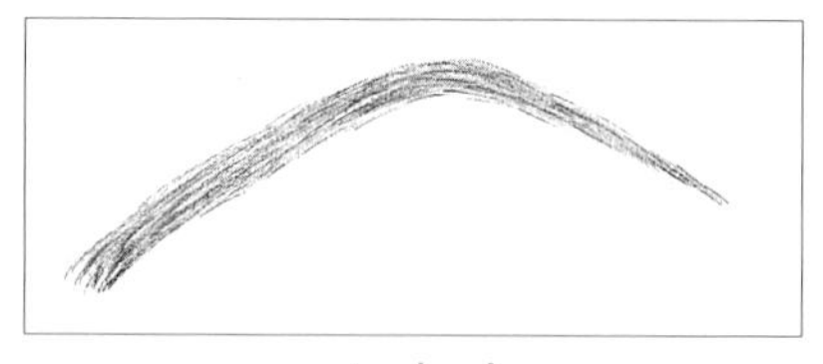

Arched

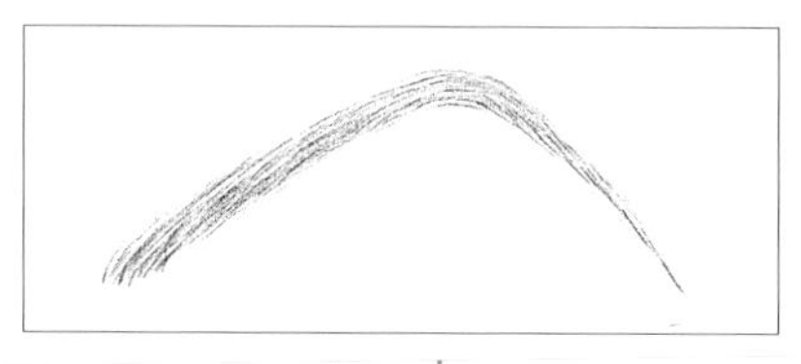

Angular

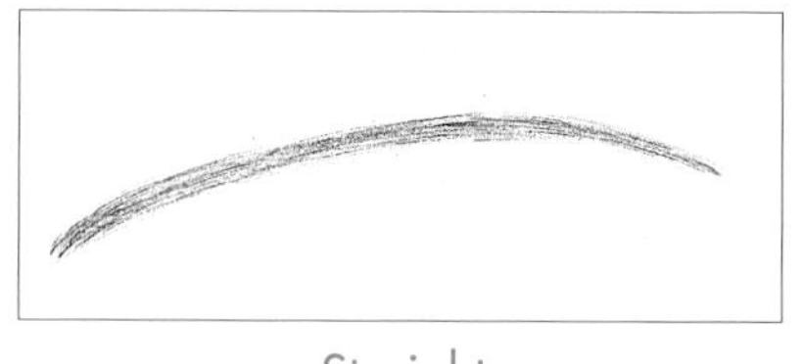

Straight

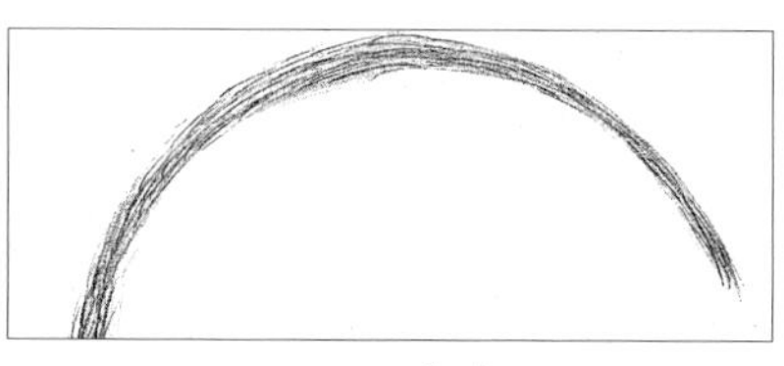

Rounded

What is achievable?

Obviously not all of these brow shapes are achievable for every client, but the skilful application to the eyebrows of temporary cosmetic colour can create the illusion of the desired brow shape.

The quantity of brow hair removed during shaping produces a thin, medium or thick final shape, as illustrated at the top of the opposite page. Other approaches, too, may be used:

- *Tattooing* This can be used to add colour permanently to the brows, for example to disguise a bald patch.
- *Hair transplants* These are available already in the USA for the client with sparse eyebrows.
- *False individual eyebrow hairs* These are applied in the same way as individual false eyelashes.

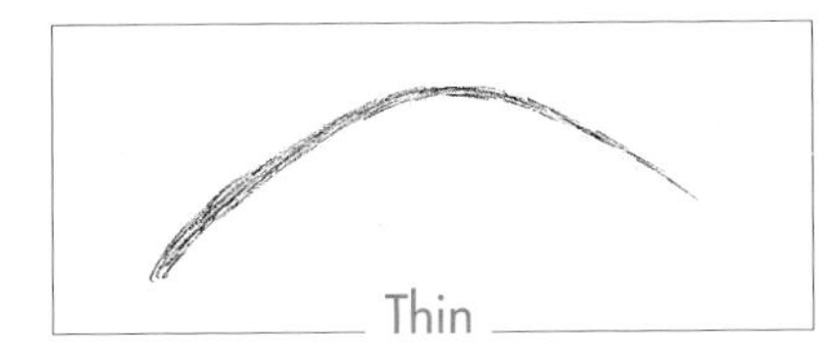

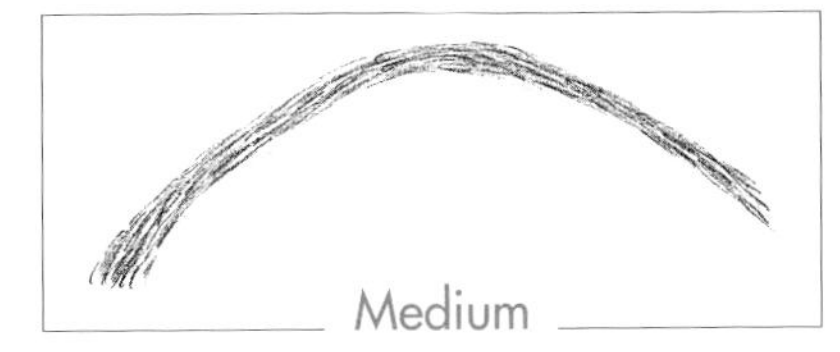

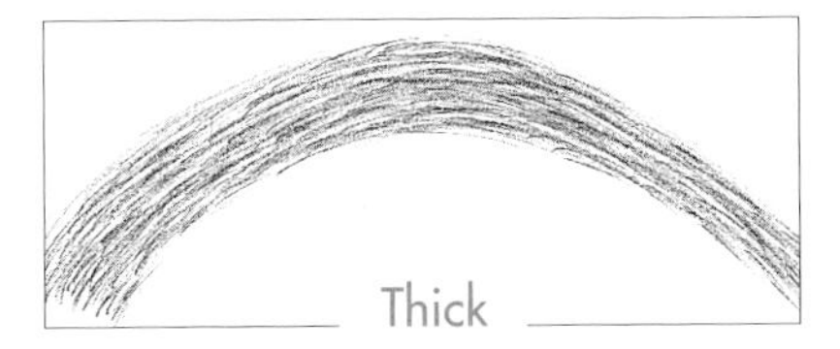

TIP

False eyebrows are available also: these are ideal for disguising any bald areas in the natural eyebrow hair.

Ideally, the distance *between* the two eyebrows should be the width of one eye. If this is not the case in fact, the illusion may be created by removing hairs or by applying temporary brow colour to create this effect.

Wide-set eyes can be made to appear closer together by extending the browline beyond the inside corner of the eye; close-set eyes can be made to look further apart by widening the distance between them.

ACTIVITY: CORRECTING BROW SHAPES (1)

You cannot change the bone structure of the facial features without cosmetic surgery, but brow shaping can create the illusion of improved facial balance and proportion. Discuss why each of the brow shapes below would complement the accompanying face shape.

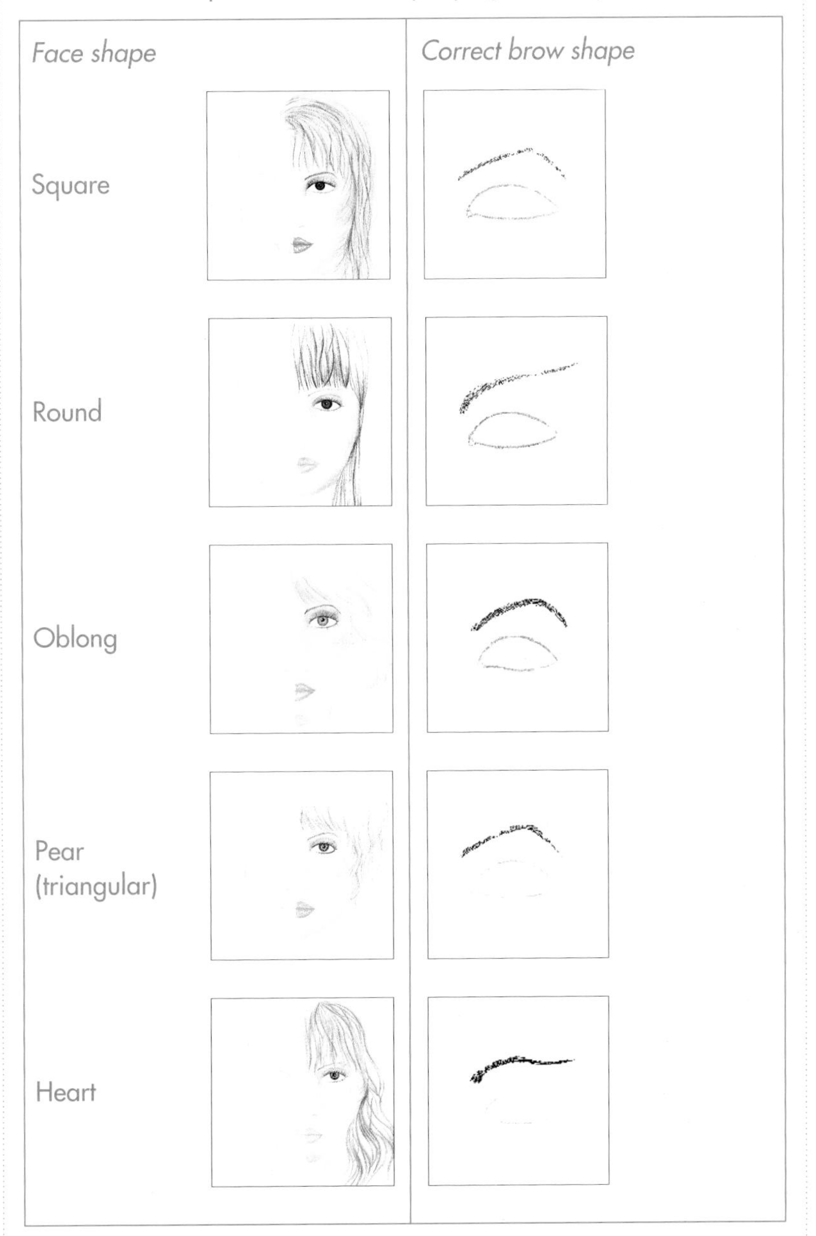

ACTIVITY: CORRECTING BROW SHAPES (2)
How would you correct the following eye shapes?

- Wide-set.
- Close-set

ACTIVITY: CORRECTING BROW SHAPES (3)
From magazines, collect pictures of faces showing different eyebrow shapes. Discuss which you think are correct and which incorrect, explaining why

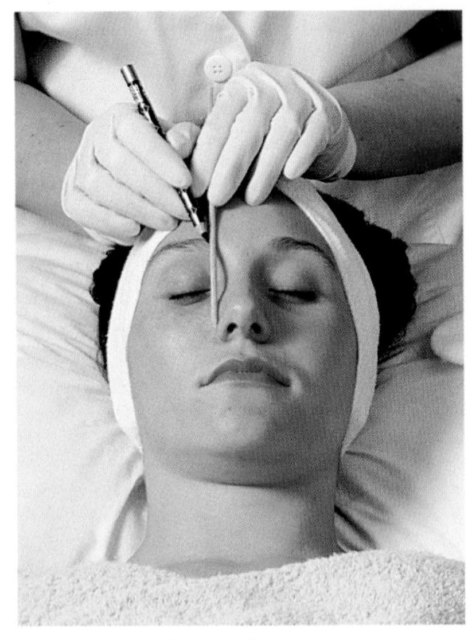
Measuring the eyebrow: the inner eye

Measuring the eyebrows to decide length

In order to determine the correct length of the client's eyebrows, there are three main guidelines:

1. Place an orange stick or spatula beside the nose and the inside corner of the eye. This is usually in line with the tear duct. Any hairs that grow between the eyes and beyond this point should be removed. If the client has a very broad nose, however, this guide is inappropriate: tweezing would commence near the middle of the brow. In this instance, use the tear duct at the inside corner of the eye as a guide.
2. Place an orange stick or spatula in a line from the base of the nose (to the side of the nostril) to the outer corner of the eye. Any hairs that grow beyond this point should be removed.
3. Place an orange stick or spatula in a vertical line from the centre of the eyelid. This is where the highest point of the arch should be.

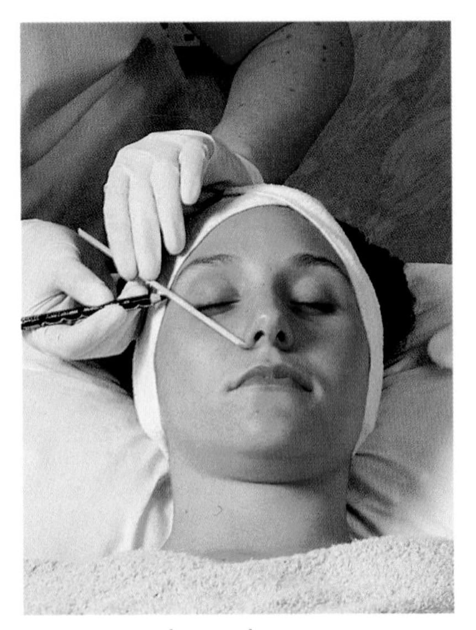
Measuring the eyebrow: the outer eye

Initially these guidelines will be needed to ensure that the correct length and brow shape are achieved, but the experienced therapist will recognise the corrective work to be carried out without the need for measuring.

PREPARING THE CLIENT

The client should be shown through to the cubicle after the record card has been completed.

Consult the record card and check the area for any contra-indications to treatment.

SHAPING THE EYEBROWS

1. Position the cold light magnifying lamp to give maximum visibility of the area.
2. Position the client on the couch (see page 183).
3. Wash your hands.

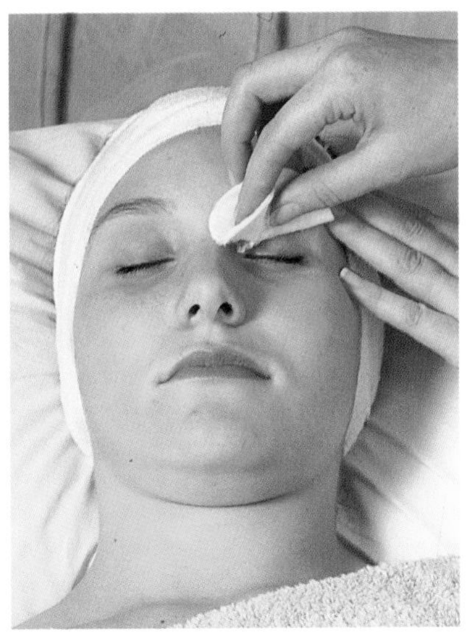

Cleansing the eyebrow

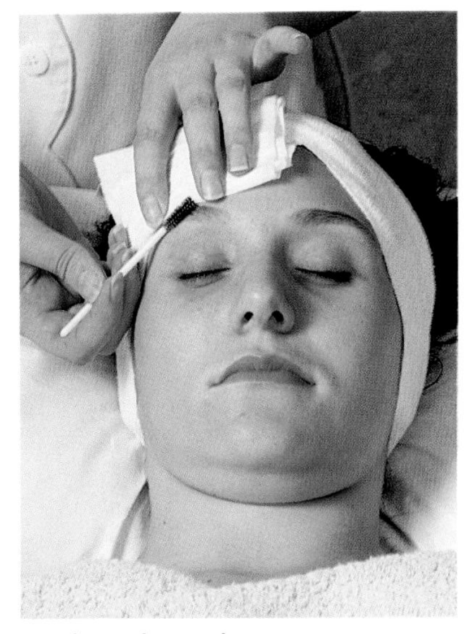

Brushing the eyebrow

HEALTH & SAFETY: CROSS-INFECTION

Use a fresh cottonwool pad for each eyebrow to avoid cross-infection.

4 Working from behind the client, cleanse the eyebrow area, using a lightweight cleansing lotion or eye make-up remover. Apply a mild antiseptic lotion to two damp cottonwool pads, then gently wipe each eyebrow. (This removes all grease from the area, so that the tweezers will not slip.) The brow area should then be blotted dry, using a folded clean facial tissue.

5 Brush the brow hair with the disposable brush, firstly *against* the natural hair growth, then with it. This enables you to define the brow shape and to observe the natural line.

6 Measure the brows (using the guidelines above).

TIP

It is often stated that before beginning shaping, the brows should be prepared with warm, damp cottonwool pads, to relax the hair follicles and soften the eyebrow tissue, thus making hair removal easier. During treatment, however, you will be wiping over the area with an antiseptic lotion which has a cooling, soothing and tightening effect on the skin, so this preparation is ineffectual.

7 Place a clean piece of cottonwool in a convenient position for collecting the removed hairs, for instance at the top of the couch next to the client's head.

8 Put on disposable gloves, as you may come into contact with tissue fluid.

TIP

Occasionally clients start sneezing when you tweeze hairs at the bridge of the nose. If this happens, leave this area till last.

9 Begin tweezing, using a sterilised pair of automatic tweezers. These are designed to remove hairs quickly and efficiently, and are therefore used for the bulk of the hair identified for removal. It is usual to start at the bridge of the nose: the skin here is less sensitive than under the brow line.

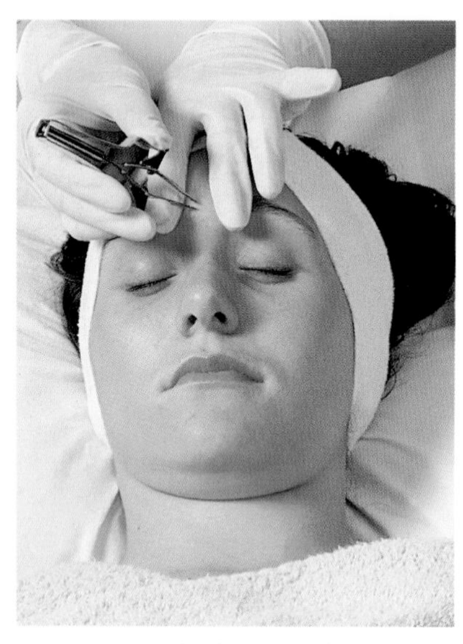
Tweezing at the bridge of the nose

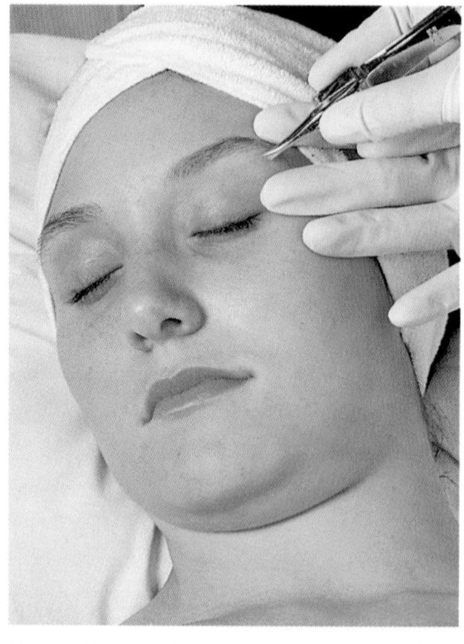
Tweezing at the outer corner of the eyebrow

10 Gently stretch the skin between the index and middle fingers, pressing lightly onto the skin. This will help you to avoid accidentally nipping the skin; it will also open the mouth of the hair follicle and minimise discomfort to the client.

11 Remove the hairs quickly, in the direction of growth. This prevents the hairs from breaking off at the skin's surface. Hair breakage can be seen to have occurred if a stubbly regrowth appears 1 or 2 days after shaping. Incorrect removal may also cause distortion of the hair follicle, or result in the hair becoming trapped under the skin as it starts to regrow (*ingrowing hair* – see page 280).

Hairs should be removed individually, and the tweezers should be wiped regularly on the pad of clean cottonwool used to collect the removed hairs.

12 Remove the hairs from underneath the outer edge of the brow, working inwards towards the nose. It is advisable to work on each brow alternately. This ensures that the brows are evenly shaped; it also reduces prolonged discomfort in any one area during shaping. Hairs should ideally be removed only from *below* the brow, otherwise the natural line may be lost. It is sometimes necessary, however, to remove the odd stray hair growing *above* the natural line.

During shaping, show the client her brows and avoid removing too much hair.

13 At regular intervals during shaping, brush the brows to check their shape. Apply antiseptic lotion to a clean dampened cottonwool pad, and wipe this gently over the eyebrow tissue to reduce sensitivity and to sanitise the area.

14 When the bulk of excess hair has been removed, manual tweezers may be used to take away any stray hairs and to define the line. Long hairs may be trimmed with scissors if

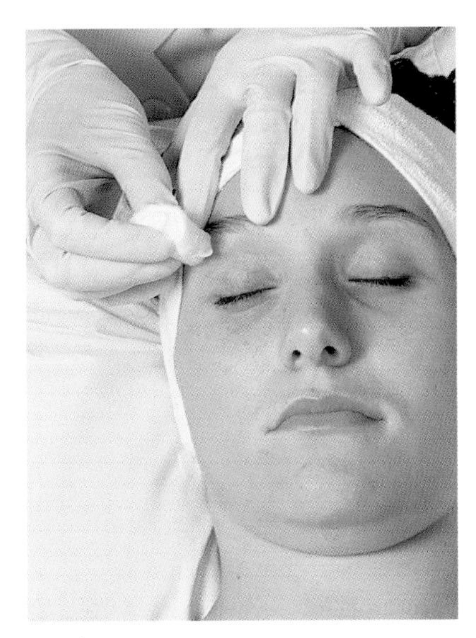

Applying antiseptic

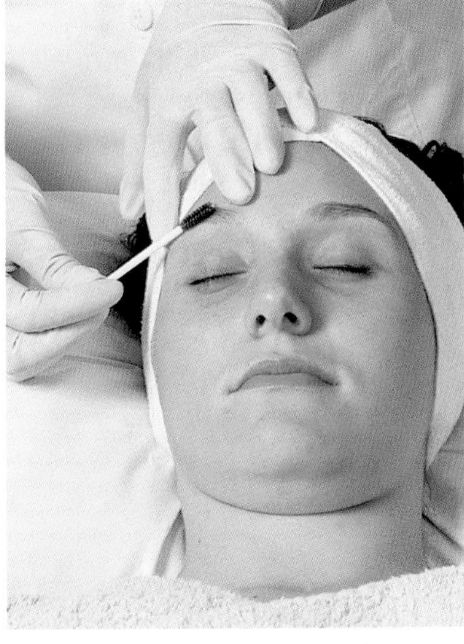

Brushing the eyebrow into shape

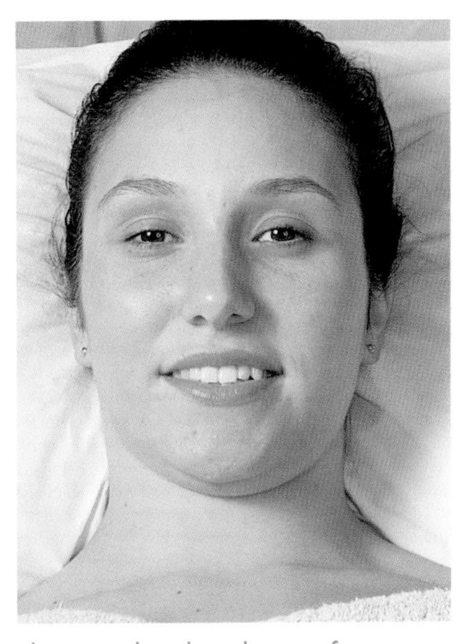

The completed eyebrow after reshaping

necessary. Any discoloured, coarse, long, curly or wavy hairs may be removed, as long as this does not alter the line or leave a bald patch.

15 On completion of brow shaping, wipe the eyebrows with the antiseptic soothing lotion, applied with clean damp cottonwool. Apply a mild antiseptic cream to the area, using clean dry cottonwool, to reduce the possibility of infection.

16 Brush the brows into shape and show the client the finished effect.

17 The hairs that have been removed should be disposed of hygienically.

TIP

Twenty per cent of hairs are not visible above the skin's surface at any one time. Explain this to the client so that she understands why stray hairs may appear shortly after treatment.

CONTRA-ACTIONS

Erythema is considered to be a contra-action to the treatment: it is recognised as a marked reddening of the skin seen over the whole area or specifically around one damaged follicle. It is usually accompanied by minor swelling of the area. If this occurs the therapist must try to reduce the redness by applying a soothing antiseptic lotion or cream to the area. In extreme cases it may be necessary to apply ice. Record details of any contra-action on the client's record card.

If the reddening reduces in response to your corrective action, you may decide in future to remove only a few stray hairs at each eyebrow-shaping treatment, to minimise the risk of this reaction recurring.

HEALTH & SAFETY: USING ICE

To maintain hygiene, place the ice cube in a new small freezer foodbag. Dispose of the bag hygienically after use.

AFTERCARE AND ADVICE

Advise the client not to wear eye make-up for at least eight hours following the eyebrow-shaping treatment. The hair follicle has been damaged where the hair has been torn out: it will be susceptible to infection unless the area is cared for whilst it heals. It should not be necessary for the client to continue using antiseptic lotion at home, but she should be advised to carry out these instructions if discomfort or continued reddening occurs.

1 Cleanse the eyebrow area using a mild antiseptic lotion or witchhazel, applied with a small piece of clean, dampened cottonwool.
2 Apply an antiseptic soothing lotion or cream, with clean, dry cottonwool.
3 Gently remove excess antiseptic lotion or cream using a clean, soft facial tissue.
4 Repeat as necessary, approximately every 4 hours. *If the reddening does not subside in the next 24 hours, contact the salon.*
5 Eye make-up may be worn as soon as the redness has gone – usually after 8 hours.

HEALTH & SAFETY: AVOIDING INFECTION

If excess antiseptic lotion or cream is left on the area, this may attract small particles of dust, which could cause infection.

KNOWLEDGE REVIEW

Eyebrow shaping treatment

Preparation

1 What factors should you consider when deciding the correct eyebrow shape for a client?
2 How can you determine the length of a client's eyebrows that will best suit her facial features?
3 How should the brow area be prepared prior to shaping treatment?

Application

1 For what purpose are each of the eyebrow tweezers illustrated used?
2 How can you minimise client discomfort during the eyebrow shaping treatment?
3 How can you ensure that hairs are removed at their roots?
4 How would a contra-action to eyebrow shaping be recognised?

Aftercare

1 Why is a soothing antiseptic lotion applied following an eyebrow shaping treatment?
2 What aftercare advice should you give to a client following an eyebrow shaping treatment?
3 When would you recommend that the client returns to the salon for an eyebrow trim, following an eyebrow shape?

Health, safety and hygiene

1 What hygiene and safety precautions should be followed when performing an eyebrow shape?

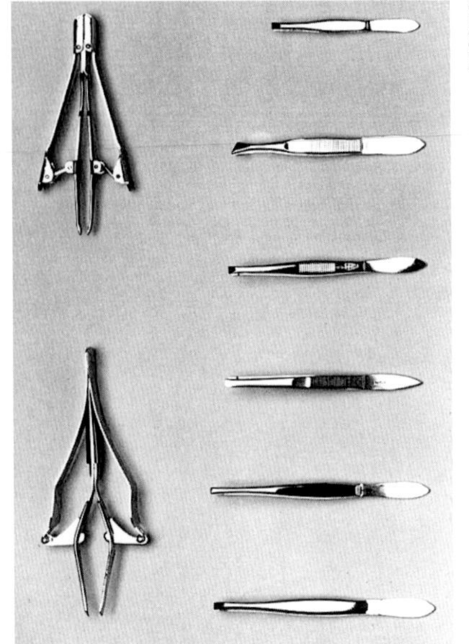
Ellisons

2 What are the acceptable methods of sterilisation for tweezers?
3 What contra-indications to eyebrow shaping should be looked for in the brow area?
4 How should consumables used during an eyebrow shape be disposed of?
5 What action would you take if a skin contra-action occurred following eyebrow shaping?

Case study

1 What factors should you consider in your treatment approach to eyebrow shaping with the following clients?
   (a) A client with excessively thick eyebrows who requests a thin eyebrow shape.
   (b) A client who has close-set eyes.
   (c) A client with a round face.
   (d) A client who has a few stray long coarse hairs.
   (e) A client with sensitive skin.

EYELASH AND EYEBROW TINTING

The hair of the eyelashes and eyebrows protects the eyes from moisture and dust, but the lashes and brows also give definition to the eye. Many clients, especially those with fair lashes and brows, feel that without the use of eye cosmetics their eyes lack this definition.

PERMANENT TINTING

Further definition of the brow and lash hair can be created if a permanent dye is applied to them. Most clients will benefit from tinting for the following reason, because the tips and the bases of these hairs are usually lighter than the body of the hairs, causing the hairs to appear shorter than they actually are. Tinting the length of the lash or brow hair makes it appear longer and bolder, yet the effect created looks natural.

Because the skin around the eye area is very thin and sensitive, dyes designed for permanently tinting the hair in this area have been specially formulated to avoid any eye or tissue reactions. *The application of any other dye materials in this area is dangerous, and may even lead to blindness.*

Permanent tints are available in different forms, including jelly, liquid and cream tints. The most popular and acceptable permanent tinting product is the cream tint: this is thicker, so it

HEALTH & SAFETY: PERMANENT TINTING

Always use a tint that is permitted for use under EU regulations and that does not contravene the latest UK Food and Drugs Act. If you use any other tint, your insurance may be ineffective.

does not run into the eye and it is easy to control during mixing, application and removal.

Several colours of permanent tint are available, including brown, grey, blue and black.

If the shade you want is not available, you can vary the shade of available tints by leaving the dye on the hair for different lengths of time, or by mixing different colours together. For example, to produce a navy blue colour, leave the tint to process for 3–5 minutes. If left to process for 10 minutes, the same tint will produce a raven blue-black colour.

TIP

When a client requests a blue eyelash tint, make clear that this will not produce an 'electric blue' fashion colour.

The development of colour

Two products are essential for the permanent tinting treatment:

- professional **eyelash** or **eyebrow tint**;
- **hydrogen peroxide** (H_2O_2).

The tint contains small molecules of permanent dye called **toluenediamine**. These need to be 'activated' before their colouring effect becomes permanent: this is achieved by the addition of hydrogen peroxide. The peroxide is said to **develop** the colour of the tint.

Chemically, hydrogen peroxide is an **oxidant**, a chemical that contains available oxygen atoms and encourages certain chemical reactions – in this case, tinting.

The hydrogen peroxide container will state either its volume or its percentage strength. To activate the tint and for safe use around the eye area, a 3% or 10-volume strength peroxide is used.

When you add the hydrogen peroxide to the tint, the small dye molecules together form large molecules which remain trapped in the cortex of the hair. The hair is thus permanently coloured, but in time, as it continues to grow, the new hair will show the natural colour.

HEALTH & SAFETY: PEROXIDE STRENGTH

Do not use a higher strength than 10-volume or 3% hydrogen peroxide. If you do, skin irritation or minor skin burning may occur.

TIP

Replace the cap of the hydrogen peroxide container *immediately* after use, or the peroxide will lose its strength.

RECEPTION

When making an appointment for this service, allow 5 minutes for a skin test, 10 minutes for an eyebrow tint, and 15 minutes for an eyelash tint. When the client is booking her treatment, ask her:

- to visit the salon 24 hours before the appointment, for a skin test;
- if she wears contact lenses, to bring her lens container so that she can place the lenses in it during the tinting treatment.

TIP

It is acceptable for the salon receptionist – provided that she has been trained to do so – to carry out the skin test.

On average, a client will need her lashes tinted every 6 weeks, or sooner if for example she takes a holiday in a climate where sun bleaches them. Eyebrow tinting, on the other hand, should be

ACTIVITY: EYELASH AND EYEBROW TINTING

List reasons why clients would benefit from this service. Discuss the reasons with your tutor.

repeated when the client feels it to be necessary, perhaps every 4 weeks, as eyebrows seem to lose colour intensity more quickly than lash hair.

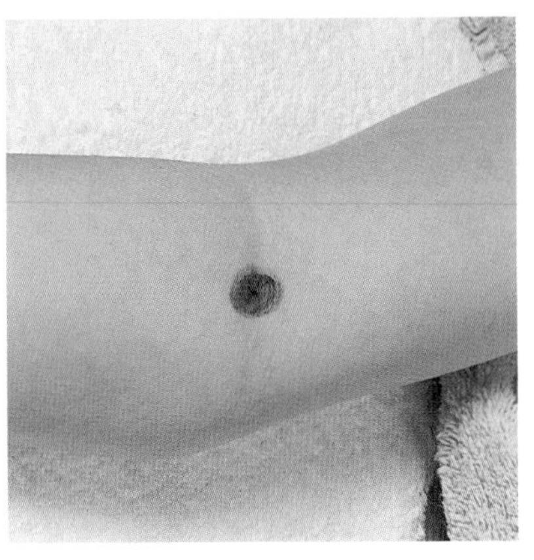
A skin test

Skin tests

Some clients are sensitive to the tint, and produce an allergic reaction immediately on contact with it; others may become allergic later. For this you therefore need to carry out a skin test before each lash- or brow-tinting treatment. (The skin test is also known as a patch test, a hypersensitivity test, or a predisposition test – see page 36.)

Two responses to the skin test are possible – positive and negative:

- a *positive* result is recognised by irritation, swelling or inflammation of the skin – if this occurs, do not proceed with the treatment;
- a *negative* result produces no skin reaction – in this case you may proceed with the treatment.

ACTIVITY: ALLERGIC REACTIONS
With colleagues, discuss the possible implications of ignoring a positive reaction to the skin test.

CONTRA-INDICATIONS

After completing the record card and inspecting the eye area, if you have found any of the following, do not proceed with the tinting treatment:

- *Inflammation or swelling* around the eye area.
- *Skin disease* in the eye area.
- *Skin disorder* in the area, such as psoriasis or eczema.
- *Cuts and abrasions* in the area.
- *Hypersensitive skin.*
- *Eye disorders*, such as conjunctivitis, blepharitis, styes or hordeola, watery eye, or cysts.
- *A positive (allergic) reaction* to the skin test.
- *Contact lenses* (unless removed).

A particularly nervous client would also be contra-indicated:

- it would be difficult for her to keep her eyes closed for 10 minutes;
- she might panic as the tint was applied, creating the possibility of tint entering the eye;
- she might blink frequently, making preparation of the eye area and application of the tint both difficult and hazardous.

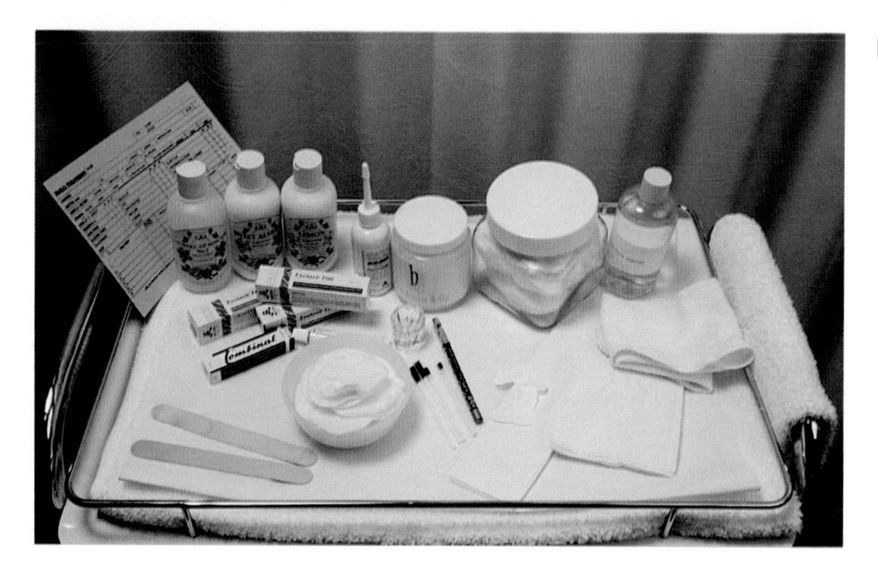

Equipment and materials

EQUIPMENT AND MATERIALS

To carry out the tinting you will need the following equipment and materials:

- *Couch* or *beauty chair* – with sit-up and lie-down positions and an easy-too-clean surface.
- *Trolley* – on which to display everything.
- *Towels (medium-sized)* – freshly laundered for each client.
- *Headband (clean)* – to protect long hair or bleached hair from the tint.
- *Cleansing milk* – used to remove make-up from the eye area.
- *Eye make-up remover (non-oily).*
- *Coloured tints (a selection).*
- *Hydrogen peroxide (10-volume/3%).*
- *Petroleum jelly* – to protect the the skin and prevent skin staining.
- *Damp cottonwool* – for cleansing the eye area and for lash and brow tint removal.
- *Eyeshields (commercial)* – to prevent skin staining during eyelash tinting.
- *Disposable brushes (2).*
- *Facial tissues (white)* – for blotting the eye area dry.
- *Disposable spatulas* – for removing the petroleum jelly from its container.
- *Non-metallic bowl* – for mixing the permanent tint (note that some metals cause immediate release of the oxygen from the hydrogen peroxide, causing ineffective processing of the tint).
- *Permanent tint (commercial).*

TIP

The brows or lash hair to be tinted must be grease-free – the grease would be a barrier to the tint.

TIP

Trimmed disposable lipbrushes can be used to apply the petroleum jelly and permanent tint.

HEALTH & SAFETY: APPLICATOR BRUSHES

Because it is impossible to sterilise applicator brushes effectively, use disposable brushes for the application of petroleum jelly and permanent tint.

- *Skin stain remover.*
- *Bottle, with a long flexible tube* – to dispense distilled water.
- *Distilled water (purified).*
- *Collodion ('new skin')* – used to cover the tint when carrying out the skin test.
- *Bowl (clean)* – to hold dampened cottonwool.
- *Swing-top bin* – lined with a disposable bin liner, for waste.
- *Hand mirror (clean).*
- *Client's record card.*

Sterilisation and sanitisation

It is necessary to use disposable applicator brushes for the application of the petroleum jelly and the permanent tint because it is impossible to sterilise brushes effectively.

Preparing the cubicle

Before the client is shown through to the cubicle, it should be checked to ensure that the required equipment and materials are available and the area is clean and tidy.

Clean and protect the couch or beauty chair as for eyebrow-shaping treatment (page 183). The couch or chair should be flat or slightly elevated.

The cubicle should be adequately lit to ensure that treatment can be given, safely, but avoid bright lighting that could cause eye irritation.

PLANNING THE TREATMENT

For the eyelashes and eyebrows, select a colour that complements the client's hair and skin colours, her age and her usual eye cosmetics. Always discuss the choice of colour carefully with the client to discover her preference. You may ask certain questions in order to help you in your selection:

- 'What colour mascara do you normally wear?'
- 'How dark would you like your eyelashes/eyebrows?'
- 'Do you normally wear eyebrow pencil? Of what colour?'

TIP

When selecting and using permanent tint for a white-haired client, note that the hair is very often resistant to colour.

ACTIVITY: CHOOSING LASH AND BROW COLOURS

Which colour do you think would be most suitable for the eyelashes and eyebrows of the following clients?

- Brunette.
- Fair.
- Red.
- Grey.
- Elderly.

PREPARING THE CLIENT

The client should be shown through to the cubicle after the record card has been completed.

1. Position the client comfortably, in a flat or slightly elevated position. If she is wearing contact lenses, these must be removed.
2. Drape a towel across the client's chest and shoulders, and protect her hair with a clean headband.
3. Wash your hands, which assures the client that the treatment is beginning in a hygienic and professional manner.
4. Consult the client's record card, then check the area for any visible contra-indications or abnormalities before proceeding.
5. Cleanse the area to be treated with a cleansing milk to dissolve facial make-up. Then use a non-oily eye make-up remover to remove eye products: apply this with clean, damp cottonwool.
6. To ensure that the area is thoroughly clean and grease-free, apply a mild toning lotion, stroked over the lash or brow hair.
7. Blot the eyelashes or eyebrows dry with a clean facial tissue. This ensures that the tint is not diluted, and also prevents the tint from being carried into the eye.
8. Prepare the pre-shaped eyeshields by applying petroleum jelly to the inner surface of each eyeshield (the surface that comes into contact with the skin).
9. Ensure that the light is not shining directly into the client's eyes. If it were, the eyes might water, carrying the tint into the eye or down the face (causing skin staining).
10. Finally, check that the client is comfortable before beginning tint application.

Cleansing the eye

TIP
Make sure that work surfaces are protected with disposable coverings, to prevent permanent staining following spillages.

TINTING THE EYELASHES

1. Remove some petroleum jelly from its container, using a new disposable spatula.
2. Working from behind the client, ask the client to open her eyes and to look upwards towards you. Using a disposable brush, apply petroleum jelly underneath the lower lashes of one eye, ensuring that it extends at the outer corner of the eye. (This is in case the client's eyes water slightly during treatment, which might otherwise lead to skin staining.) The petroleum jelly must not come into contact with the lash hair, where it would create a barrier to the tint.
3. Place the prepared eyeshield on the skin under the lower lashes, ensuring that it adheres to the petroleum jelly and fits 'snugly' to the base of the lower lashes.
4. Repeat the above process for the other eye.
5. Ask the client to close her eyes gently. Instruct her not to open them again until you advise her to do so, in about 10 minutes' time.

HEALTH & SAFETY: MAINTAINING HYGIENE
Do not use the same spatula in the jar again, or you might contaminate the product.

TIP
Speak to your client as you apply the eyelash tint. Check that she is comfortable and understands what you are doing. Remember: the eyes are very sensitive and will water readily.

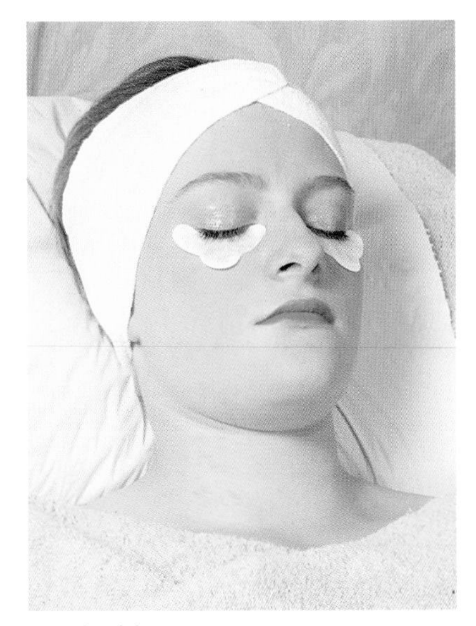
Eyeshields

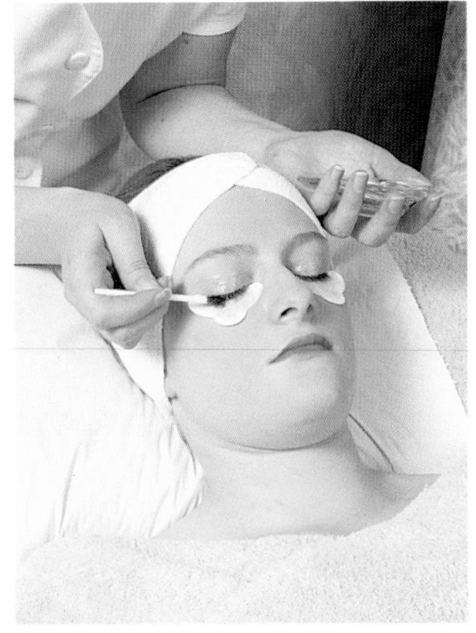
Applying the tint

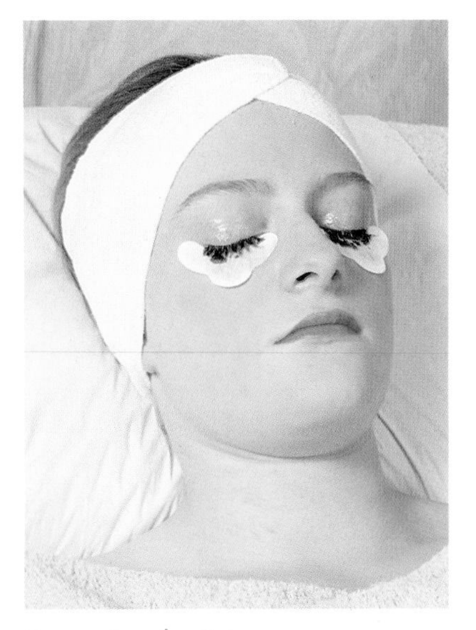
Processing the tint

6 Apply petroleum jelly to the upper eyelid, in a line on the skin at the base of the lashes.

> **HEALTH & SAFETY: USING TINTS**
> Always read the manufacturer's instructions carefully before using a permanent tint.

7 Considering the length and density of the client's eyelashes, mix the required amount of tint with 10-volume (3%) hydrogen peroxide. As a guide, a 5 mm length of tint from the tube, mixed with 2 or 3 drops hydrogen peroxide, is usually sufficient. Mix the products to a smooth cream in the tinting bowl, using the disposable brush. Always re-cap bottles and tubes tightly after use, to avoid deterioration of materials.

8 Wipe excess tint off the brush onto the inside of the bowl. Apply the tint thinly to each hair. Work from the base of the lash to the tip, ensuring that each hair is evenly covered. Press down gently with the applicator to ensure that the lower lashes also are covered. The inner and outer few lashes should also be covered, down to the base.

9 Allow the tint to process, for approximately 5–10 minutes from the completion of application. Discard any unused mixture as soon as the tint has been applied.

10 On completion of processing, remove the eyelash tint by applying clean damp cottonwool pads over each eye, wiping away most of the tint and removing the protective eyeshield in one movement, an outward sweep. Using fresh dampened

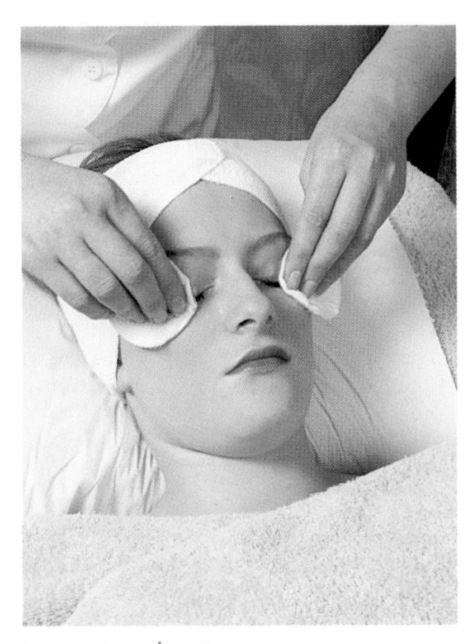
Removing the tint

> **HEALTH & SAFETY: CLIENT COMFORT**
> The client should not be left during the lash-tinting treatment. You must be available, both to offer reassurance and to take the necessary action if the eyes water.
>
> If the eyes do water whilst the eyes are closed, hold a tissue at the corner of the eye to soak up the moisture. Take further action if watering persists and the eyes begin to sting.

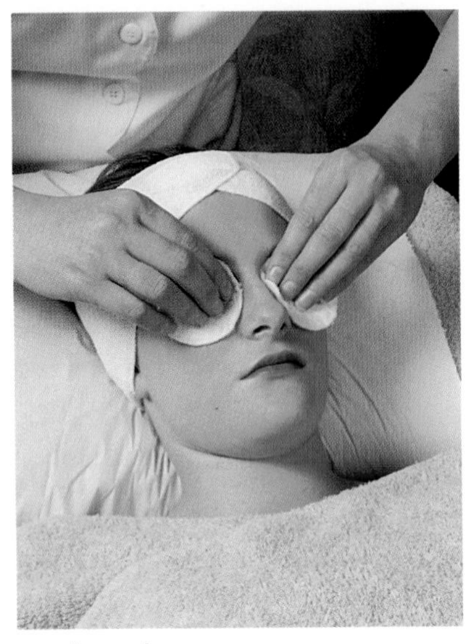

Soothing the eyes

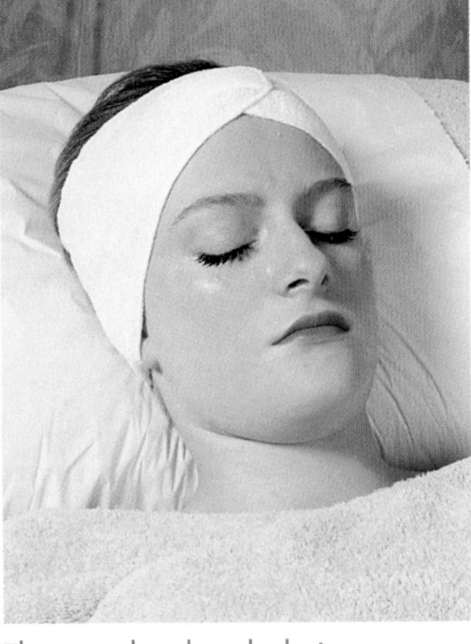

The completed eyelash tint

cottonwool pads, gently stroke down the lashes from roots to tips, until all excess tint has been removed. With a sweeping action on each eye and using one cottonwool pad, wipe from the side to the middle against the lash growth, whilst the other hand supports the eye tissue. *All tint must be thoroughly removed before the client opens her eyes.*

11 Ask the client to open her eyes. If removal has been correctly carried out, the lashes and their bases will be free from tint. (While training, if any tint remains at the base of the lashes after the eyes have been opened, ask the client to close her eyes again and finish the removal process using clean damp cottonwool.) Check that every lash has been tinted, especially the base of each lash and the inner and outer corner lashes.

12 Once you are satisfied that all tint has been removed, place a cool damp cottonwool pad over each eye for 2–3 minutes to soothe the eye tissue.

13 Show the client the result, ensuring that the colour is dark enough and that she is satisfied with the final effect.

14 Complete the client's record card.

15 Take the client to reception to book her next appointment.

TINTING THE EYEBROWS

1 Remove some petroleum jelly from its container, using a new disposable spatula.

2 Brush the brow hair away from the skin, using a disposable brow brush.

3 Using a disposable brush, surround each eyebrow with petroleum jelly, as close as possible to the brow hair (to avoid skin staining).

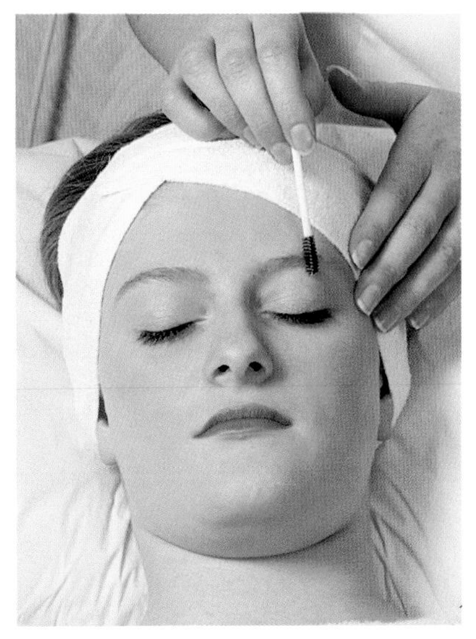
Brushing the eyebrow against the growth

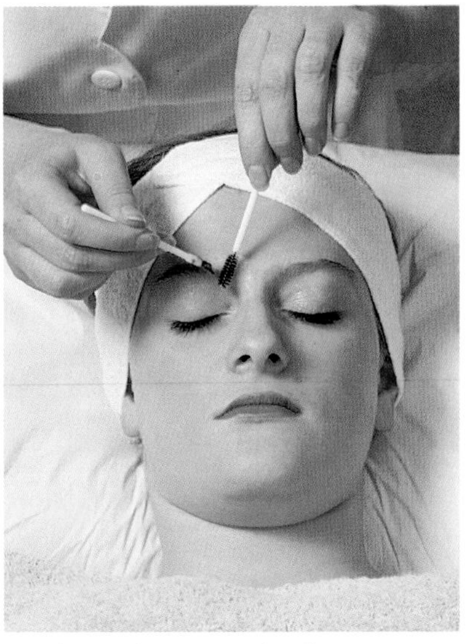
Applying the tint to the first eyebrow

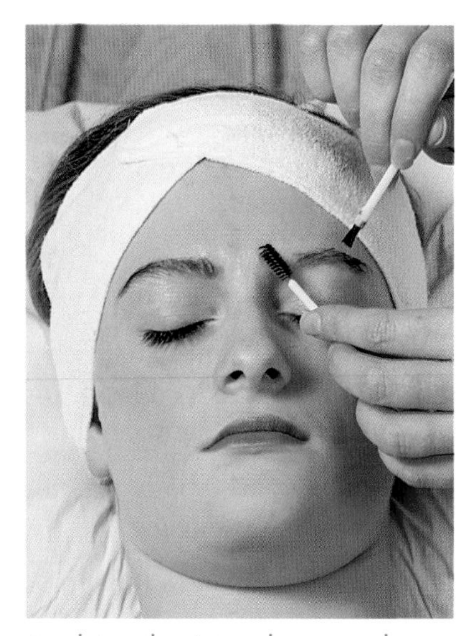
Applying the tint to the second eyebrow

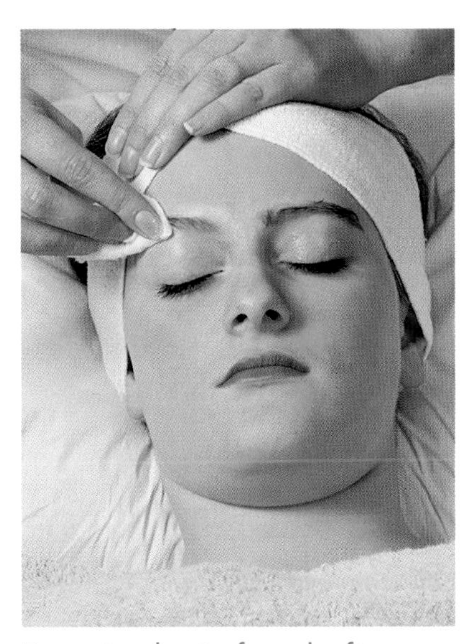
Removing the tint from the first eyebrow

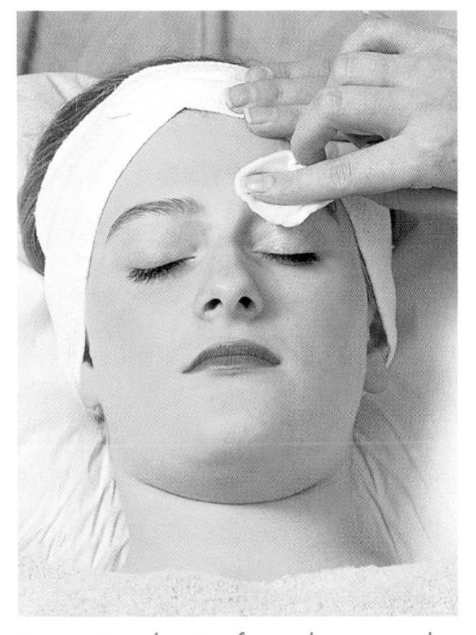
Removing the tint from the second eyebrow

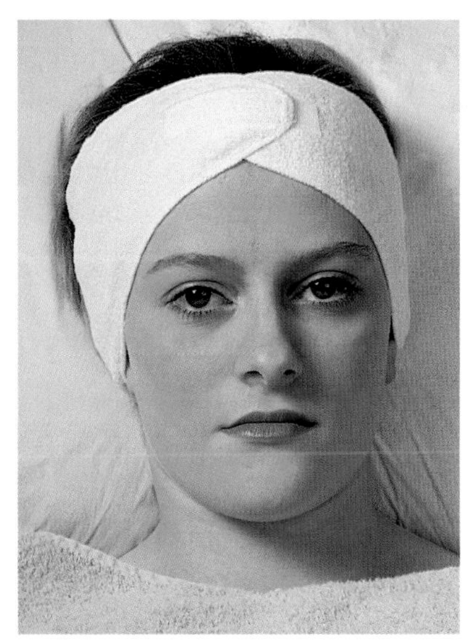
The completed effect, showing both eyelash and eyebrow tint

4 Mix approximately 5 mm of the chosen tint colour with 2 or 3 drops of 10-volume (3%) hydrogen peroxide in a tinting bowl. Ensure that the tint is mixed thoroughly to a creamy consistency.

5 Wipe excess tint off the brush onto the inside of the tinting bowl. Apply the tint neatly and economically to the brow hair of the first eyebrow, ensuring that the brow hairs, from the base to the tips, are evenly covered.

6 Apply the tint to the second eyebrow, following the same procedure.

7 Immediately after application of the tint to the second eyebrow, remove the tint from the first eyebrow. Use a clean

dampened cottonwool pad. Place it on the eyebrow, then wipe it across the eyebrow in an outward sweep, removing the excess tint. Ensure that all traces of excess tint have been removed, to prevent skin staining.

Never leave tint on the eyebrows for longer than 2 minutes. Eyebrow hair colour develops much more quickly than lash hair.

8. Remove the tint from the second eyebrow in the same way.
9. Show the client the effect of the tinted eyebrows. If the brow hair is not dark enough, re-apply the tint to the eyebrow hair, following the same application and removal procedure.
10. When you are both satisfied with the colour of the tinted eyebrows, complete the client's record card.
11. Take the client to reception to book her next appointment.

The effect of tinting depends on the natural colour:

- *Blonde hair* develops colour rapidly – if the tint is left on the eyebrow too long, a harsh appearance will be created.
- *Red hair* is more resistant to the tint, and developing will take a little longer – allow 15 minutes' processing time when tinting lash hair.
- *Dark hair* requires tinting to increase the intensity of the natural eyebrow colour, giving a glossy conditioned appearance.

HEALTH & SAFETY: SKIN STAINS
If skin staining accidentally occurs, use a professional skin stain remover designed for this purpose. Afterwards, use plenty of clean dampened cottonwool pads to avoid skin irritation.

ACTIVITY: EXPLAINING POOR RESULTS
What reasons can you think of to explain why a permanent tint applied to the eyelashes or eyebrows has not coloured the hair successfully? Discuss your answers with your tutor.

CONTRA-ACTIONS

If the client complains of discomfort during the treatment, tint may have entered the eye. Take the following action:

1. Remove the tint immediately from the eye area, using clean damp cottonwool pads in an outward sweep.
2. When you are satisfied that all excess tint has been removed (that is, when the cottonwool shows clean), flush the eye with distilled water, dispensed from a bottle. Repeat the rinsing process until discomfort has been relieved.

If there is a noticeable sensitivity of the eye tissue after eyebrow or eyelash tinting, recommend that the client does not receive the treatment again. Record this on her record card.

KNOWLEDGE REVIEW

Permanent tinting

Preparation

1. What products are required to produce the permanent colouring of eyelash hair and brow hair?
2. Why should brow or lash hair be grease-free prior to permanent tinting treatment?
3. What is the purpose of petroleum jelly when used in a permanent tinting treatment?
4. How do you select the colour when carrying out a permanent tinting treatment?

Application

1. How can you ensure that tint application will colour the hair evenly?
2. Why is the quantity of tint applied to the area important in relation both to efficiency and to the final result?
3. What is the difference in processing time between a brow tint and an eyelash tint?

Aftercare

1. When would you recommend that the client returned to the salon for a permanent tinting treatment:
   (a) for the eyelashes?
   (b) for the eyebrows?

Health, safety and hygiene

1. What should you ensure has been carried out before a permanent tinting treatment?
2. How and why should this be carried out?
3. How would an adverse (positive) reaction to a skin test be recognised?
4. Name *four* contra-indications that would prevent permanent tinting treatment from being carried out.
5. If a client complained of irritation during an eyelash tinting treatment, what action would you take?
6. How would a contra-action to permanent tint be recognised?

Case study

1. How would permanent tinting be performed when treating the following?
   (a) A very nervous client, to ensure a safe, efficient eyelash tinting treatment.
   (b) A client who requires both an eyebrow shape and an eyebrow tint. In which order should these treatments be given, and why?
2. For how long would you allow the tint to process when treating a client with:
   (a) blonde hair?
   (b) dark hair?
   (c) red hair?

ARTIFICIAL EYELASHES

False eyelashes are made from small threads of nylon fibre or real hair. They are attached to the client's natural lash hair, imitating the natural eyelashes and making the lashes appear longer and thicker, and thereby drawing attention to the eye. There are two main types: **semi-permanent individual lashes** and **strip lashes**.

False lashes are applied for the following reasons:

- to create shape and depth in the eye area, when completing a corrective eye make-up;
- simply to add definition to the eye area;
- to enhance an evening or fantasy make-up;
- to provide thick long lashes for photographic make-up;
- to provide an alternative eyelash-enhancing effect for a client who is allergic to mascara.

False eyelashes

RECEPTION

When making an appointment for false-eyelash application, find out why the client wants false lashes and determine which type would be more appropriate.

Strip lashes

Artificial **strip lashes** are designed to be worn for a short period, either for a day or an evening. They are attached to the natural eyelashes with a soft, weak adhesive. After removal the strip must be cleaned before re-application. Allow 15 minutes for the application of strip lashes; if applying them in conjunction with a make-up, allow an additional 30 minutes for the make-up.

Individual lashes

Artificial *individual* lashes are attached to the natural lashes with a strong adhesive. They may be worn for approximately 4–6 weeks, and are therefore known as **semi-permanent lashes**. Allow 20–30 minutes for the application of individual eyelashes;

Salon System

Strip and individual lashes

and again allow a further 30 minutes if applying them in conjunction with a make-up.

Although individual lashes can be worn for up to 6 weeks they look effective only for approximately 3 weeks. After this time, the appearance of the artificial lashes begins to deteriorate, the lash adhesive becomes brittle, and the eyelash area may become irritated.

Due to the cyclic nature of hair replacement some individual lashes will be lost when the natural lash falls out. These lashes may be replaced each week, as necessary; the client is usually charged a price for each individual lash replaced. The client must be told of this service as part of the aftercare advice.

Booking the treatment

When a client makes an appointment for an artificial lash treatment, she should be asked the following questions:

- *Has she had a similar treatment before in this salon?* If she has not, she should visit the salon beforehand for a skin test to assess her sensitivity to the adhesive (see page 36).
- *Is she having the false eyelashes applied for any particular reason, such as a holiday or a special occasion?* In deciding which type of false eyelash would be most appropriate, take into consideration the effect required and for how long the lashes are to be worn.

If the client wears glasses, the artificial lashes must not be so long as to touch the lenses. Also, if the lens magnifies the eye, this must be taken into account. Ask the client to bring her glasses with her to the salon.

CONTRA-INDICATIONS

If following completion of the record card or inspection of the eye area you have found any of the following, do not apply false eyelashes:

- *Skin disease.*
- *Skin disorder* in the eye area, such as psoriasis or eczema.
- *Inflammation or swelling* around the eye.
- *Hypersensitive skin.*
- *Any eye disorder*, such as styes or hordeola, conjunctivitis, blepharitis, watery eye, or cysts.
- *A positive (allergic) reaction* to the adhesive skin test.
- *Contact lenses* (unless removed).

An unduly nervous client with a tendency to blink could prove hard to treat in this way. Use your discretion in deciding on the suitability of a client for treatment.

EQUIPMENT AND MATERIALS

To apply the false eyelashes you will need the following equipment and materials:

- *Couch or beauty chair* – with sit-up and lie-down positions and an easy-to-clean surface.
- *Trolley* – on which to display everything.
- *Towels (2, medium-sized)* – freshly laundered for each client.
- *Headband* – freshly laundered for each client.
- *Disposable tissue roll* – for example bed roll.
- *Cleansing milk* – used to remove facial make-up in the eye area.
- *Eye make-up remover (non-oily).*
- *Damp cottonwool* – for removing cleansing product from the eye area.
- *Facial tissues (white)* – for blotting the eyelashes dry.
- *Disposable mascara brush.*
- *Manual tweezers (2 pairs, sterilised)* – special tweezers are available, designed specifically to assist in attaching individual eyelashes.
- *Surgical spirit* – for wiping the points of the tweezers to remove adhesive.
- *Plastic palette (sanitised)* – on which to place the artificial lashes prior to application.
- *Eyelash adhesive* – for strip lashes.
- *Strip eyelash lengths (a selection)* – in a choice of colours.
- *Eyelash adhesive* – for individual lashes.
- *Individual eyelash lengths (a selection)* – in a choice of colours.
- *Eyelash adhesive solvent* – for removing and cleaning artificial lashes.
- *Sterilised dish (small)* – lined with foil, in which to place the eyelash adhesive during lash application.
- *Sterilised scissors (1 pair)* – used for trimming the length of strip lashes.
- *Hand mirror (clean).*
- *Client's record card.*

Sterilisation and sanitisation

In order that you can effectively clean the dish after the treatment, you need to line the dish with a disposable lining.

When preparing to apply artificial *strip* lashes, clean the surface of the palette onto which you will stick the lashes once you have removed them from their packet. Use surgical spirit, applied with clean cottonwool. The palette may be stored in the ultra-violet light cabinet until ready for use. *Individual* lashes come in a special 'contoured' package: you can hold this securely whilst removing individual lashes, so the lashes can be kept hygienically until required.

Always have a spare pair of tweezers available during

application of the individual or strip false lashes. Should you accidentally drop the tweezers with which you are working, you will need a clean, sterile pair.

Scissors, used to trim strip lashes, should be sterilised before use.

Preparing the cubicle

Before the client is shown through to the cubicle, it should be checked to ensure that the required equipment and materials are available and the area is clean and tidy. The plastic-covered couch should be clean, having been thoroughly washed with hot soapy water, or wiped thoroughly with surgical spirit or a professional alcohol-based cleaner. The couch or chair should be protected with a long strip of disposable tissue-paper bed roll, or a freshly laundered sheet and a bath towel. A small towel should be placed neatly at the head of the couch, ready to be draped across the client's chest for protection during treatment. (The paper tissue will need changing and the towels will need to be laundered for each client.)

The couch or beauty chair should be in a slightly elevated position, to give the optimum position for the therapist when applying the false lashes. In this position, too, the client will not be staring into the overhead light (which might cause the eyes to water).

ACTIVITY: POSITION OF CLIENT

Can you think of further disadvantages of having the client flat when applying false lashes?

PLANNING THE TREATMENT

A variety of false lashes is available, including lashes intended for corrective work as well as those simply intended to enhance the natural lashes. Lashes may be short, medium or long; their texture may be, fine, medium or thick, with some having a feathered effect. Strips designed for use on the lower lashes are called **partial lashes**: here small groups of hairs are placed intermittently along the length of the false-lash base.

In a commercial salon, the most popular colours are usually black and brown. For special effects, however, strip lashes are available in fantasy colours, complete with glitter and jewels!

Factors when choosing false eyelashes

Before applying artificial lashes the beauty therapist should consider the following points, and advise the client accordingly.

The client's age

Artificial lashes create a very bold, dramatic effect, which can make an older client look too hard. Remember that the skin colour and the natural hair colour change with age: the lash chosen must enhance the client's appearance.

> **TIP**
> In America streaked lashes are available, to give a more subtle effect for the mature client.

The client's natural lashes

Does the client have short or long, sparse or thick, very curly or straight lashes? Choose an artificial lash to complement the natural lash. Here are some guidelines:

- *Short and stubby lashes* These are commonly seen on older clients who have overhanging eyelids. Choose a medium lash length in a medium thickness at the outer corner of the eyelid; the lashes should become gradually shorter from the centre of the eyelid to the inner corner. Brush the artificial and natural lashes together after application to ensure that they blend.
- *Sparse lashes* Place individual short lashes along the natural lashline; or, to give a more natural appearance, you may wish to apply partial strip lashes to the upper eyelid.
- *Curly eyelashes* These are very common on Afro-Caribbean clients. Choose a longer, sweeping strip or individual false lash, in black. The chosen lash and colour should give emphasis and depth to the eye.

Short and stubby lashes

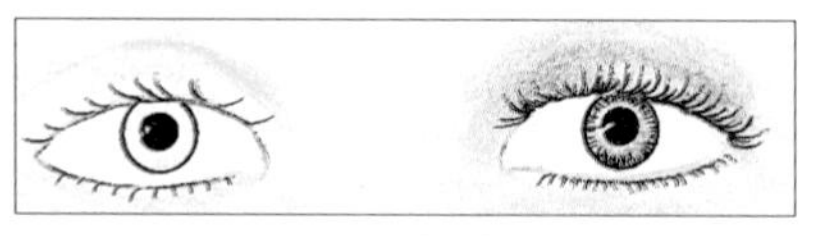

Sparse lashes

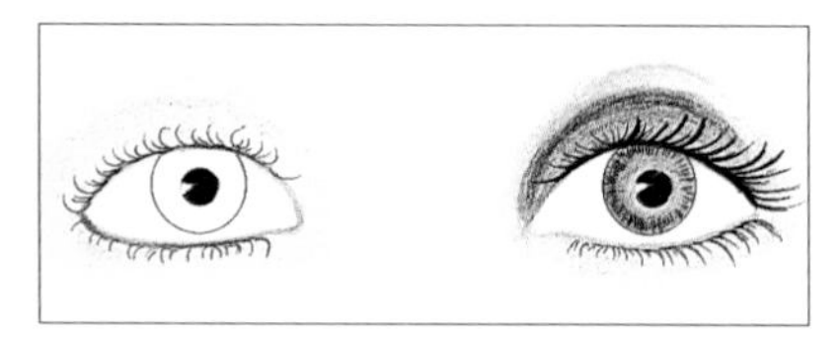

Curly lashes

The natural eyelash colour

Select false lashes that complement the hair and skin tone. Natural-hair false eyelashes offer the greatest choice of colour, but these are expensive and may be difficult to purchase.

> **ACTIVITY: CHOOSING EYELASH COLOURS**
> Suggest a choice of false eyelash colour for the clients below:
>
> - a mature grey-haired client;
> - a young red-haired client;
> - a mature client with bleached hair;
> - a mature Afro-Caribbean client.

> **TIP**
> Permanent waving is available as a cosmetic treatment, using a specialised perming lotion to curl the eyelashes.

Using false eyelashes for corrective purposes

Here are some outlines of corrective techniques for various eye shapes.

- *Small eyes* Place false lashes at the outer corners of the upper eyelid. These should be longer than the natural lashes.
- *Close-set eyes* Place fine, long, individual or partial lashes at

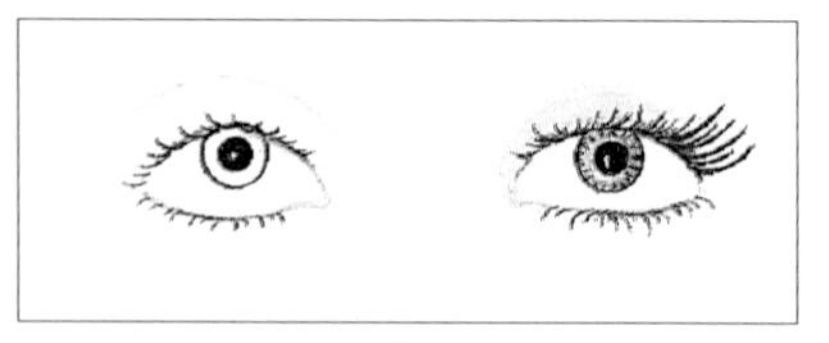

Small eyes

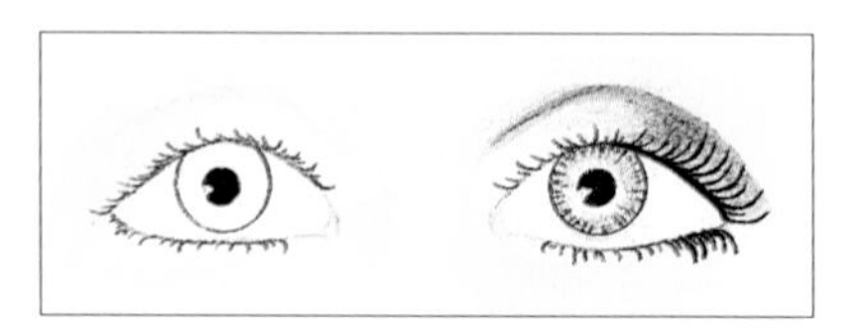

Close-set eyes

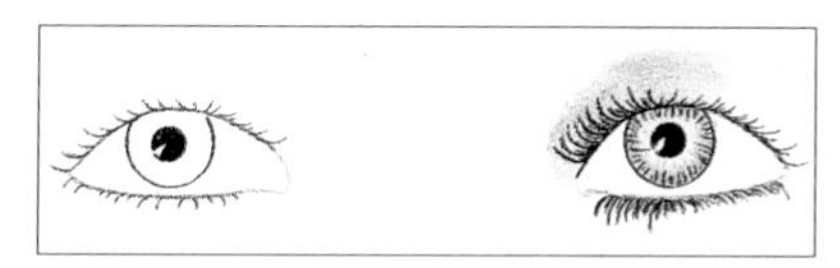

Wide-set eyes

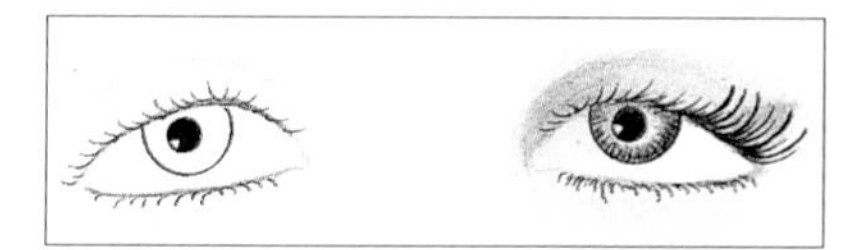

Downward-slanting eyes

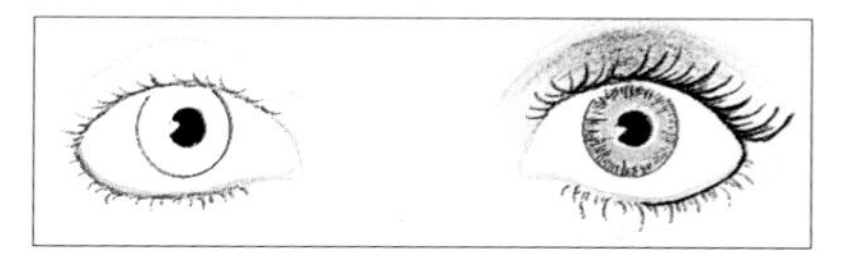

Round eyes

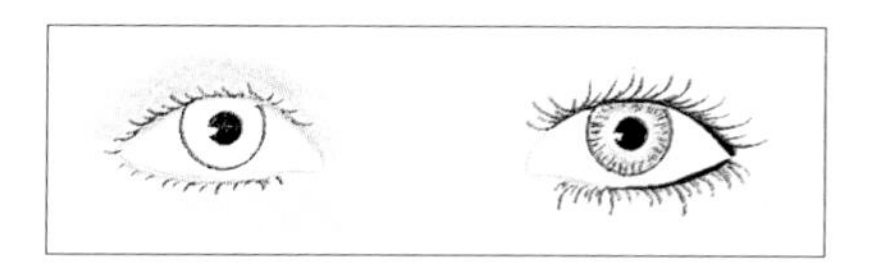

Deep-set eyes

Overhanging lids

the outer third of the eye. They may be applied to the lower lashes as well as to the upper.

- *Wide-set eyes* Apply medium-length lashes at the inner corner of the eye and to the centre, becoming slightly shorter towards the outer corner of the upper eyelid. This may be repeated on the lower lash also.
- *Downward-slanting eyes* Apply longer lashes (individual or partial strip lashes) to the outer corners of the upper eyelid.
- *Round eyes* Individual or strip lashes should be used to lengthen the lashline. Apply the false lashes from the centre of the upper eyelid outwards.
- *Deep-set eyes* Apply fine lashes to the upper and lower lashes. The upper-lid false lashes should be longer, to draw attention to the eye.
- *Overhanging lids* Apply longer lashes to the upper eyelid, from the outer corner and tapering to a shorter length at the centre of the lid and toward the inner corner.

PREPARING THE CLIENT

Show the client through to the prepared treatment area after filling in the record card at reception.

1. Position the client comfortably on the treatment couch or beauty chair (which should be slightly elevated). If the client wears contact lenses, these must be removed before the treatment begins.
2. Drape a clean towel across the client's chest and shoulder. Protect her hair with a clean headband.
3. Wash your hands, which indicates to the client that treatment is beginning and in a hygienic and professional manner.
4. Consult the client's record card, then check the treatment area for visible contra-indications or abnormalities before proceeding.
5. It is usual to carry out a full facial cleanse (rather than cleansing only the eye area) as make-up is usually applied to complement the false eyelashes. Use a cleansing milk to dissolve facial make-up, followed by a non-oily eye make-up remover to cleanse the eye area. Both products should be removed with clean damp cottonwool.
6. To ensure that the eye tissue and eyelashes are thoroughly clean and grease-free, apply a mild toning lotion: stroke this over the skin using clean damp cottonwool.
7. Blot the lashes dry, using a fresh facial tissue for each eye. (Any moisture left on the natural lashes would reduce the effectiveness of the eyelash adhesive.)

APPLYING SEMI-PERMANENT LASHES

1. Check that everything you need is on the trolley.
2. Check that the back of the couch or beauty chair is slightly raised, at a height that is comfortable for you.
3. Discuss the treatment procedure with the client. Explain that she will be required to keep her eyes open during the treatment. Reassure her that she may blink during application. Very often clients feel that they shouldn't, and their eyes begin to water.
4. Ask her to tilt her head downwards very slightly. This tends to lower the upper eyelids, making application easier.
5. Depending on the effect required, you may start application of the individual lashes at different positions along the natural lashline. In general, apply shorter lashes to the inner corners of the eyelid, and longer lashes to the outer corners; this creates a realistic effect and ensures client comfort. If you are applying individual lashes to the entire upper lid, it is practical to start application at the inner corner of the eyelid and work outwards: this follows the natural contour of the eye.
6. With the sterile tweezers, select a lash from the package, holding it near its centre. Brush the underside of the individual lash, at the root, through the adhesive. The adhesive should extend slightly beyond the root. You need sufficient adhesive, but not too much – excess adhesive should be removed by wiping the lash against the inside of the adhesive container.

TIP

Use a clean pair of tweezers to remove the individual lashes from their container when required. Holding the tweezers firmly, grip each lash near its base. (This avoids misshaping the outer lash hairs.)

7. Working from behind the client, hold the tweezers at the angle at which the false lash will be applied to the natural lashline. Hold the brow tissue with your other hand, to steady the eyelid.

TIP

Do not prepare the lash adhesive until you are ready to use it. It tends to dry on contact with the air.

TIP

When applying the individual lashes to the inner portion of the eyelid, hold the skin taut, stretching the skin slightly. This will enable you to position the false lash more easily.

Salon System

The natural lashes

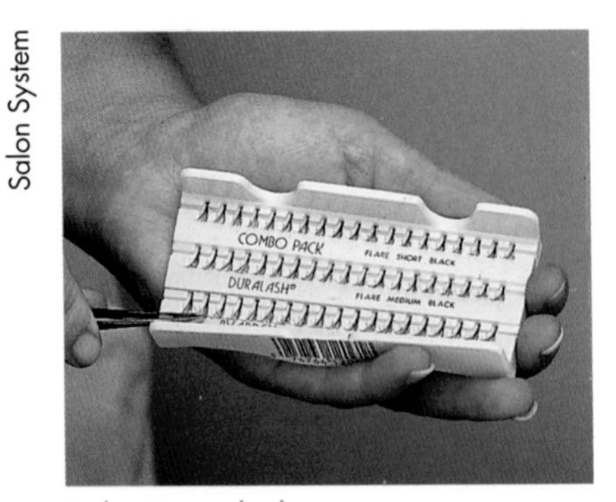

Salon System

Selecting a lash

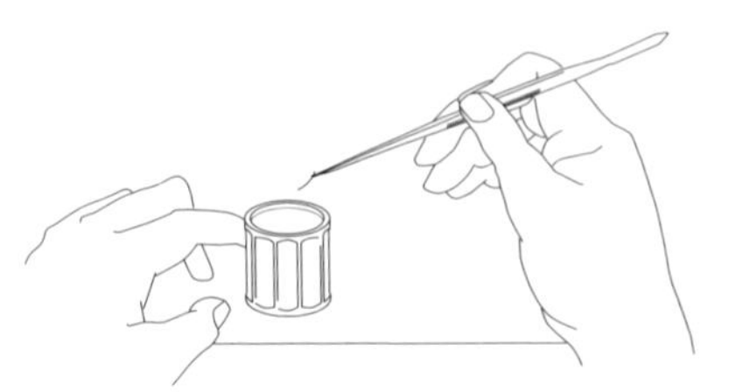
Brushing through the adhesive

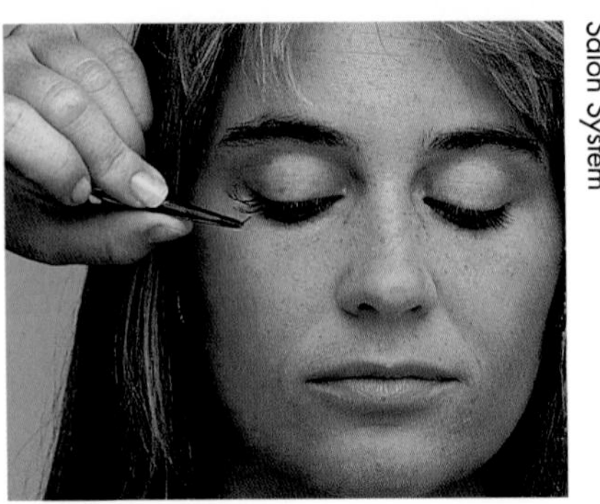
Salon System

Applying a lash to the upper lid

TIP

Positioning the individual lashes correctly requires experience. It is a good idea initially to practise application without adhesive.

TIP

To ensure efficient application, don't get adhesive on the points of the tweezers.

Using a stroking movement, place the underside of the false lash on top of the natural lash. Stroke the adhesive along the length of the natural eyelash. Guide the false lash towards the base of the natural lash, so that the false lash rests along the length of the natural lash. Wait a few seconds to allow the adhesive to dry (to prevent the lashes from sticking together). Continue placing further false lashes side by side until the desired effect is achieved.

During application, keep checking your work. If a lash is out of line, remove it while the adhesive is still soft. (If the adhesive has set, the lash will need to be removed using adhesive solvent – see the photograph on page 213).

8 Apply the false lashes one at a time, to each eye alternately. This avoids sensitising the eye, and makes it easier for you to create a balanced effect.

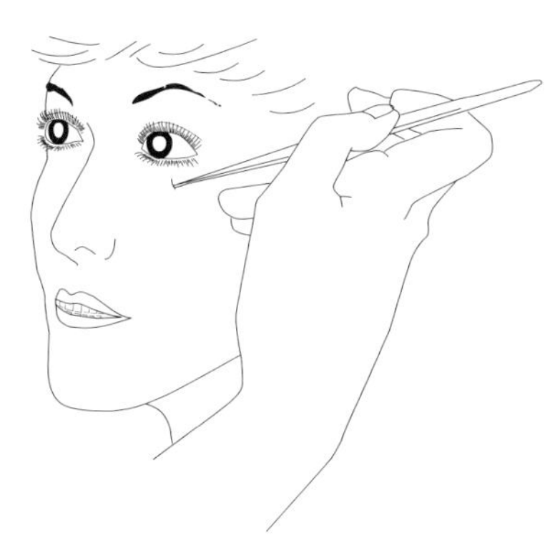

Applying a lash to the lower lid

9 If the client requires false lashes to be applied to the *lower* lid, the application technique is slightly different.

Work facing the client, with the client looking upwards, her eyes slightly open. Follow the same general procedure for applying the false eyelashes; here, however, the lashes curve downwards and the adhesive is applied to the *upper* surface of the lash.

Lashes applied to the lower eyelid are usually shorter than those chosen for the upper lid; and more adhesive is required for the lashes to be secure and have maximum durability.

TIP

Tell the client that artificial lashes applied to the lower lashes tend to fall off after one week. (This is probably due to the natural watering of the eye affecting the adhesive.)

The completed effect

Salon System

10 When you have completed the lash application, ask the client to sit up, and show her the completed effect.

11 If the client is satisfied with the result, you can apply a water- or powder-based eye make-up if desired. Do not apply mascara, as this will reduce the adhesion to the natural lash. Mascara also clogs the lashes together, and is difficult to remove without affecting the eyelash adhesive. On completion, the lashes can be gently brushed – using a disposable brush – to remove particles of eyeshadow.

APPLYING STRIP LASHES

Strip false eyelashes are applied *before* carrying out the eye make-up – this avoids the eye make-up being spoilt if the eyes water slightly during application.

1 Carefully apply moisturiser and foundation, taking care not to get any cosmetic products on the lashes. (If you do, gently wipe over the lashes with the non-oily eye make-up remover, and blot the eyelashes dry again with a clean facial tissue.)

2 Brush the lashes to separate them, using a clean disposable mascara brush. This makes false-lash application easier, and removes any fine particles of loose powder.

3 Remove the strip lashes from their container, and place them on a clean sanitised palette. Each strip is designed to fit either the left or the right eye: remember which is which when placing them on the palette.

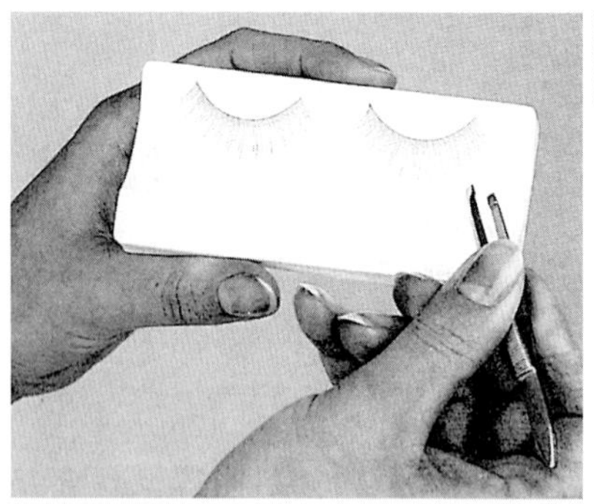
Bellitas

Removing the strip lashes from their container using tweezers

4 Check the length of the strip against the client's eyelid. The strip should never be applied directly from one corner of the eyelid to the other, but should start about 2mm from the inner corner of the eye, and end 2mm from the outer corner. This ensures a natural effect and maximises the durability of the false lash.

When you remove the strip lash from the package you will find that there is adhesive on the backing strip, which fixes the lash in the container: this adhesive is sufficient to hold the lash onto the client's natural lash while you measure the length.

5 To trim the false lashes you require a sharp pair of scissors.

First correct the length of the *strip* if necessary. Hold the lashes securely with one hand, and then trim the strip at the outer edge (as shown below).

Then trim the lashes themselves, if necessary. Never reduce the length of the lashes by cutting straight across them: the result this would not look natural. Natural eyelashes are of varying lengths, due to the nature of the hair growth cycle; it is this effect that you must simulate. To shorten the lash, 'chip' into the lash. Use the *points* of the scissors to shorten the lash length (see below). Cut the lashes so that the shorter lashes are at the inner corner of the eyelid, gradually increasing toward the outer corner.

HEALTH & SAFETY: LASH LENGTH

If the lashes were not shorter at the inner corner of the eyelid, they would irritate the client's eye.

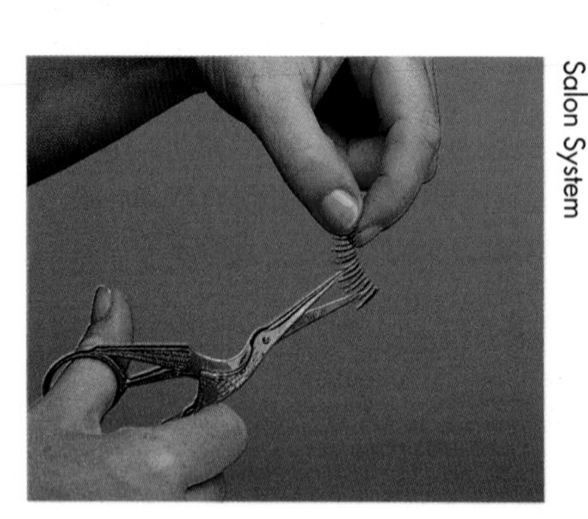
Salon System

Trimming the lashes

TIP

Do not apply adhesive directly from the tube to the eyelash base – you would apply too much adhesive.

Salon System

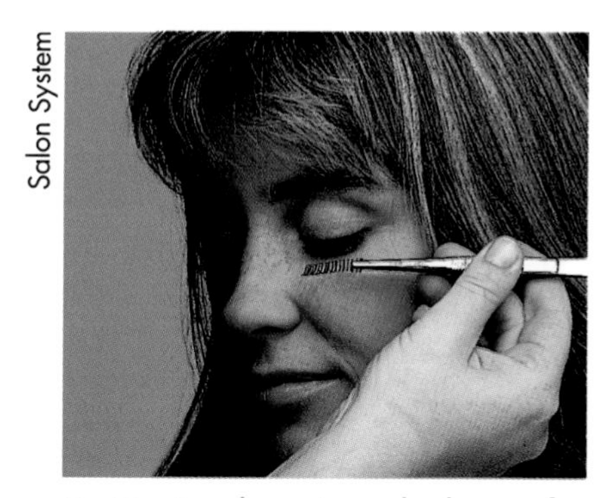

Positioning the strip to the base of the natural lashline

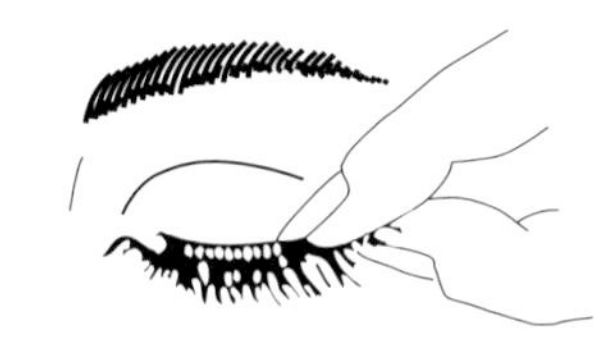

Pressing the strip lash and the natural lash together

Salon System

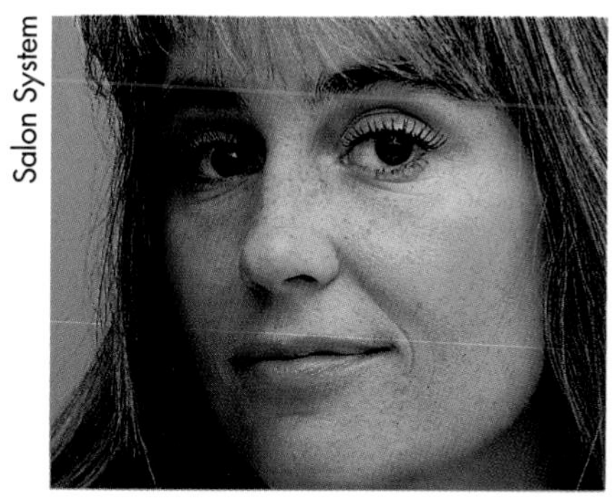

The completed effect

TIP

If a client has straight eyelashes that grow downwards, curl them slightly using eyelash curlers (page 168). If you don't a gap will be visible between the real lashes and the false strip lash.

6 Check that everything required for the false eyelash application is available on your trolley.

7 The couch or beauty chair should be in a slightly raised position. During the treatment you will be working from behind the client: the height must be comfortable for you.

8 Discuss the treatment procedure with the client. Explain to her that she will be required to keep her eyes open during the application. Ask her to tilt her head downwards very slightly – this lowers the upper eyelids, making application easier.

9 Using the sterile tweezers, remove one of the eyelash strips from the palette. Handle it very carefully, as it can easily become misshapen.

Remembering that the strip is designed to fit either the right or the left eye, place it against the appropriate eyelid and check the length (with the client's eyes closed).

10 Once satisfied that the length of the strip lash is correct, remove the adhesive tape used to hold it in the container.

11 Place a small quantity of strip-lash adhesive on the sanitised palette.

12 Ask the client to look down slightly, with her eyes half open. With one hand lift her brow to steady the upper eyelid.

Holding the strip lash with the sterile tweezers at its centre, drag it at its base through the adhesive. (The adhesive must be moist.) It is usually white, but when it dries it becomes colourless.)

Position the base of the strip lash as close as possible to the base of the natural eyelash, ensuring that it is about 2 mm in from the inner and outer corners of the eye. *Gently* press the false and natural eyelashes together with your fingertips, along the length of the lash and at the outer corners.

13 When you are sure that the first strip lash is secure, apply the second in the same way.

14 If strip lashes are to be applied to the bottom lashes also, apply these now, in the same way as the upper lashes. (Strip lashes for the lower lids are fine, with an extremely thin base. These lashes should be trimmed as before to ensure comfort and durability in wear.)

15 Allow 3–5 minutes for the adhesive to dry.

16 Gently brush the lashes from underneath the natural lashline, using a clean disposable mascara brush. This will blend the natural and false lashes together. Check that both sets of lashes are correctly positioned, and that a balanced look has been achieved.

17 Artificial lashes look more realistic if eyeliner is applied to the client's eyelid: this disguises the base of the strip lash.

CONTRA-ACTIONS

If during application of false lashes the eye starts to water, blot the tears with the corner of a clean tissue. The tears can cause the adhesive to take on an unsightly white crystallised appearance. Any possible irritation of the eyes should therefore be avoided, during both preparation of the eye area and application itself.

Never place eyelashes *underneath* the natural eyelashes – eye irritation would occur.

While practising individual eyelash application you may find at some point that you have accidentally glued a couple of the lower and upper natural lashes together. Apply adhesive solvent to a cottonwool-tipped orange stick, and gently roll this over the lash length to dissolve the adhesive.

If solvent or adhesive should accidentally enter the eye, rinse the eye thoroughly and immediately, using distilled water in a sanitised eyebath. Repeat this until discomfort is no longer experienced.

HEALTH & SAFETY: CLIENT COMFORT
Check the eyelash application as you work. Ensure that the lower and upper lashes are not stuck together, and that the false eyelashes are accurately and evenly applied.

AFTERCARE AND ADVICE

The following aftercare instructions should be given to the client after false eyelash application:

- Avoid rubbing the eyes, or the lashes may become loosened.
- Do not use an oil-based eye make-up remover, as its cosmetic constituents would dissolve the adhesive.
- Use only dry or water-based eye make-up (as these may readily be removed with a non-oily eye make-up remover).
- If the lashes are made of a synthetic material, heat will cause them to become frizzy. Advise the client to avoid extremes of temperature, such as a hot sauna.
- Do not touch the eyes for 1½ hours after application, while the adhesive dries thoroughly.

If the client has had *individual* artificial lashes applied, the following home-care advice should be given on caring for the lashes:

- Use a non-oily eye make-up remover daily to cleanse the eyelids and eyelashes. Avoid contact with oil-based preparations in the eye area, such as moisturisers and cleansers: the oil content would dissolve the adhesive, and the lashes would become detached.
- Do not attempt to remove the artificial lashes – pulling at the artificial lash would pull out the natural eyelashes also.
- After bathing, gently *pat* the eye area dry with a clean towel.

Clients with individual false eyelashes should have the false lashes maintained by regular visits to the salon. Lost individual lashes can be replaced as necessary; this is often described as an eyelash **infill** service.

If the client wishes to have the individual false lashes removed, this should be done professionally.

HEALTH & SAFETY: SOLVENTS

Although solvents were originally formulated to remove artificial lashes from the natural lash, great care must be taken to avoid skin/eye irritation. Ideally this solution should be used to clean the artificial lashes after removal. Removal of artificial lashes should be with an oil-based eye make-up remover product.

TIP

Disposable cotton buds may be used to apply eyelash adhesive solvent.

Removing individual eyelashes

1. Position the client lying on the couch.
2. Wash your hands.
3. Remove make-up from the eye area, cleansing the skin with a suitable eye make-up remover.
4. While the client's eyes are open, place a pre-shaped eyeshield underneath the lower lashes of each eye. (This will protect the eye tissue from the solvent.) Position the eyeshields so that they fit snugly to the base of the lower lashes.
5. Ask the client to close her eyes gently, and not to open them again until you tell her to do so.
6. Prepare a new disposable orange stick by covering it at the pointed end with clean dry cottonwool.
7. Moisten the cottonwool with the artificial eyelash adhesive solvent.

HEALTH & SAFETY: EYE CARE

Do not allow eyelash adhesive solvent to come into contact with the eye tissue – it could cause irritation of the skin. Ensure that there is sufficient solvent only on the cottonwool: it should be moist but not dripping wet, or the solvent might enter the eye.

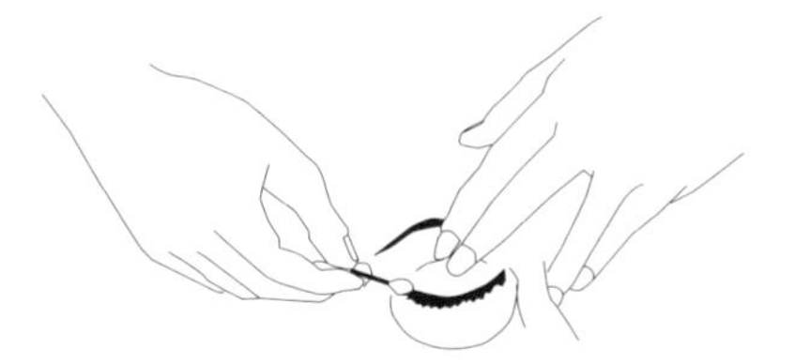

Applying adhesive solvent to remove false eyelashes

8. Treating one eye at a time, gently stroke down the false eyelashes with the adhesive solvent until the adhesive dissolves and the false eyelash begins to loosen.
9. When you are satisfied that the eyelash adhesive has dissolved, gently attempt to remove the false eyelash. Support the upper eyelid with the fingers of one hand, using the other hand to remove the eyelash with a sterile pair of manual tweezers. If the adhesive has been adequately dissolved, the eyelash will lift away easily from the natural eyelash. If there is any resistance, repeat the solvent application until the eyelash comes away readily.

HEALTH & SAFETY: CLIENT COMFORT

Never attempt to remove the artificial eyelashes until they have begun to loosen – if you do, the client's natural eyelashes will be removed also, causing her discomfort.

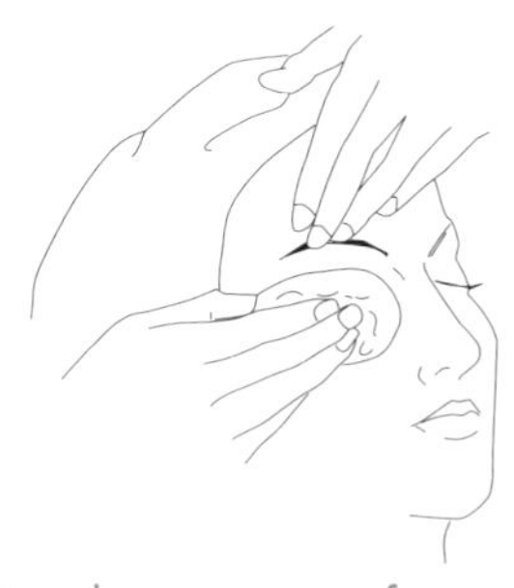

Soothing the eye area after removal of the lashes

10. As the artificial eyelashes are removed, collect them on a clean white facial tissue or a clean pad of cottonwool.
11. Having removed all of the artificial eyelashes from one eye, soothe the area by applying damp cottonwool pads soaked in cool water. (This will also remove any remaining solvent.) A

damp cottonwool pad may be placed over the eye, while you remove the false lashes from the other eye.

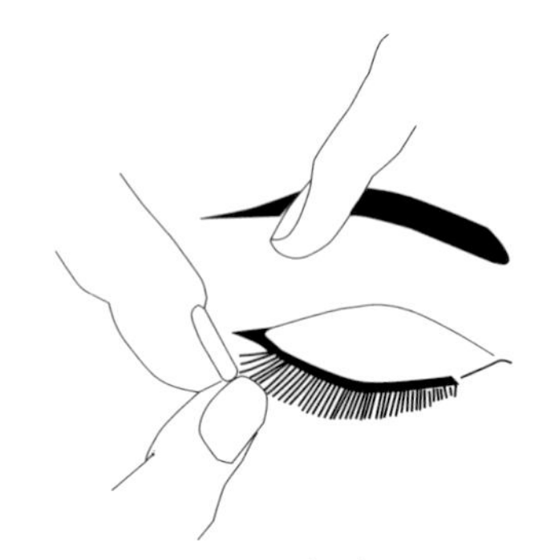
Removing strip eyelashes

Removing strip false lashes

If a client wears strip eyelashes she will need instructions on how to remove and care for the false lashes herself.

1. Use the fingertips of one hand to support the eyelid at the outer corner. With the other hand, lift the lash strip base at the outer corner of the eye. Gently peel the strip away from the natural lash, from the outer edge towards the centre of the eyelid.

> **HEALTH & SAFETY: CLIENT COMFORT**
> When removing the strip lash, avoid pulling the natural lashes with the false lashes.

2. Peel the adhesive from the backing strip, using a clean pair of manual tweezers. Take care to avoid stretching the strip lash.
3. Clean the strip lash in the appropriate way.
   - *Strips made from human hair* Clean with a commercial lash cleaner or 70% alcohol. This removes the remaining adhesive and the eye make-up.
   - *Synthetic strips* Place in warm soapy water for a few minutes, to clean the lashes and remove the remaining adhesive. Rinse in tepid water.
4. After cleaning the strip lashes should be recurled.

Recurling strip lashes

For reasons of hygiene and because of the time involved, this service will not be offered by the salon. The client, however, will need advice on how to recurl the lashes at home.

1. On removal from the water, place the lashes side by side, ensuring that the inner edges are together inside a clean facial tissue.
2. Wrap a tissue around an even, barrel-shaped object such as a felt-tip pen, and secure it with an elastic band.
3. The false lashes, inside the facial tissue, should then be rolled

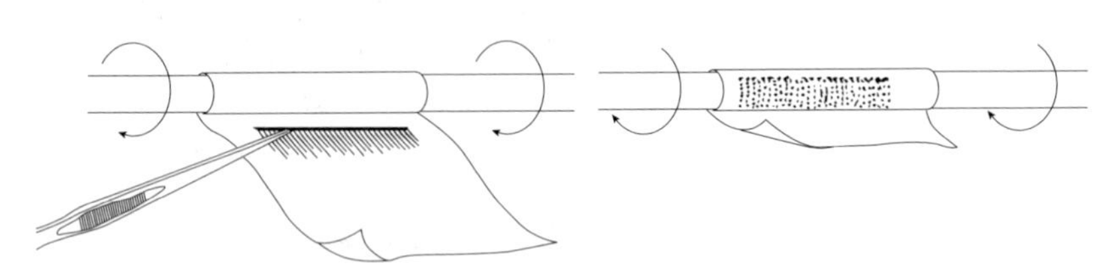
Recurling strip lashes

> **HEALTH & SAFETY: SEMI-PERMANENT LASHES**
> The client should be advised to return to the salon for semi-permanent lashes to be removed professionally.

around this object. Keep the base of the lash straight, so that the whole lash length curls around the object.

Once recurled, the strip lashes can be returned to the contoured shelves in their original container, and stored for further use.

KNOWLEDGE REVIEW

Artificial eyelash application

Preparation

1. What should be considered in the choice of artificial eyelash for a client?
2. How should the eye area be prepared to ensure that the eyelashes adhere securely to the natural eyelashes?
3. How would you explain the treatment application for (a) strip lashes, and (b) individual eyelashes, to ensure client comfort during application?

Application

1. In what position should the client be while you apply artificial eyelashes?
2. When applying false eyelashes along the length of the eyelid, how can you avoid irritating the corners of the eye?
3. What differences are there between the application of strip and individual eyelashes?

Aftercare

1. What aftercare advice should be given to the client following the application of individual eyelashes?
2. For how long should a client be advised to wear:
   (a) strip eyelashes?
   (b) individual eyelashes?
3. Explain the correct removal of:
   (a) strip eyelashes;
   (b) individual eyelashes.

Health, safety and hygiene

1. If a client has a known skin sensitivity, what should be carried out before the application of semi-permanent eyelashes?
2. What safety precautions should be taken during the application of false eyelashes?

Case studies

1. Which eyelashes would you select for a client who is to attend an evening function, and whose occupation prohibits her from wearing artificial lashes?
2. When applying semi-permanent eyelashes, what colour and length would you select for:
   (a) a client with short, stubby, fair lashes?
   (b) a client with small eyes?

EYELASH PERMING

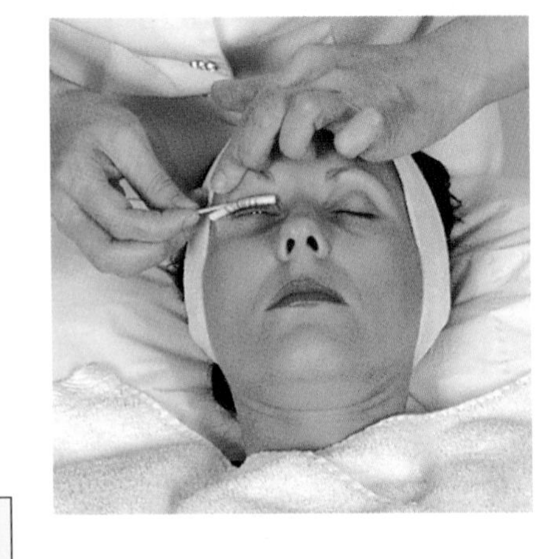

Eyelash perming entails permanently curling the lashes, which enhances the appearance of the eyes. The lashes immediately appear longer, which suits most clients. The treatment also suits:

- short sparse lashes, to make them appear longer and denser;
- downward-slanting eyes, as the eyes appear lifted when the outer lashes are curled;
- special occasions and holidays;
- clients who are unable to wear eye make-up at work.

The effect lasts as long as the hair growth cycle of the eyelashes.

> **HEALTH AND SAFETY: EYELASH CURLING**
> Repeated mechanical eyelash curling is time-consuming and can weaken the lashes, causing breakage of the natural lash. Permanent curling is the ideal alternative.

RECEPTION

When making an appointment for this service, recommend that the client has a skin test. Ask her to visit the salon for this, 24 hours before the appointment.

If she wears contact lenses, ask her to bring her lens container so that she can place the lenses in it during the perming treatment.

Advise the client that the treatment requires that she keep her eyes closed for a long time. Some clients may find this uncomfortable, and may choose not to have the treatment.

The client should not have an eyelash tint immediately before perming, as the colour of the lashes will be lightened by the treatment process. However, tinting is effective *following* eyelash perming and should be promoted. Allow a minimum of 24 hours following the eyelash perming service before tinting.

On average, a client will need her eyelashes perming every 2–3 months. Treatment takes approximately 45 minutes.

> **TIP**
> Permanent curling is not advisable if:
> - the lashes are naturally curly;
> - the lashes are very short and sparse;
> - the lashes are fragile.

Skin tests

Some clients may be sensitive to the perm solution, and may produce an allergic reaction immediately on contact with it; others may become allergic later. For this reason, a skin test must be carried out before *each* eyelash perming treatment. This is essential not just for new clients, but also if there has been a lengthy interval since the last eyelash perming service.

1. Apply a small amount of perm solution to the inside of one elbow, and a small amount of fixing lotion to the other.
2. After 15 minutes, remove the lotion with cool water.

TIP

To maximise revenue and to enhance the client's treatment, offer the client another service while the lashes are being permed. Suitable treatments include a manicure or hand treatment.

HEALTH AND SAFETY: ALLERGIES

Some perming lotions contain lanolin. If using such lotions, ask the client before application whether she is allergic to lanolin.

3 Advise the client how to recognise a positive skin reaction: skin reddening, itching and swelling. Advise her to apply a soothing agent if this occurs and to notify you. This reaction can then be recorded on the client's record card.

CONTRA-INDICATIONS

After completing the record card and inspecting the eye area, if you have found any of the following do not proceed with the eyelash perming treatment:

- *Inflammation or swelling* around the eye area.
- *Skin disease* in the eye area.
- *Skin disorder* in the area such as psoriasis or eczema.
- *Cuts and abrasions* in the eye area.
- *Hypersensitive skin.*
- *Eye disorders*, such as conjunctivitis, blepharitis, styes or hordeola, watery eye, or cysts.
- *Positive (allergic) reaction* to the skin test.
- *Contact lenses* – unless removed.
- *Considerable nervousness* – a particularly nervous client may make application, fixing and removal hazardous.

EQUIPMENT AND MATERIALS

To carry out the perming you will need the following equipment and materials:

- *Couch or beauty chair* – with sit-up and lie-down positions and an easy-to-clean surface.
- *Trolley* – on which to place everything.

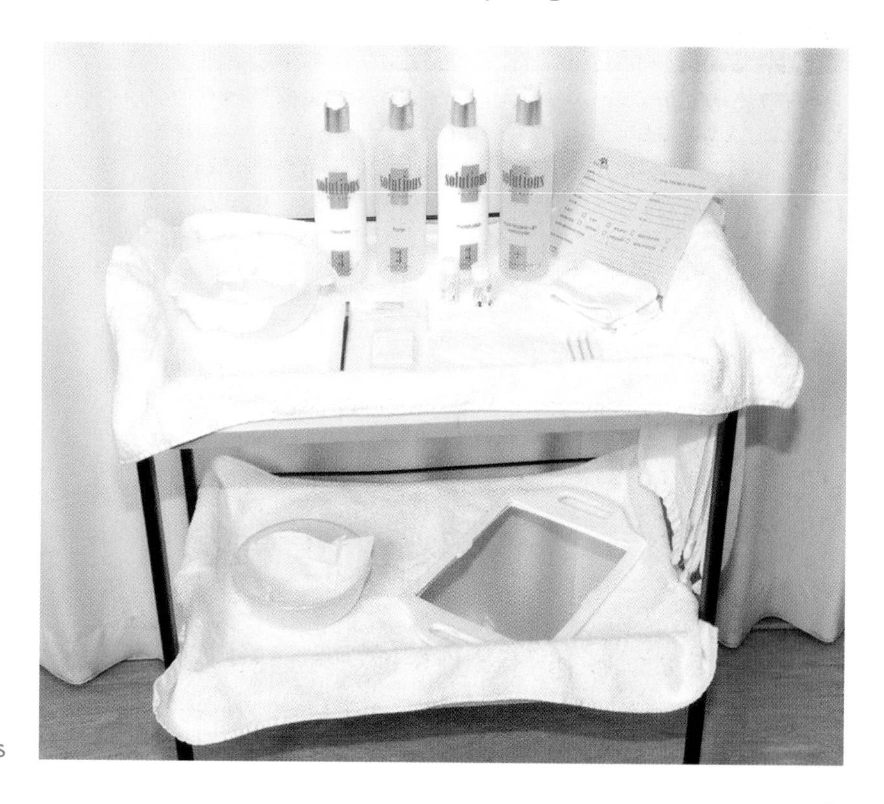

Equipment and materials

Salon System

Perm lotions

- *Towels (medium-sized)* – freshly laundered for each client.
- *Headband (clean)* – to protect the hair.
- *Eye make-up remover (non-oily).*
- *Mild perm solution (usually 6% thioglycollate)* – especially designed for use in the eye area. (It is this solution that curls the lash into the desired new shape.)
- *Fixing lotion* – sodium bromate. (It is this which makes the new curl permanent.)

HEALTH AND SAFETY: STORAGE
Store the perm and fixing lotion in a well-ventilated area, away from direct sources of heat and light. If a spillage occurs, wear gloves while you clean it up, and increase ventilation in the area.

TIP
Perm lotions for the eye area are usually of *gel* formulation: this makes the lotion easy to control. Bear this in mind when selecting this product. The bottles are small to reduce the risk of oxidation, which would make the lotion ineffective.

- *Eyelash curlers (a selection of different sizes)* – small curlers are used for fine hair, larger curlers for thicker hair.
- *Eyelash adhesive* – to secure the lashes to the curlers.
- *Lint-free pads* – to remove the perm and fixing lotion.
- *Disposable wooden cocktail or orange sticks* – to secure the natural lashes to the curlers.
- *Disposable brushes* – to apply the perm lotion and fixing lotion.
- *Damp cottonwool* – for cleansing the eye area and to remove excess products.
- *Bottle, with a long, flexible tube* – to dispense distilled water.
- *Distilled water (purified).*
- *Bowls (2, clean)* – to hold damp cottonwool or lint-free pads.
- *Moisturising agent* – to facilitate removal of the curlers.
- *Swing-top bin* – lined with a disposable bin liner, for waste.
- *Hand mirror (clean).*
- *Record card.*

TIP
A specialised conditioning agent may be applied to the lashes. This would be professionally applied as part of the perming service to rehydrate the lashes.

Sterilisation and sanitisation

Ideally, applicators used for the perming chemical agents should be disposable. Alternatively, several brushes must be available to allow effective sanitisation between clients.

Preparing the cubicle

Before the client is shown through to the cubicle it should be checked to ensure that the required equipment and materials are available and the area is clean and tidy.

The couch or beauty chair should be flat or slightly elevated. Clean and protect the couch or chair as for eyebrow-shaping treatment (page 183).

The cubicle should be adequately lit to ensure that the treatment can be given safely, but avoid bright lighting which might cause eye irritation.

PREPARING THE CLIENT

The client should be shown through to the cubicle after the record card has been completed.

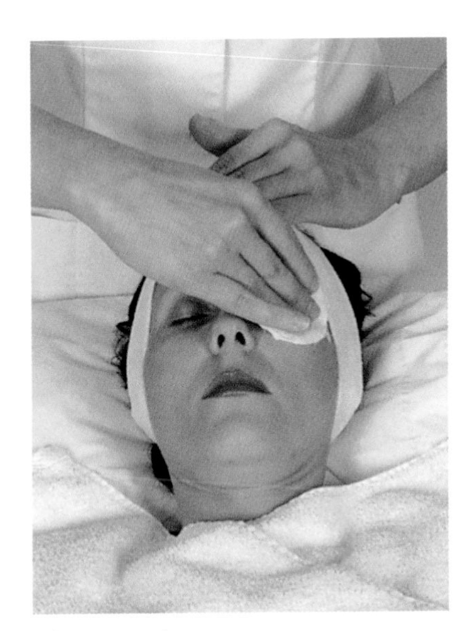

Cleansing the eye area

1 Position the client comfortably, in a flat or slightly elevated position. If she wears contact lenses, these must be removed.

2 Drape a towel across the client's chest and shoulders, and protect her hair with a clean headband.

3 Wash your hands. This assures the client that the treatment is beginning in a hygienic and professional manner.

4 Consult the client's record card, then check the area for any visible contra-indications or abnormalities before proceeding.

5 Cleanse the surrounding eye area with a cleansing milk to dissolve facial make-up. Then use a non-oily eye make-up remover to remove eye products: apply this with clean damp cottonwool.

6 Blot the eyelashes dry with a clean facial tissue. This ensures that the perm lotion is not diluted, and also prevents the perm lotion being carried into the eye.

TIP

If possible, after cleansing leave the lashes to dry for 10 minutes before perming, to ensure that the lashes are completely dry.

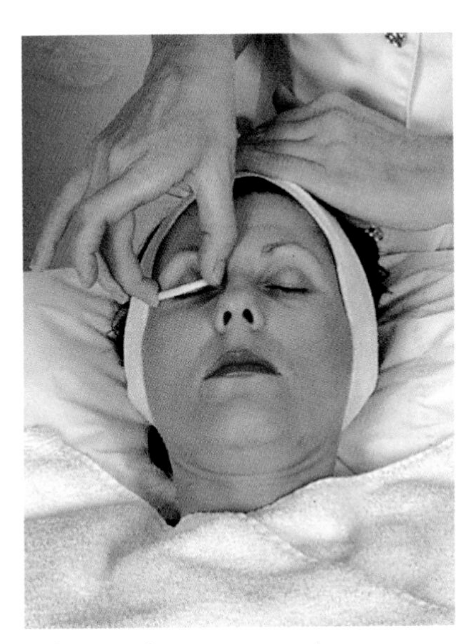

Selecting the correct curler

7 Explain the procedure to the client and tell her about the sensations she may experience during eyelash perming. Warn her that she will need to keep her eyes closed for a long period. Some clients may find this difficult, while others will enjoy this as a time of relaxation!

8 Comb through the lashes to ensure they are evenly separated and dry. This will optimise the results.

PERMING THE EYELASHES

1 Select the correct size of curler for the client's lash length. Bend it so that it fits the contour of the eye. If the curler is too long it may be trimmed.

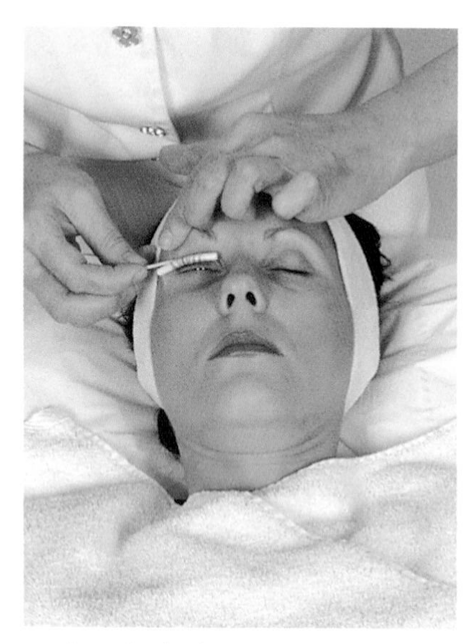

Curling the lashes

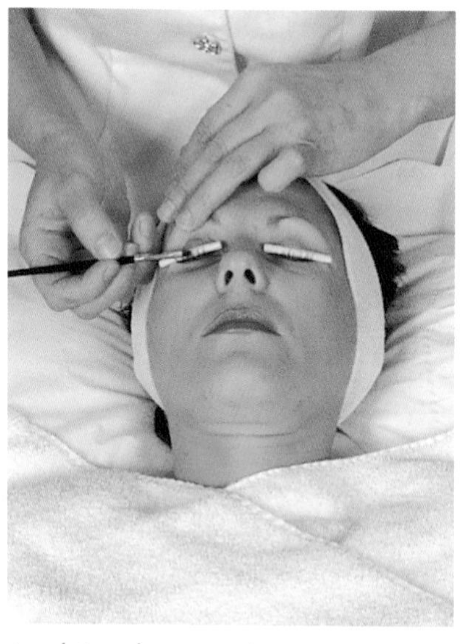

Applying the perm lotion

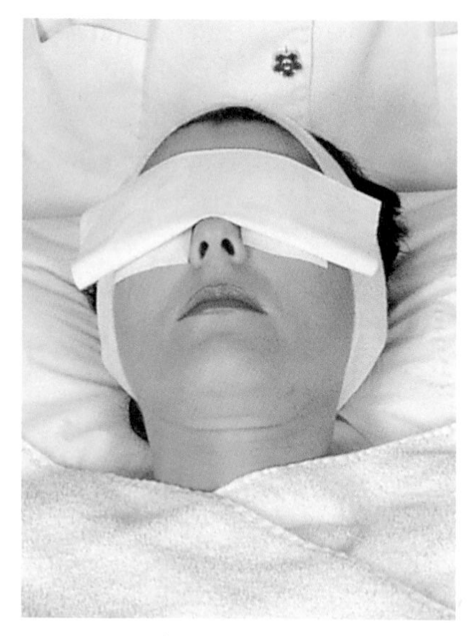

Processing the perm lotion

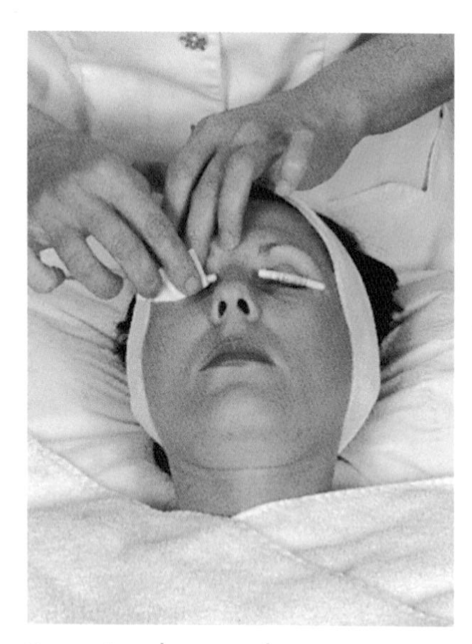

Removing the perm lotion

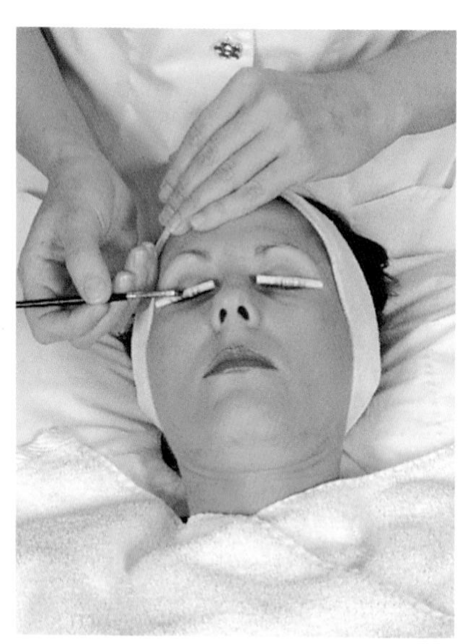

Applying the fixing lotion

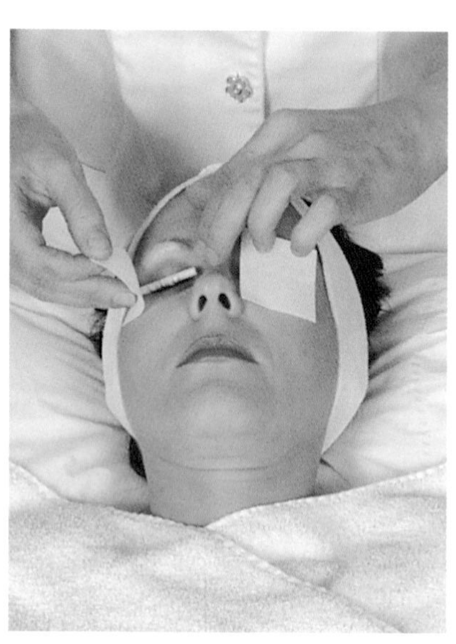

Covering the lashes with lint-free pads

2 Place the curler at the base of the upper lashes, near the inner tear duct.

3 Gently curl the natural lashes around the curler, using the disposable stick. Ensure that the lashes do not overlap each other and are straight or the ends will be 'crooked', spoiling the overall effect and appearance.

4 When you are satisfied that the lashes are straight and that

TIP

Avoid overhandling the curler as this will affect the natural adhesion of the curler to the skin.

TIP

If necessary, apply non-permanent eyelash adhesive using a fine wooden or plastic cocktail stick to secure the lashes to the curler and to keep them even and straight so that the lower and upper lashes do not stick together.

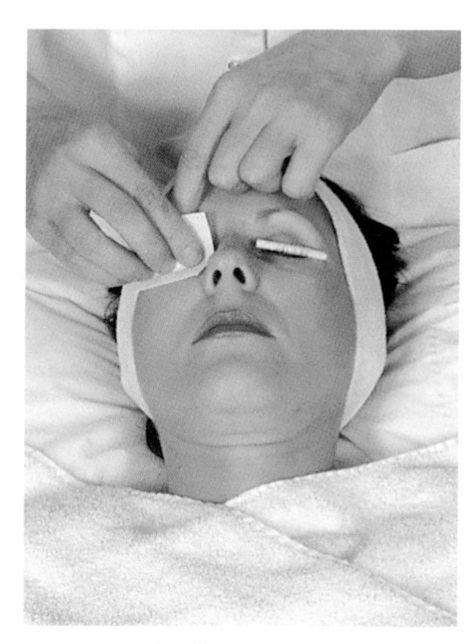

Removing the fixing lotion

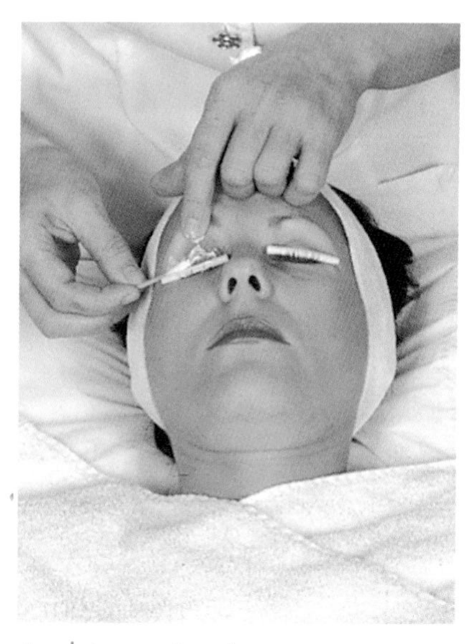

Applying moisturiser

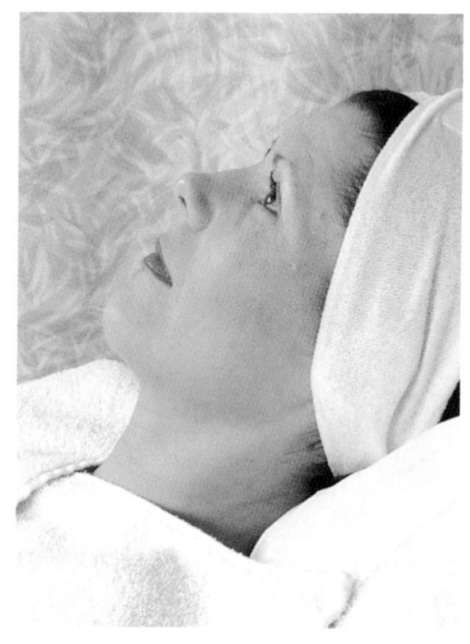

The completed effect

the client is comfortable, apply the perm lotion evenly to the upper lashes, using a disposable brush.

5. Cover the lashes with plastic film or dry lint-free pads. This creates warmth which aids the perming process. Allow:
   - 10 minutes for fine hair;
   - 15 minutes for coarse hair.
6. Remove the perm lotion using dry lint-free pads, gently blotting the lashes.
7. Apply the fixing lotion with a clean disposable brush.
8. Cover the lashes with plastic film or dry lint free pads for 10–15 minutes.
9. Gently remove the fixing solution from the lashes, using dry lint-free pads.
10. When all fixing lotion has been thoroughly removed, gently remove each curler, rolling downwards. A moisturising agent may be applied to the lashes to aid the removal of the curlers. Warn the client that she will feel a gentle pulling movement as the curlers are removed.
11. Wipe excess product from the lashes with damp clean cottonwool.
12. Show the client the finished result in a mirror.
13. Wash your hands again.

HEALTH AND SAFETY: PROTECTIVE GLOVES
Protective gloves may be worn to reduce the risk of chemical contact with the skin.

CONTRA-ACTIONS

If the client complains of discomfort during the treatment, the chemical lotion may have entered the eye. Take the following action:

1. Remove the perm or fixing lotion immediately, using clean damp cottonwool pads.
2. When you are satisfied that all excess lotion has been removed, flush the eye if necessary with distilled water, dispensed from the water bottle. Repeat the rinsing process for at least 15 minutes, until discomfort has been relieved. If the irritation does not cease after rinsing, dry the eye area, protect the eye with a sterile pad, and advise the client to seek medical attention.

If there is a noticeable sensitivity or a burning sensation in the eye area after eyelash perming, recommend that the client does not receive the treatment again. Record this on her record card.

AFTERCARE AND ADVICE

Advise your client not to apply make-up for at least 24 hours. Similarly, she should not receive any other eye treatments for at least 24 hours.

Subsequently, however, eyelash tinting will enhance the effect of the newly permed lashes, so you can promote this service to the client.

KNOWLEDGE REVIEW

Eyelash perming

Preparation

1. What should be discussed with the client before an eyelash perming treatment?
2. What materials do you need to perform eyelash perming?
3. How should you prepare the treatment area for eyelash perming?

Application

1. What chemical agents produce the curled effect?
2. What are the two treatment processes carried out? How long should be allowed for each?
3. How are the curlers removed following treatment?

Aftercare

1. What aftercare advice should be given to the client following eyelash perming?
2. What other complementary eye treatment could be recommended following eyelash perming, and why?
3. How often should this service be carried out?

Health, safety and hygiene

1. What is the procedure for an eyelash perm skin test?
2. Why should client consultation records be current, accurate and complete?
3. What are the possible contra-actions to eyelash perming? How should each be dealt with?

Case studies

1. Which clients would be most suitable to receive this service?
2. How could you promote this service in the salon? Design a suitable promotional package identifying all the benefits.

CHAPTER 8

Manicure and pedicure

MANICURE

THE PURPOSE OF A MANICURE

The word **manicure** is derived from the Latin words *manus*, meaning 'hand', and *cura*, meaning 'care'. A manicure is carried out on the hands for the following reasons:

- to improve the hands' appearance;
- to keep the nails smooth;
- to keep the cuticles attractive and healthy;
- to keep the skin soft.

SKIN AND NAIL DISORDERS OF THE HANDS AND FEET

When a client attends for a manicure of pedicure treatment, the manicurist should always look at the client's skin and nails to check that no infection or disease is present which might contra-indicate treatment. Below is a list of common diseases and disorders that may be seen on the hands and feet. Not all of these contra-indicate treatment.

Disorder	*Cause*	*Appearance*	*Salon treatment*	*Home-care advice*
Onychophagy	Excessive nail-biting	Very little nail plate; bulbous skin at the fingertip; nail walls often red and swollen, due to biting of the skin surrounding the nails	Regular weekly manicures Cuticle treatment to maximise the visible nail-plate area	Bitter-tasting preparations painted onto the nail plate Wear gloves to avoid biting

Dr A. L. Wright

Onychorrhexis 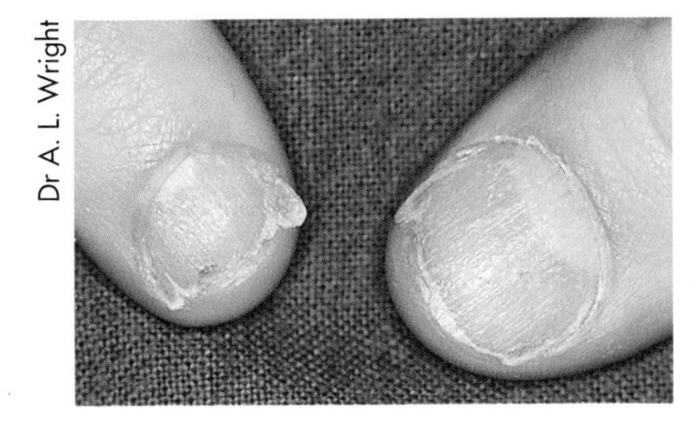 Dr A. L. Wright	Using harsh detergents without wearing gloves Poor diet Not wearing gloves in cold weather	Split, flaking nails	Warm-oil manicures on a weekly basis	Regular use of a rich handcream Always wear rubber gloves when cleaning or washing up Always wear warm gloves in cold weather Ensure a balanced diet
Hangnail Judith Ifould	Biting the skin around the nails Cracking of the dry skin or cuticle	Epidermis around the nail plate cracks and a small piece of skin protrudes between the nail plate and the nail wall, sometimes accompanied by redness and swelling: this condition can be extremely painful	Warm-oil treatments to soften the skin and cuticles Remove the protrusion of dead skin with cuticle nippers: do not cut into live tissue	Regular use of a rich handcream Wear rubber gloves when cleaning and washing up Wear warm gloves in cold weather Ensure a balanced diet
Longitudinal ridges in the nail plate (corrugated nails) 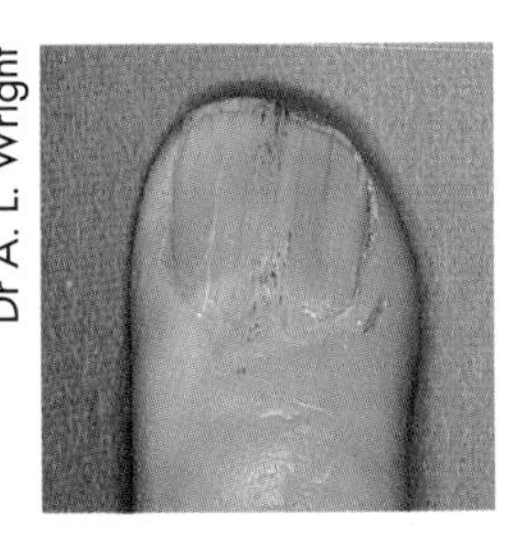 Dr A. L. Wright	Illness Damage to the matrix Age	Grooves in the nail plate running from the cuticle to the free edge: may affect one or all nails	Abrasive buffing to smooth out the ridges Use of a ridge-filling base coat prior to nail enamel application	General manicure advice
Pterygium 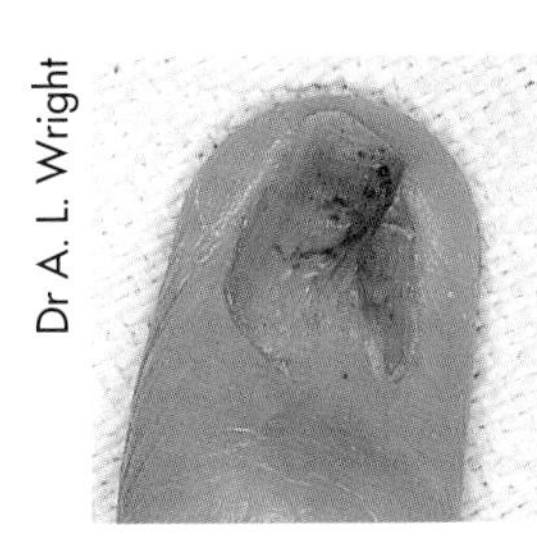 Dr A. L. Wright	Neglect of the nails	Overgrown cuticles, often tightly adhered to the nail plate: if left untreated, this may lead to splitting of the cuticle and subsequent infection	Warm-oil treatments weekly Once softened, remove excess cuticle with cuticle nippers	Regular use of a rich cuticle cream Wear rubber gloves when cleaning and washing up

Transverse furrows in the nail plate (Beau's lines) 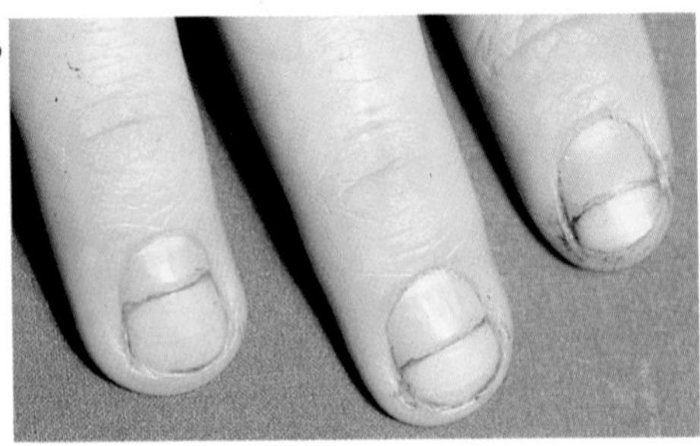 Dr A. L. Wright	Temporary arrested development of the nail in the matrix, due to illness or trauma of the nail	Groove in the nail plate, often on all nails simultaneously, running from side to side: this will grow out with the nail	General manicure regularly	General manicure advice
Leuconychia 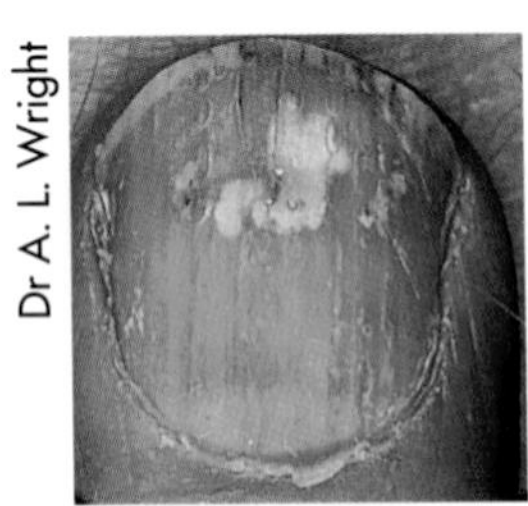 Dr A. L. Wright	Trauma to the nail plate or matrix, due to pressure or hitting with a hard object	White spots or marks on the nail plate: will grow out with the nail	General manicure treatment	Be careful with the hands Wear protective gloves when doing housework or gardening
Bruised nails Dr A. L. Wright	Trauma to the nail (e.g. trapping it in a door); severe damage can result in loss of the nail	Part of the nail plate may appear blue or black	Although this disorder does not contra-indicate treatment, it is advisable to postpone manicuring the nails until the condition is no longer painful Nail enamel may be used to disguise the damaged nail	Be careful with the hands and nails Seek medical advice if swelling is present or if pain persists
Calluses  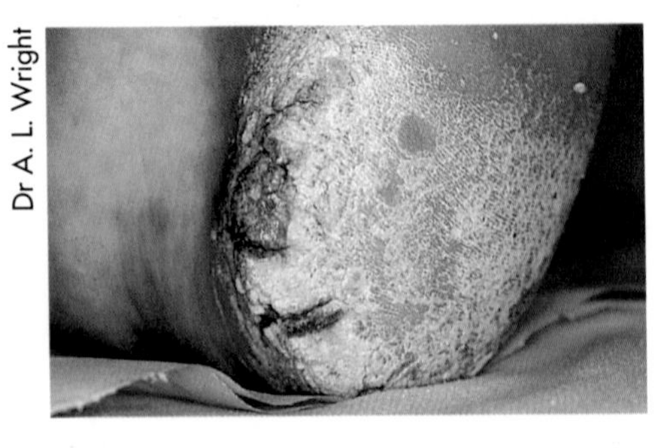 Dr A. L. Wright	Incorrect footwear	Thick, yellowish, hardened patches of skin, usually found on prominent areas of the foot such as the heel and the ball of toe: may be painful	Use a rasp or pumice stone gently to remove any build-up of hard skin: painful calluses should be treated by a chiropodist	Ensure that shoes fit correctly Avoid standing for long periods Alternate style of footwear regularly

Corns 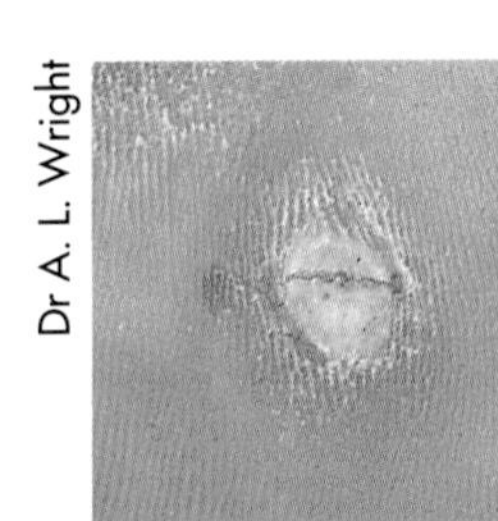 Dr A. L. Wright	Incorrect footwear (corns are often found on toes which have been squeezed together by tight shoes)	Similar to calluses except that the affected area is smaller and more compact; corns often look white, and may be extremely painful	Small corns may be treated in the same way as a callus, but if the client has large or painful corns she should be treated by a chiropodist	Ensure that shoes fit correctly Avoid standing for long periods Alternate style of footwear regularly
Bunions	Long-term wear of ill-fitting shoes, especially those with high heels or pointed toe areas A weakness in the arches of the feet	The large joint at the base of the big toe protrudes, forcing the big toe inwards towards the other toes	None – refer the client to a chiropodist if the bunion is painful; gentle massage may help to ease any pain or discomfort	Try to keep pressure off the affected area
Chilblains	Poor blood supply to the hands and feet, aggravated in cold weather	Fingers and toes may be red, blue or purple in colour; the client may complain of painful or itchy areas	Regular manicures or pedicures, with special attention paid to massage which will help to improve the circulation	Keep affected areas warm and dry Avoid tight footwear, which might restrict the circulation If the condition is severe, seek medical advice
Ingrowing toenails 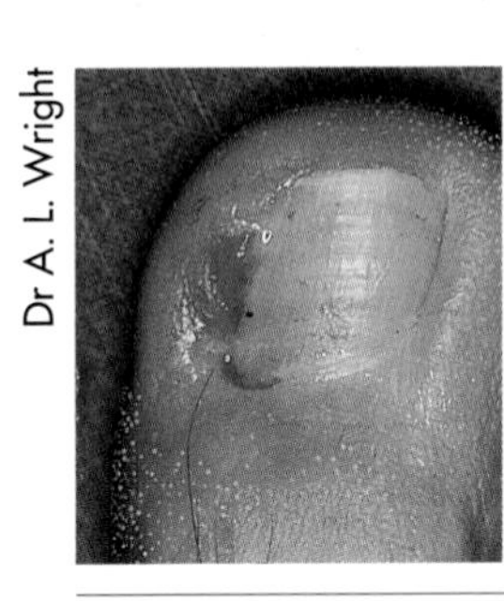 Dr A. L. Wright	Ill-fitting shoes Cutting toenails too far down at the sides	The side of the nail penetrates the nail wall: redness, inflammation and pus may be present, depending on the severity of the condition	None – refer client to a chiropodist	Avoid tight footwear Do not cut the toenails too short, particularly at the sides

THE NAIL

The structure and function of the nail

Nails grow from the ends of the fingers and toes and serve as a form of protection. They also help when picking up small objects.

The nail plate

The **nail plate** is composed of compact translucent layers of keratinised cells: it is this that makes up the main body of the nail. The layers of cells are packed very closely together, with fat but very little moisture.

The nail gradually grows forward over the nail bed, until finally it becomes the free edge. The underside of the nail plate is grooved by longitudinal ridges and furrows, which help to keep it in place.

In normal health the plate curves in two directions:

- transversely – from side to side across the nail;
- longitudinally – from the base of the nail to the free edge.

There are no blood vessels or nerves in the nail plate: this is why the nails, like hair, can be cut without pain or bleeding. The pink colour of the nail plate derives from the blood vessels that pass beneath it.

Function: To protect the nail bed.

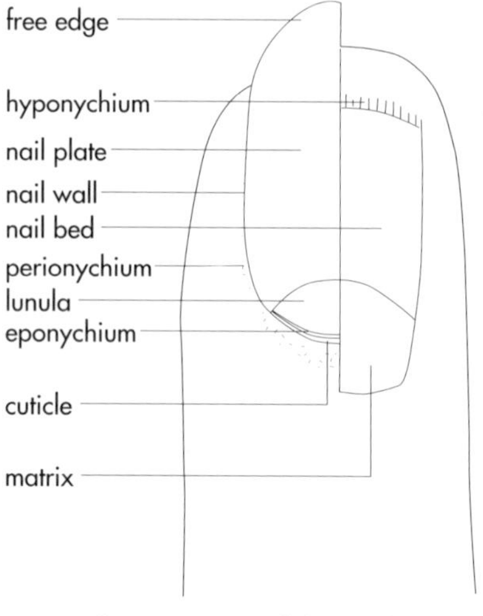

The structure of the nail

The free edge

The **free edge** is the part of the nail that extends beyond the fingertip; this is the part that is filed.

Function: To protect the fingertip and the hyponychium (see below).

The matrix

The **matrix**, sometimes called the **nail root**, is the growing area of the nail. It is formed by the division of cells in this area, which is part of the germinating layer of the epidermis. It lies under the eponychium (see below), at the base of the nail, nearest to the body. The process of keratinisation takes place in the epidermal cells of the matrix, forming the hardened tissue of the nail plate.

Function: To produce new nail cells.

The nail bed

The **nail bed** is the portion of skin upon which the nail plate rests. It has a pattern of grooves and furrows corresponding to those found on the underside of the nail plate; these interlock, keeping the nail in place, but separate at the end of the nail to form the free edge. The nail bed is liberally supplied with blood vessels, which provide the nourishment necessary for continued growth; and sensory nerves, for protection.

Function: To supply nourishment and protection.

The nail mantle

The **nail mantle** is the layer of epidermis at the base of the nail, before the cuticle.

Function: To protect the matrix from physical damage.

The lunula

The **lunula** is located at the base of the nail, lying over the matrix. It is white, relative to the rest of the nail, and there are two theories to account for this:

- newly formed nail plates may be more opaque than mature nail plates;
- the lunula may indicate the extent of the underlying matrix – the matrix is thicker than the epidermis of the nail bed, and the capillaries beneath it would not show through as well.

Function: None.

The hyponychium

The **hyponychium** is part of the epidermis under the free edge of the nail.

Function: To protect the nail bed from infection.

The nail grooves

The **nail grooves** run alongside the edge of the nail plate.

Function: To keep the nail growing forward in a straight line.

The perionychium

The **perionychium** is the collective name given to the nail walls and the cuticle area.

Function: To protect the nail.

The nail walls

The **nail walls** are the folds of skin overlapping the sides of the nails.

Function: To protect the nail plate edges.

The eponychium

The **eponychium** is the extension of the cuticle at the base of the nail plate, under which the nail plate emerges from the matrix.

Function: To protect the matrix from infection.

ACTIVITY: RECOGNISING NAIL STRUCTURE

With a colleague, try to identify the structural parts of each other's nails. Write down both the parts that you can see and the parts that you cannot.

The cuticle

The **cuticle** is the overlapping epidermis around the base of the nail. When in good condition, it is soft and loose.

Function: To protect the matrix from infection.

Nail growth

Cells divide in the matrix and the nail grows forward over the nail bed, guided by the nail grooves, until it reaches the end of the finger or toe, where it becomes the free edge. As they first emerge from the matrix the translucent cells are plump and soft, but they get harder and flatter as they move toward the free edge. The top two layers of the epidermis form the nail plate; the remaining three form the nail bed.

The nail plate is made up of a protein called keratin, and the hardening process that takes place in the nail cells is known as **keratinisation**.

The nail bed has a pattern of grooves and furrows corresponding to those found on the underside of the nail plate: the two surfaces interlock, holding the nail in place.

Fingernails grow at approximately twice the speed of toenails. It takes about 6 months for a fingernail to grow from cuticle to free edge, but about 12 months for a toenail to do so.

THE HAND AND THE FOREARM

The bones of the hand

The wrist consists of 8 small **carpal** bones, which glide over one another to allow movement. This is called a **condyloid** or **gliding joint**.

There are then 5 **metacarpal** bones that make up the palm of the hand.

The fingers are made up of 14 individual bones called **phalanges** – 2 in each of the thumbs, and 3 in each of the fingers.

ACTIVITY: IDENTIFYING BONES IN THE HAND
Look very closely at your hand. Can you identify where the bones are? Try feeling the bones with your other hand. How many can you feel?

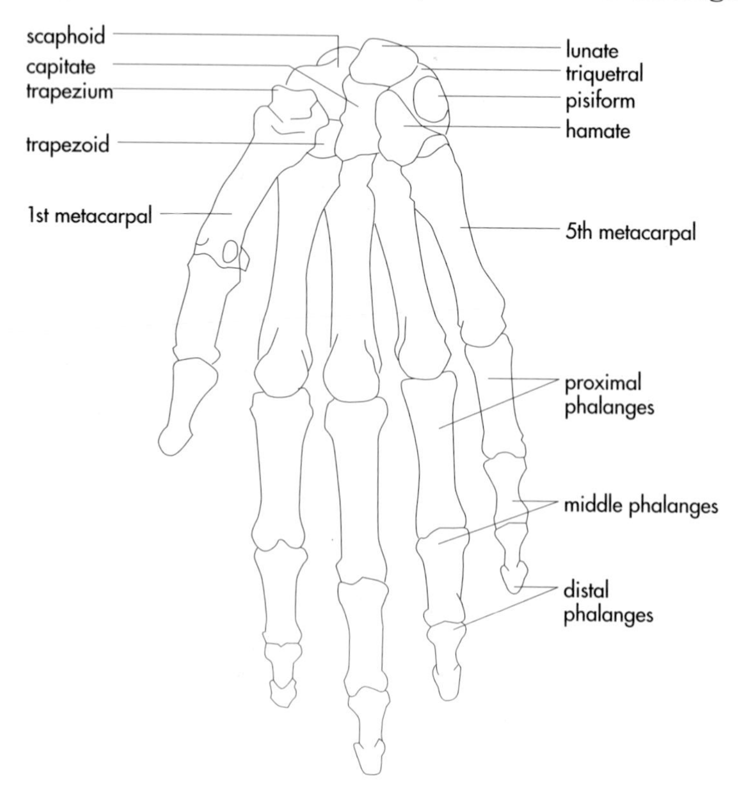

Bones of the hand

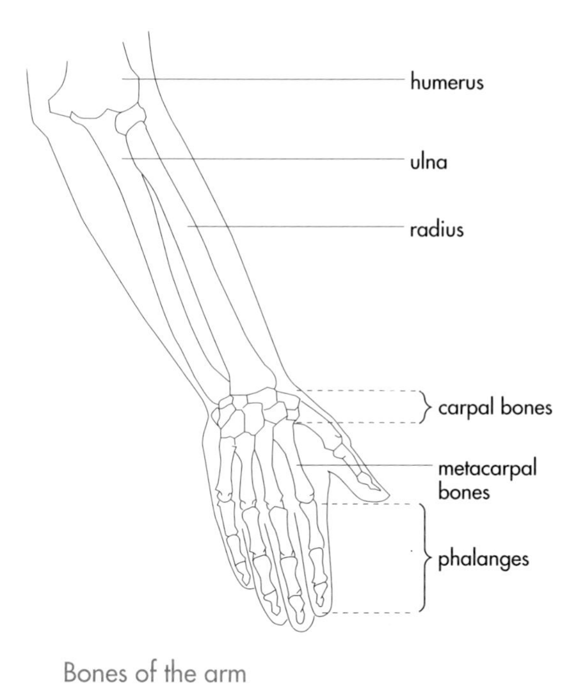

Bones of the arm

The bones of the arm

The arm is made up of 3 long bones: the **humerus** is the bone of the upper arm, from the shoulder to the elbow; the **radius** and **ulna** lie side by side in the lower arm, from the elbow to the wrist.

Having two bones in the lower arm makes it easier for your wrist to rotate. This movement that causes the palm to face downwards is called **pronation**; the movement that causes it to face upwards is called **supination**.

The muscles of the hand and arm

The hand and fingers are moved primarily by muscles and tendons in the forearm. These muscles contract, pulling the tendons, and thereby move the fingers much as a puppet is moved by strings.

The muscles that bend the wrist, drawing it towards the forearm, are **flexors**; other muscles, **extensors**, straighten the wrist and the hand.

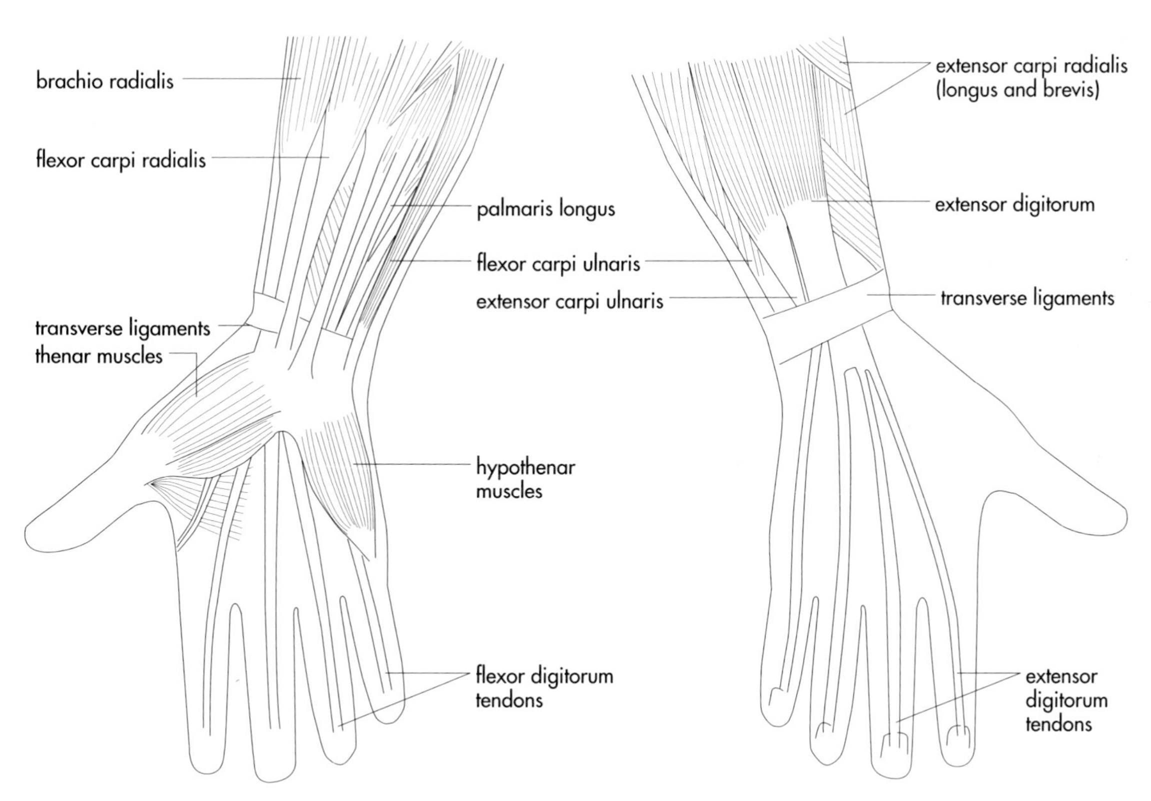

Muscles of the arm and hand

ACTIVITY: OBSERVING THE TENDONS
Hold your palm face upwards, with your sleeve pulled back so that you can see your forearm. Move the fingers individually towards the palm. Can you see the tendons moving?

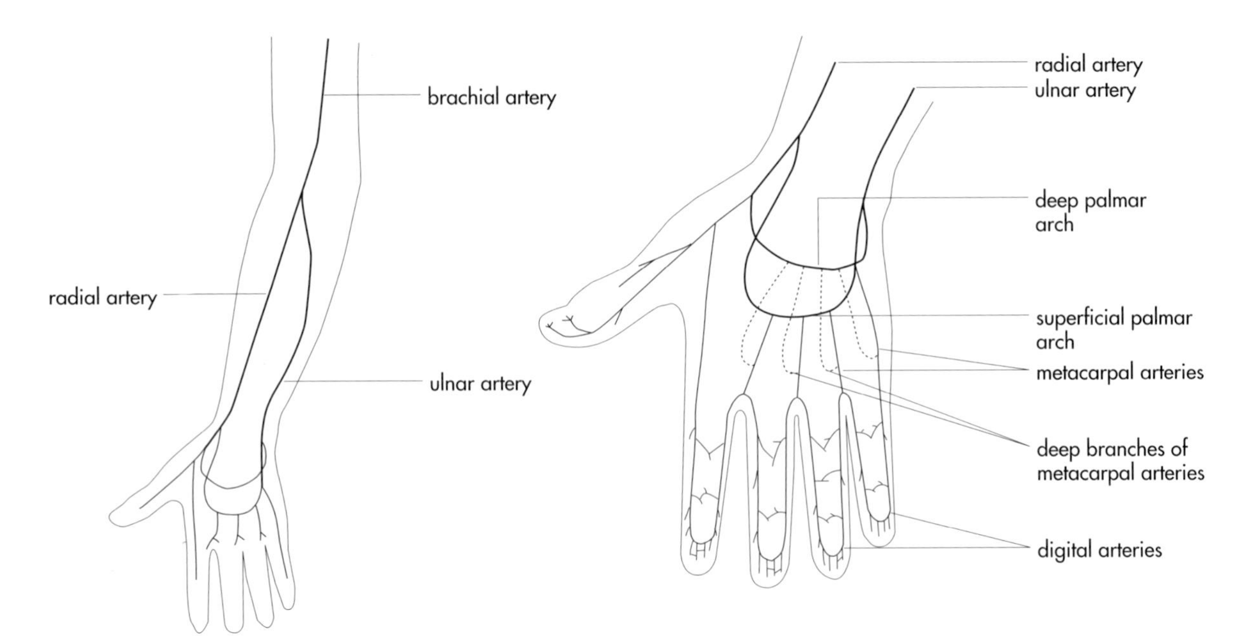

Arteries of the arm and hand

The arteries of the arm and hand

The arm and hand are nourished by a system of arteries which carry blood to the tissues. You can see the colour of the blood from the capillaries beneath the nail: it is these that give the nail bed its pink colour.

RECEPTION

When a client makes an appointment for a manicure treatment, the receptionist should ask a few simple questions that will save time when she arrives for treatment.

- Does she require nail polish? With drying time, this part of the treatment can last up to 20 minutes, so the appropriate length of time needs to be booked out in the appointment book.
- Are any nails damaged or in need of repair? Again extra time will need to be allowed for this work.
- Does she require any services in addition to her nail treatment, such as paraffin wax or warm oil? Allow extra time accordingly.

Allow 30–45 minutes for a manicure.

ACTIVITY: MAKING APPOINTMENTS
What other questions could the receptionist ask in order to make the business run more efficiently? Write down your answers.

CONTRA-INDICATIONS

If a client has any of the following conditions, a manicure treatment must not be carried out:

- *Tinea unguium.*
- *Paronychia.*
- *Cuts or abrasions* on the hands or arms.
- *Infectious skin diseases,* such as ringworm.
- *Warts* on the hands or arms.

Equipment and materials

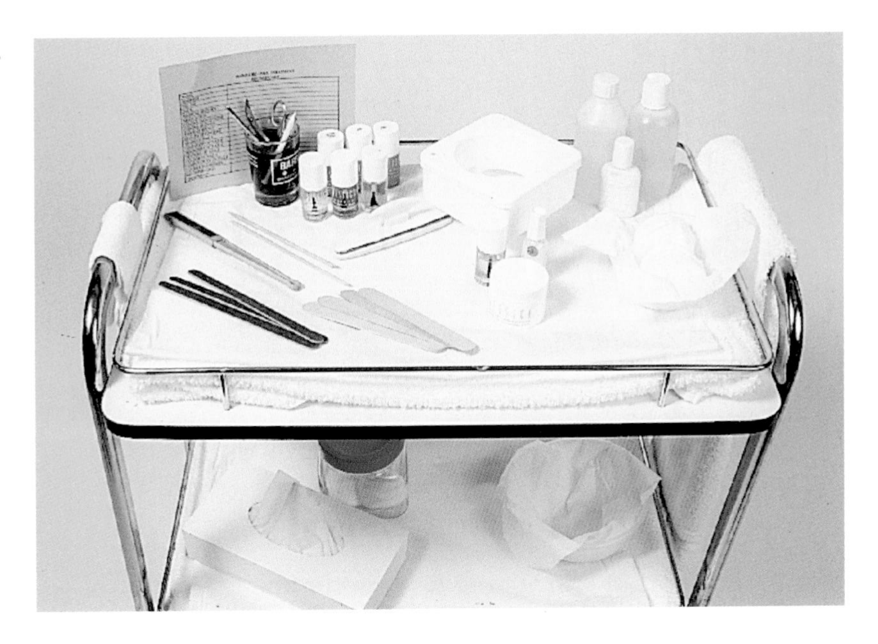

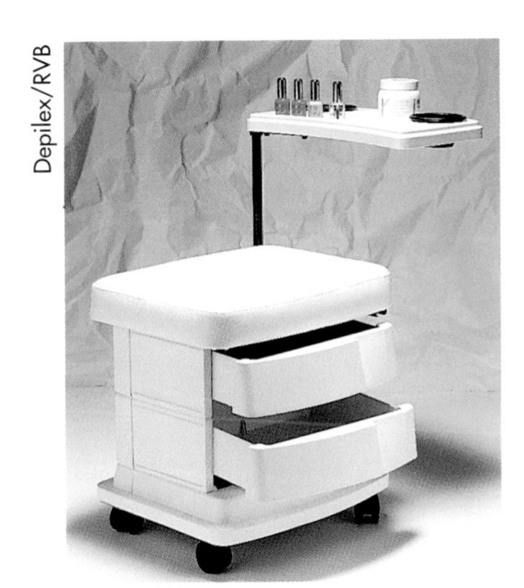

Depilex/RVB

A manicure station

EQUIPMENT AND MATERIALS

- *Manicure table or trolley.*
- *Medium-sized towels (3).*
- *Small bowls lined with tissues (3).*
- *Finger bowl.*
- *Emery board.*
- *Orange sticks* – tipped at either end with cottonwool. (Orange sticks should be disposed after each client, as they cannot be effectively sterilised.)
- *Hand cream, oil or lotion.*
- *Top coat.*
- *Base coat.*
- *Coloured nail enamels* – a selection.
- *Cuticle knife.*

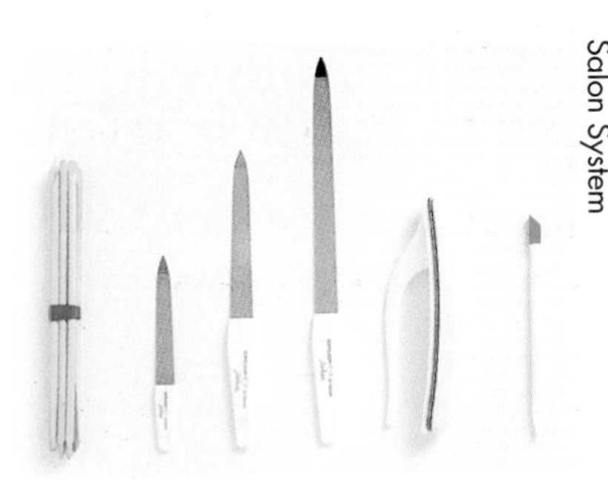

Salon System

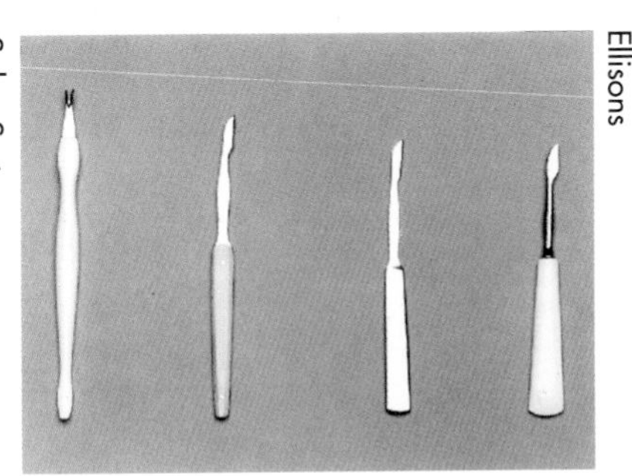

Ellisons

left Hoof sticks and buffers
right Cuticle knives

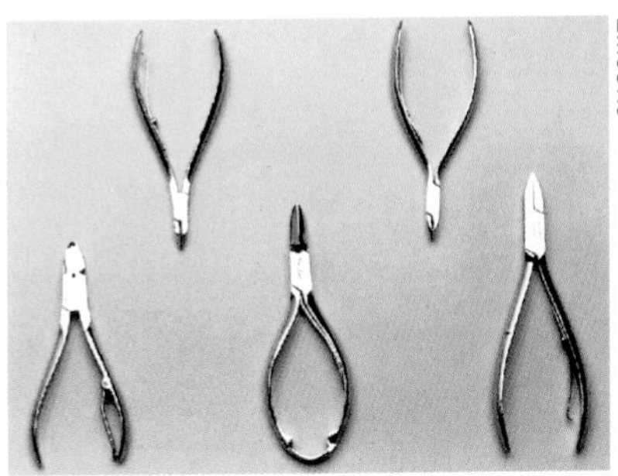

Ellisons

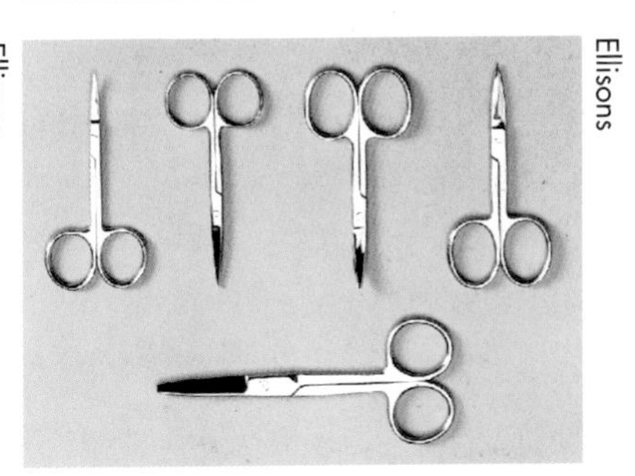

Ellisons

left Cuticle nippers
right Scissors

- *Hoof stick.*
- *Cuticle nippers.*
- *Nail scissors.*
- *Buffers.*
- *Buffing paste.*
- *Cuticle remover.*
- *Cuticle oil.*
- *Cuticle massage cream.*
- *Client's record card.*
- *Cottonwool.*
- *Tissues.*
- *Jar of sanitising solution.*
- *Antiseptic.*
- *Four-sided buffer.*

Products used in manicure and pedicure treatments

Product	*Ingredients*	*Use*
Nail enamel remover	Acetone or ethyl acetate Perfume Colour Oil	To remove nail enamel To remove grease from the nail plate prior to enamelling
Handcream	Vegetable oils (e.g. almond oil) Perfume Emulsifying agents (e.g. beeswax or gum tragacanth) Emollients (e.g. glycerine or lanolin) Preservatives	To soften the skin and cuticles To provide slip during hand massage
Nail bleach	Citric acid or hydrogen peroxide Glycerine Water	To whiten stained nails and the surrounding skin
Nail enamel	Formaldehyde Solvent Colour pigments Resin Nitrocellulose Plasticisers Pearlised particles	To colour nail plates To provide some protection
Cuticle cream	Emollients (e.g. lanolin or glycerine) Perfume Colour	To soften the cuticles
Nail strengthener	Formaldehyde	To harden soft nails
Cuticle remover	Potassium hydroxide Glycerine	To soften dry cuticles
Buffing paste	Perfume Colour Abrasive particles (e.g. pumice, talc or silica)	To shine the nail plate (used with a buffer)

Sterilisation and sanitisation

(Note: These instructions apply to pedicure also.)

Hygiene must be maintained in a number of ways:

- ensure that tools and equipment are clean and sterile before use;
- disinfect work surfaces regularly;
- use disposables wherever possible;
- always follow hygienic working practices;
- maintain a high standard of personal hygiene.

Manicure and pedicure tools and equipment can be sterilised or sanitised by the following methods:

TIP

Any metal to be placed in the autoclave should be of a high-quality stainless steel, to prevent rusting.

Tool/Equipment	*Method*	*Term used*
Cuticle knife	Autoclave	Sterilisation
Cuticle nippers	Autoclave	Sterilisation
Orange stick	Throw away after use	Disposable
Callus file	Autoclave	Sterilisation
	Chemical (e.g. disinfectant)	Sanitisation
Overall	Wash in hot soapy water (60°C)	Sanitisation
Bowl	Chemical (e.g. disinfectant)	Sanitisation
Emery board	Throw away after use	Disposable
Buffer	Wipe handle with surgical spirit Wash buffing cloth in hot (60°C) soapy water	Sanitisation
Towel	Wash in hot soapy water (60°C)	Sanitisation
Spatula	Throw away after use	Disposable
Nail clippers	Autoclave	Sterilisation
Scissors	Autoclave	Sterilisation
Hoof stick	Immerse in chemical (e.g. disinfectant)	Sanitisation
Trolley	Wipe with chemical (e.g. disinfectant)	Sanitisation

Preparing the working area

Ensure that all manicure tools and equipment are sanitised, and that all necessary materials are neatly organised on the trolley. Keeping the working area tidy promotes an organised and professional image, and prevents time being wasted as you try to find materials.

Place a towel over the work surface, then fold another towel into a pad and place it in the middle of the work surface. The pad helps to support the client's forearm during treatment. Place the third towel over the pad, with more of the towel on the manicurist's side – this is used to dry the client's hands during treatment.

A tissue or disposable manicure mat should then be placed on top of the towels, to catch any nail clippings or filings. This can be thrown away later, avoiding irritation to the client from filings.

PLANNING THE TREATMENT

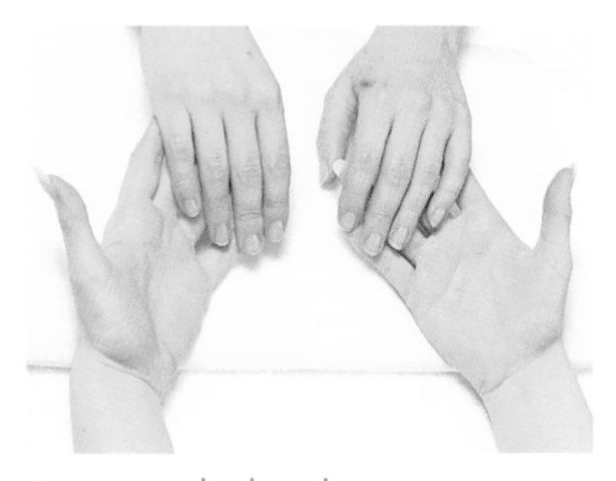

Assessing the hands

Before carrying out a manicure or pedicure treatment, it is necessary to assess the condition of the client's skin, nails and cuticles. This is done in order that the most appropriate equipment and products may be chosen. Also, by correctly assessing and analysing the client's hand or foot condition and writing this on her record card, you will be able to see over a period of time how the condition is progressing.

The parts to assess are these:

- *The cuticles* Are they dry, tight or cracked, or are they soft and pliable?
- *The nails* Are they strong or weak, brittle or flaking? Are they discoloured or stained? What shape are they – square, round, oval? Are they long or short? Are they bitten?
- *The hands* Is the skin dry, rough or chapped, or is it soft and smooth? Is the colour even?
- *The feet* Is there any hard skin?

Whilst assessing the client's hands or feet for treatment, you should also be looking for any *contra-indications* to treatment.

ACTIVITY: ASSESSING THE HANDS AND FEET

Look closely at your own hands and feet. Assess their condition and make notes about everything you see.

Then assess the hands and feet of a colleague. How do they differ from your own?

PREPARING THE CLIENT

Offer the client a lightweight gown to cover her clothing. This will prevent damage to her clothes from accidental spillage of products during treatment. Ask her to remove any jewellery from the area to be treated, to prevent the jewellery being damaged by creams and to avoid obstructing massage movements. Put the jewellery in a tissue-lined bowl where the client can see it. Ensure that the client is seated at the correct height, and close enough to the manicurist to avoid having to lean forward.

When the client is comfortably seated, wash your hands – preferably in view of the client, who will then observe hygienic procedures being carried out. This will assure her that she is receiving a professional treatment.

Consult the client's record card, and begin the treatment.

Removing nail enamel

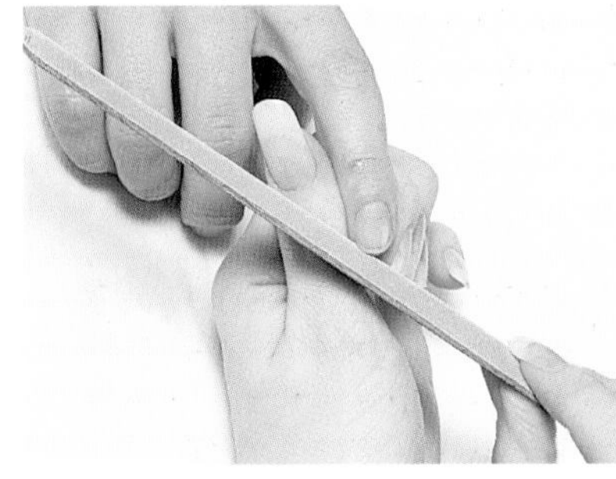
Filing

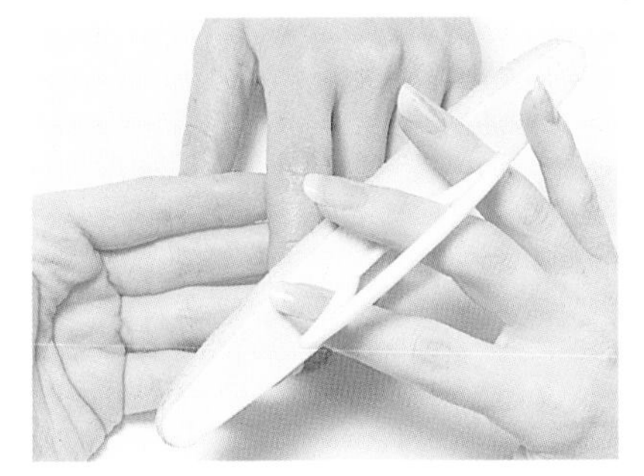
Buffing

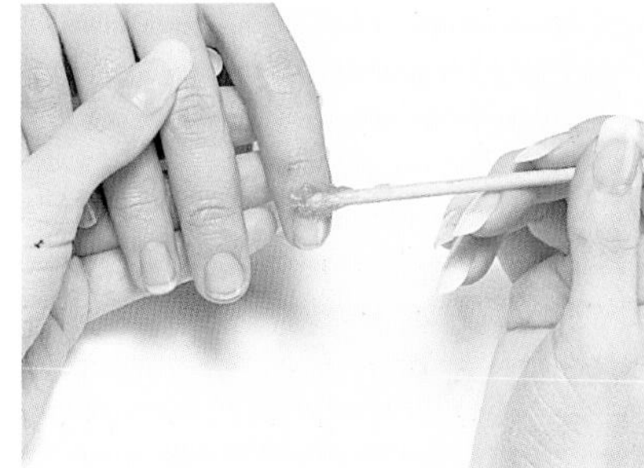
Applying cuticle cream

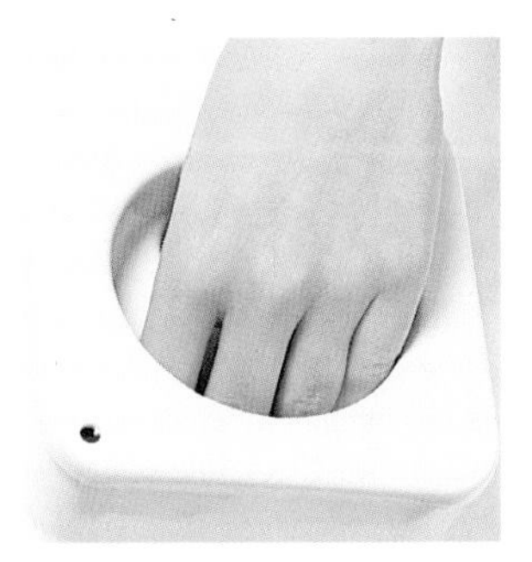
The hand in a bowl

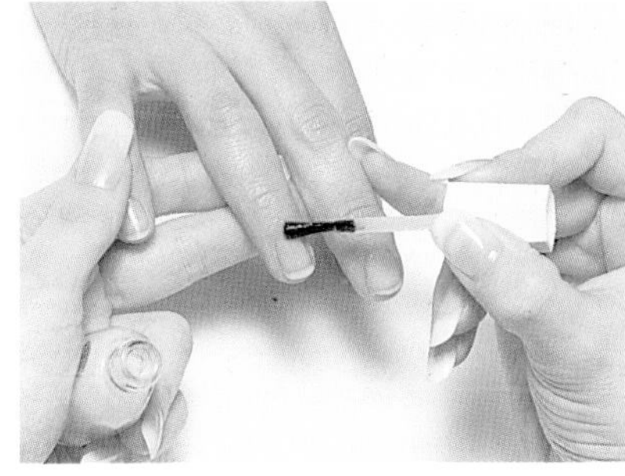
Applying cuticle remover

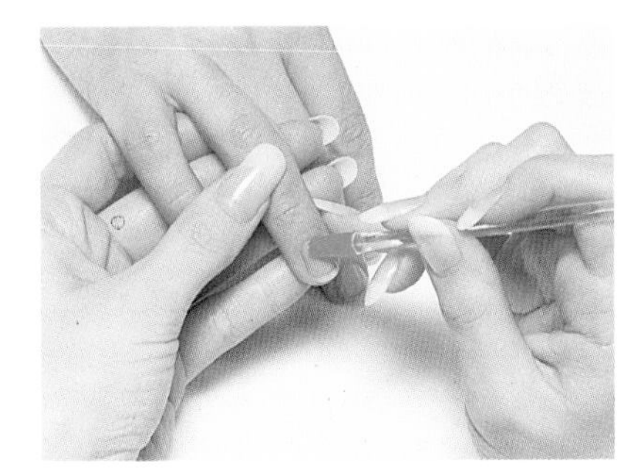
Pushing back cuticles

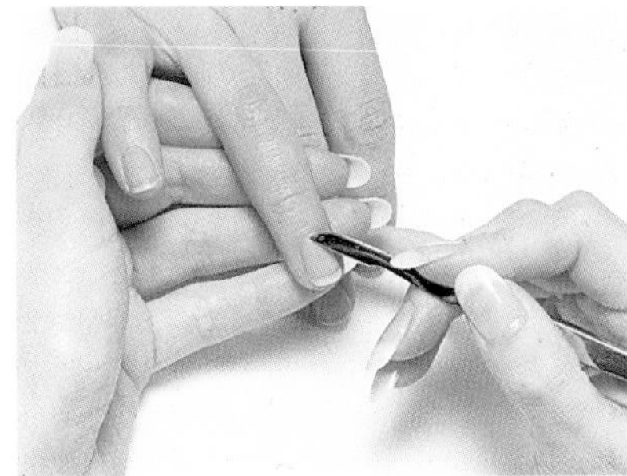
Using a cuticle knife

TIP

Keep the client's jewellery in full view throughout the treatment, so that she doesn't forget it when she leaves.

THE MANICURE PROCEDURE

This procedure shows briefly the stages in the manicure. Each step is discussed in detail later in the chapter.

1. Remove any existing nail enamel, using fresh cottonwool for each hand.
2. File and buff the nails of the left hand.
3. Apply cuticle cream.
4. Place the left hand in a manicure bowl containing warm water and a little antiseptic liquid soap.
5. Repeat steps **2** and **3** for the right hand.
6. Remove the left hand from the manicure bowl and dry with a towel.
7. Place the right hand into the manicure bowl.
8. Apply cuticle remover to the left hand, and push back the cuticle with a cottonwool-tipped orange stick or hoof stick.
9. Remove excess cuticle with nippers, and excess eponychium with a knife.
10. Wipe the nails with cottonwool to remove the cuticle remover.
11. Apply cuticle oil and massage it in with your thumbs.
12. Repeat steps **8–11** for the right hand.
13. Massage both hands.
14. Remove grease from the nail plate with a cottonwool pad soaked in nail enamel remover.

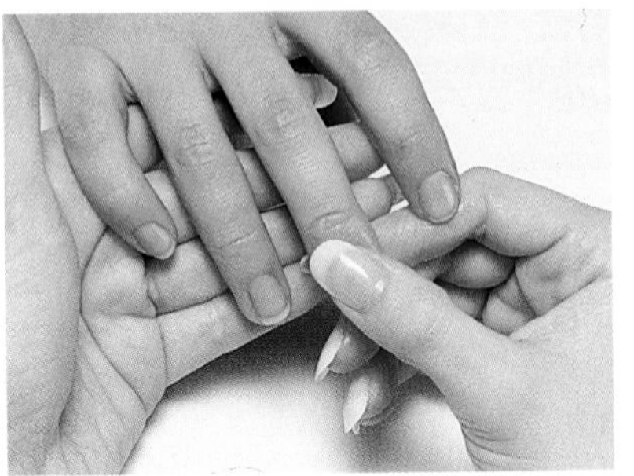

Massaging in cuticle oil

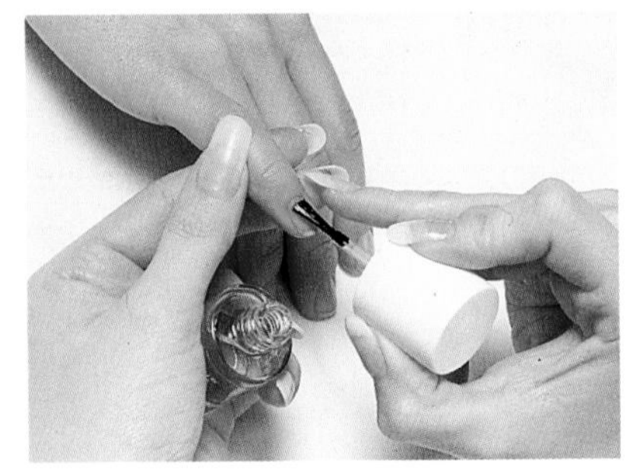

Applying the base coat

Applying nail enamel

15 The client may find it convenient to pay for her treatment at this stage, to avoid smudging her enamel later.

16 Apply the enamel: base coat (once); enamel (twice); and top coat (once). If the client doesn't want enamel, buff to a shine with paste or use a four-sided buffer.

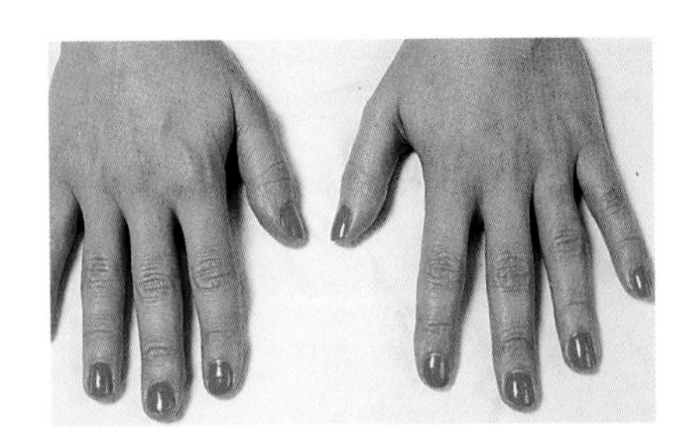

The completed manicure

Filing

The part of the nail that is filed is the free edge. When filing the natural nail, use a fine emery board. Very often emery boards have different degrees of coarseness on either side, indicated by different colours. Use the darker, rougher side to remove excess length, and the lighter, smoother side for shaping and removing rough edges. A flexible emery board is preferable to a stiff one as it generates less friction.

Always file the nails from the side to the centre, with the emery board sloping slightly under the free edge. Use swift, rhythmical strokes. Avoid a sawing action – this would generate friction and might cause the free edge to split.

Never file completely down the sides of the nail, as strength is required here to balance the free edge. Always allow about 4 mm of nail growth to remain at the sides of the nail.

Oval

The ideal nail shape is oval. This is the shape that offers the most strength to the free edge.

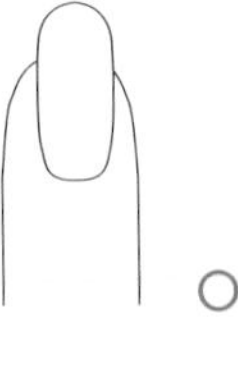

Oval

Square

A recent trend from America is to have the free edge square. The client should be informed, however, that if she has severe corners on the nails she will be more likely to catch and break them.

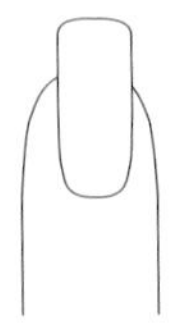

Square

Pointed

One nail shape that should never be recommended is the pointed nail. This leaves the nail tip very weak and likely to break.

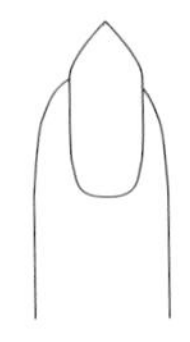

Pointed

Buffing

In manicure, **buffing** is used for these reasons:

- to give the nail plate a sheen;
- to stimulate the blood supply in the nail bed, increasing nourishment and encouraging strong, healthy nail growth;
- to smooth any surface irregularities.

An assortment of buffers

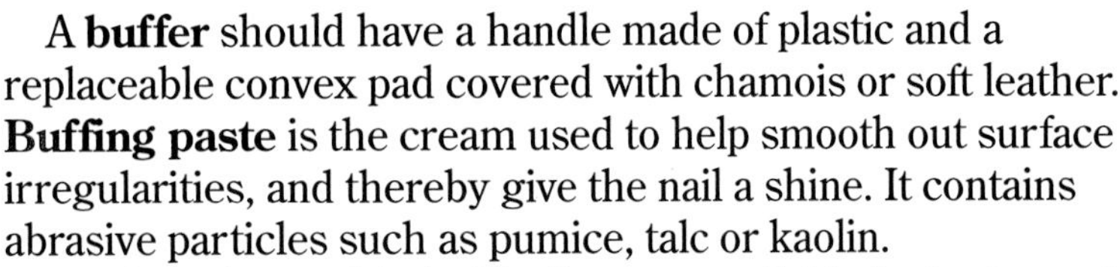

A **buffer** should have a handle made of plastic and a replaceable convex pad covered with chamois or soft leather. **Buffing paste** is the cream used to help smooth out surface irregularities, and thereby give the nail a shine. It contains abrasive particles such as pumice, talc or kaolin.

A relatively new kind of buffer now available is the **four-sided buffer**: this is shaped like a thick emery board and has four types of surface, ranging from slightly abrasive to very smooth. It can be used to bring the nail to a shine without the need for buffing paste. It cannot be effectively sterilised, however, and must therefore be discarded after use on one client.

Buffing is carried out after filing and before the nails are soaked in the finger bowl. It could also be used instead of polish at the end of the manicure, or on a man's nails.

If it is being used, **buffing paste** is applied by taking a small amount out of the pot with a clean orange stick and applying this to each nail plate. With the fingertip, use downward strokes from the cuticle to the free edge to spread the paste without getting it under the cuticle (which would cause irritation). With the buffer held loosely in the hand, buff in one direction only from the base of the nail to the free edge, using smooth, firm, regular strokes. Use approximately six strokes per nail.

Cuticle work

Cuticle work is carried out to keep the cuticle area attractive and also to prevent cuticles from adhering to the nail plate, which could lead to splitting of the cuticle as the nail grows forward, and subsequently to infection of the area.

The work is carried out after soaking the nails in warm soapy water. This step loosens dirty particles from the free edge and softens the skin in the cuticle area.

Procedure

1. Take the fingers from the soapy water and pat them dry with a soft towel.
2. Apply cuticle remover to the cuticle and nail walls, using the applicator brush. (**Cuticle remover** is a slightly caustic solution that helps soften and loosen the cuticles and the eponychium from the nail plate.)
3. Gently push back the cuticle with a cottonwool-tipped orange stick. This is tipped with cottonwool to avoid splinters from the wood, and also so that the cottonwool may be replaced if

necessary.) Use a gentle, circular motion to push back the cuticle, holding the orange stick like a pen.

Use a fresh orange stick for each part of the manicure treatment, and when working on different hands, to prevent cross-infection.

4. Hold the cuticle knife at 45° to the nail plate and stroke it in one direction only, gently loosening any eponychium that has adhered to the nail plate: do not scratch it backwards and forwards. The cuticle knife should have a fine-ground flat blade which can be resharpened when necessary. Dampen it regularly in the manicure bowl to prevent scratches occurring on the nail plate.
5. Hold the nippers comfortably in the palm of the hand, with the thumb resting just above the blades – this gives firm control over what can be a dangerous instrument. Use the cuticle nippers to remove any loose or torn pieces of cuticle, and to trim excess dead cuticle. *Do not cut into live cuticle*: if you do, it will bleed profusely and will be very uncomfortable for the client. Not every client will require the use of cuticle nippers – use them only when needed. (Cuticle nippers should have finely-ground cutting blades to give a clean cut and to avoid tearing the cuticle.)

Hand massage

Hand massage is generally carried out near the end of the manicure treatment, just prior to enamelling. It can also be carried out on its own if the client wants the effects of the massage but does not need or want treatment to her nails.

The reasons for offering a hand massage during a manicure are as follows:

- to moisturise the skin with handcream;
- to increase circulation;
- to keep a range of movement in the joints;
- to ease discomfort from arthritis or rheumatism;
- to relax the client;
- to help remove any dead skin cells (desquamation).

Hand massage sequence

1. **Effleurage to the whole hand and forearm** Use long sweeping strokes from the hand to the elbow, moving on both the outer and the inner sides of the forearm.
   Repeat step **1** *a further 5 times.*
2. **Thumb kneading to the back of the hand and the forearm** Use the thumbs, one in front of the other, and move backwards and forwards in a gently sawing action. Move from the hand to the elbow, then slide the thumbs back down to the hands.
   Repeat step **2** *a further 2 times.*

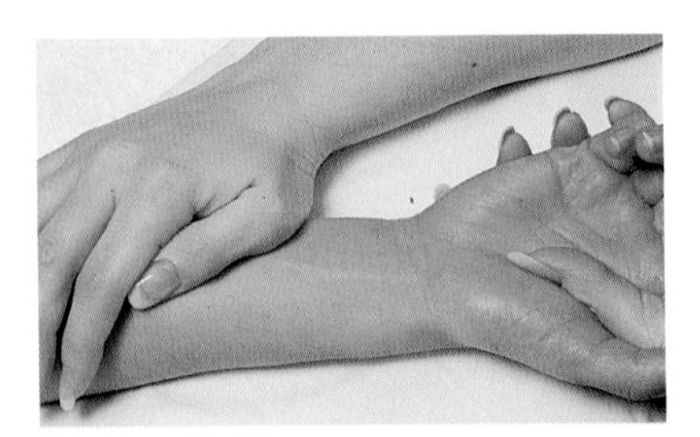
Effleurage to the hand and forearm

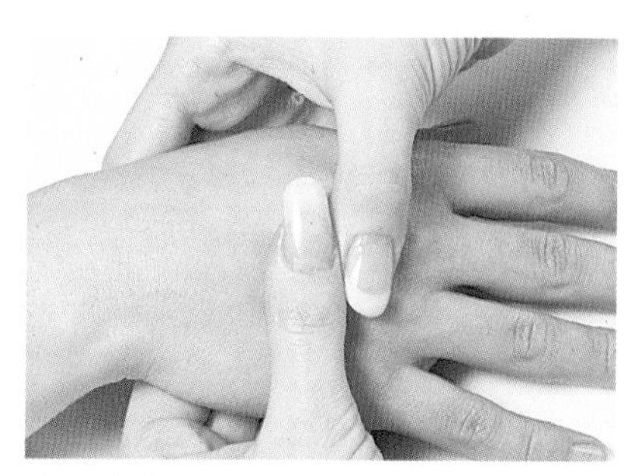
Thumb kneading to the back of the hand

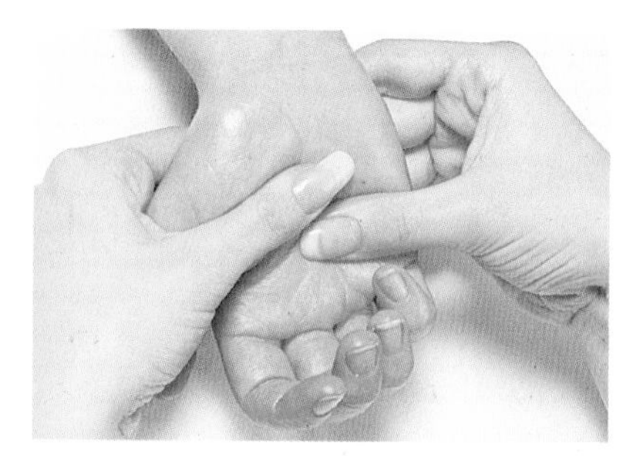
Thumb kneading to the palm

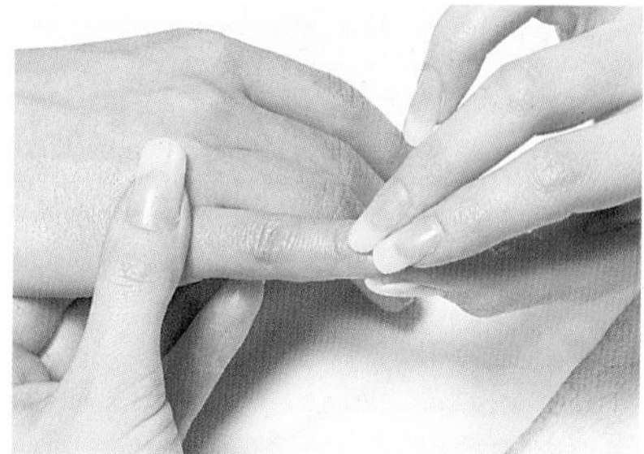
Finger circulations

3 **Thumb kneading to the palm and the inner forearm** Use the same movements as in step **2**.

4 **Finger circulations, supporting the joints** Supporting the knuckles with one hand, hold the fingers individually and gently take each through its full range of movements, first clockwise and then anticlockwise. Move from the little finger to the thumb.
Repeat step **4** *a further 2 times.*

5 **Wrist circulations, supporting the joints** Support the wrist with one hand and put your fingers between the client's gently grasping her hand. Move the wrist through its full range of movement, first clockwise and then anticlockwise.
Repeat step **5** *a further 2 times.*

6 **Effleurage to the whole hand and forearm** Use the same movement as in step **1**.
Repeat step **6** *a further 5 times.*

Treatment plan

After analysing the client's nails and adjacent skin, a treatment plan should be considered. In order to correct nail problems the client should attend the salon weekly. She should also be advised of the appropriate treatment preparations to use at home, so as to support the salon treatment.

Specialist nail treatments include:

- **Nail strengthener** This is used on brittle, damaged nails, to strengthen, condition and protect them against breaking, splitting or peeling.
- **Ridge filler** This is used on nails with ridges to provide an even surface, allowing a smoother application of enamel.

Nail enamelling

Nail enamel is used to coat the nail plate for a number of reasons:

- to adorn the nail;
- to disguise stained nails;
- to add temporary strength to weak nails;
- to coordinate with clothes or make-up.

Types of enamel

- **Cream** This has a matt finish, and requires a top coat application to give a sheen.
- **Crystalline** This is pearlised by the addition of natural fish scales or synthetic ingredients such as bismuth oxychloride.
- **Base coat** This protects the nail from staining by a strong-coloured nail enamel; it also gives a good grip to enamel, and smooths out minor surface irregularities.
- **Top coat** This gives a sheen to cream enamel, and adds longer wear as it helps to prevent chipping.

Contra-indications to nail enamel

Do not apply enamel in these circumstances:

- if there are diseases and disorders of the nail plate and surrounding skin;
- if the client is allergic to nail enamel.

In addition, crystalline nail enamel should not be applied to excessively ridged nails as it may appear to exaggerate the problem. Short or bitten nails should be painted only with pale enamels, to avoid attracting attention.

How to apply nail enamel

Before nail enamel is applied the client's hand jewellery may be replaced, to avoid smudging afterwards.

1. After ensuring that the nail plate is free from grease, start with the thumb and apply three brush strokes down the length of the nail from the cuticle to the free edge, beginning in the centre, then down either side close to the nail wall.
   Take care to avoid touching the cuticle or the nail wall. If flooding occurs, remove the varnish immediately with an orange stick and enamel remover.
2. Apply one base coat, two coats of coloured enamel, and one top coat. Top coat is required only after using cream enamel; pearl enamel does not need a top coat.

TIP
A nail 'fast-drying' product may be applied to reduce the time taken for the enamel to dry.

Styles of application

- **Traditional application** This style is the one most commonly requested by clients: the entire nail plate is covered with polish.
- **French application** This style involves enamelling the nail plate of the nail bed pink or pale beige, and the free edge white.
- **Free lunula application** This style involves applying enamel over the whole nail plate except the area of the lunula.

- **Application to give the appearance of longer nails** This style creates an optical illusion that the nails are longer then they really are. The whole nail plate is enamelled, leaving a slightly larger gap than usual along the nail walls.

TIP

For short nails, select a pale, neutral colour. Darker, more dramatic colours suit healthy, long nails, especially on clients with darker skin tones.

Reasons for peeling and chipping nail enamel

Chipping may be explained by any of the following:

- The nail enamel was not thick enough because of over-thinning with solvent.
- No base coat was used.
- Grease was left on the nail plate prior to enamelling.
- The nail plate is flaking.
- The enamel was dried too quickly by artificial means.

Peeling may have the following explanations:

- No top coat was used.
- Successive coats were not allowed to dry between applications.
- The nail enamel was too thick, due to evaporation of the solvent.
- Grease was left on the nail plate prior to enamelling.

Enamels should be stored in a cool dark place, to avoid separation and fading. The caps and the rims of bottles *must* be kept clean, not only for appearance but also to ensure that the bottle is airtight.

If enamel does thicken, **solvent** may be added to restore the correct consistency. This should be done twenty minutes prior to use, to ensure an even consistency.

Nail art

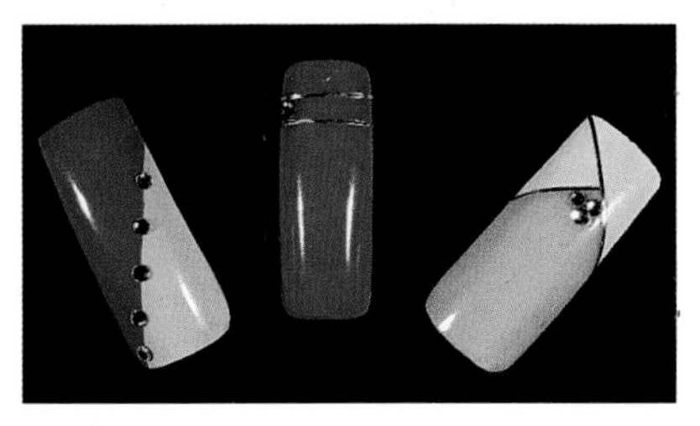

Nail art

Nail art is an exciting and elaborate form of nail decoration. It is becoming increasingly popular, especially for parties and special occasions; styles are limited only by the manicurist's imagination. The artwork can be as simple as a single stripe across a painted nail, or as intricate as a detailed desert island scene, painted in different colours.

Nail art may make use of:

- nail polish;
- specially designed paints;
- transfers;
- gemstones;
- foil strips;
- feathers;

- snakeskin (yes, really!);
- any other item you choose!

Nail jewellery

Nail jewellery deserves a mention here, too. Although cumbersome and a little impractical for everyday wear, jewellery on the nails can be an unusual accessory to an outfit. Nail jewellery ranges from a simple stud through the nail tip to rings, chains and tassels. Special equipment is required to carry out this nail service, but such work would certainly attract attention to the client, particularly at a party.

CONTRA-ACTIONS

The client – or the manicurist – may at some time develop an allergy to a manicure product that has been successfully used previously. This could be for a number of reasons, including new medication being taken or illness.

The symptoms of an allergic reaction could be as follows:

- redness of the skin;
- swelling;
- itching;
- raised blisters.

The symptoms do not necessarily appear on the hands. In the case of nail enamel allergy, the symptoms often show up on the face, which the hands are continually touching.

In the case of an allergic reaction:

- Remove the offending product immediately, using water or, in the case of enamels, solvent.
- If symptoms persist, seek medical advice.

Always record any allergies on the client's record card, so that the offending product may be avoided in future.

AFTERCARE AND ADVICE

It is important when carrying out a manicure that the client knows how to care for her nails at home. It is your duty as a professional manicurist to ensure that the correct aftercare advice is given. If it isn't, the client may unwittingly undo all the good work you have done during the treatment.

When giving aftercare advice you have a good opportunity to recommend retail products, such as nail enamel or handcream, thereby enhancing retail sales and the salon's profit.

Aftercare advice will differ slightly for each client, according to individual needs, but generally it will be as follows:

- Wear rubber gloves when washing up.
- Wear protective gloves when gardening or doing housework.
- Always wear gloves in cold weather.

Ellisons

Retail products

- Dry the hands thoroughly after washing, and apply handcream.
- Avoid harsh soaps when washing hands.
- Do not use the fingernails as tools (for instance, to prise lids off tins).

ACTIVITY: DESIGNING AN AFTERCARE LEAFLET
Devise an aftercare leaflet for clients, advising a suitable home-care routine.

Exercises for the hands

Hand exercises play an important role in the home-care advice given to clients, for the following reasons:

- They keep the joints supple, allowing greater movement.
- Circulation is increased, encouraging healthy nail and skin growth.
- Good circulation helps to prevent cold hands.
- Exercises keep the client interested in her hands, and so more likely to keep regular salon appointments.

Exercise routine

1. Rub the palms together, back and forth, until warm.
2. Make a tight fist with each hand, then slowly stretch out all the fingers as far as possible.
   Repeat step **2** *a further 3 times.*
3. With the fingers extended, rotate the wrists slowly in large clockwise circles.
   Repeat step **3** *a further 3 times.*
4. With the fingers extended, rotate the wrists slowly in large anticlockwise circles.
   Repeat step **4** *a further 3 times.*
5. Play an imaginary piano vigorously with the fingers for 10 seconds.
6. With the hands together as if praying, gently widen the fingers as far as possible, then relax.
   Repeat step **6** *a further 3 times.*

MANICURE FOR A MALE CLIENT

A man's hands differ slightly from a woman's so slight adaptations to the basic manicure are necessary if the treatment is to be effective:

- File the nails to a shorter length.
- Omit the coloured nail enamel.
- Shape the nails square rather than oval.
- Buff the nails with paste, if a shine is required.

- Use unperfumed lotion for massage.
- Use a lotion or an oil rather than cream for massage.
- Use deeper movements during hand and arm massage.

> **ACTIVITY: COMPARING HANDS**
> Carefully look at and touch a man's hands, then look at and touch those of a woman. Write down as many differences as you can.
> From your observations, can you think of any further adaptations that may be necessary in manicuring a man's hands?

HAND AND NAIL TREATMENTS

In addition to a manicure, further treatments may be added as appropriate. Here are some examples.

Warm-oil treatment

Warm-oil treatment involves gently heating a small amount of vegetable oil (such as almond oil) and soaking the cuticles in it for five minutes. This softens the cuticles and the surrounding skin, and is an excellent treatment for clients with dry cracked cuticles.

Exfoliating treatment

Exfoliating treatment is carried out as part of the massage routine. The massage is performed as usual, using a mildly abrasive cream. This offers the following benefits:

- the removal of dead skin cells;
- improvement of the skin texture;
- improvement of the skin colour;
- increased circulation.

The abrasive particles must be thoroughly removed with hot damp towels before continuing with the rest of the manicure.

Ellisons

An exfoliating treatment

Hand treatment mask

An appropriate **treatment mask** may be applied, according to the client's treatment requirements. This may be either stimulating and rejuvenating, or moisturising. The hands may be placed inside warm hand gloves for 10 minutes to enable the mask to penetrate the epidermis. The mask is then removed, and followed with treatment massage cream.

Ellisons

A hand treatment mask

> **ACTIVITY: RESEARCHING TREATMENTS**
> Research other types of hand and nail treatments. Write down the details of your research, and try out the treatments on clients.

KNOWLEDGE REVIEW

Manicure

Anatomy

1. What is the function of the matrix?
2. What type of joint is found in the wrist?
3. Why does the nail bed look pink?
4. How do the fingers move?

Preparation

1. Why should the manicurist consult the client's record card prior to treatment?
2. List *three* contra-indications to manicure.
3. Why is it important to keep the working area tidy?
4. Why is it important to assess the condition of the client's hands and nails before commencing treatment?

Application

1. How would you adapt your hand massage technique when treating a male client?
2. If a client had badly bitten nails, what type and colour of nail enamel would you suggest that she tried?
3. Why is it important to avoid a sawing action when filing the nails?
4. Why are the nails soaked at the beginning of the treatment?

Aftercare

1. What aftercare advice would you give to a client with very dry hands and dry cuticles?
2. List *three* retail products that you would recommend to a manicure client.
3. Why should you recommend that the client wears rubber gloves when washing up?

Health, safety and hygiene

1. What precautions should be taken to avoid cross-infection?
2. List the materials used in the manicure that are disposable.
3. Why is it important not to get buffing paste under the cuticle?

Case study

1. For each of the imaginary clients below, suggest a treatment routine. Include details of the treatment, the products used, the home-care advice, the aims of the treatment, and the probable cause of the condition.
   (a) A florist whose nails are very soft and weak.
   (b) An elderly woman with very strong but ridged nails.
   (c) A male junior doctor who is trying to stop biting his nails.

THE PURPOSE OF A PEDICURE

The word **pedicure** is derived from the Latin word *pedis*, meaning 'foot' and *cura*, meaning 'care'. The treatment is very similar to manicure except that it is carried out on the feet instead of the hands. The reasons for carrying out a pedicure are these:

- to make the feet more attractive;
- to reduce the amount of hard skin;
- to relax tired, aching feet.

THE FOOT AND THE LOWER LEG

The bones of the foot

The foot is made up of 7 **tarsal** bones, 5 **metatarsal** bones, and 14 **phalanges**. These bones fit together to form arches which help to support the foot and to absorb the impact when we walk, run and jump.

Bones of the foot

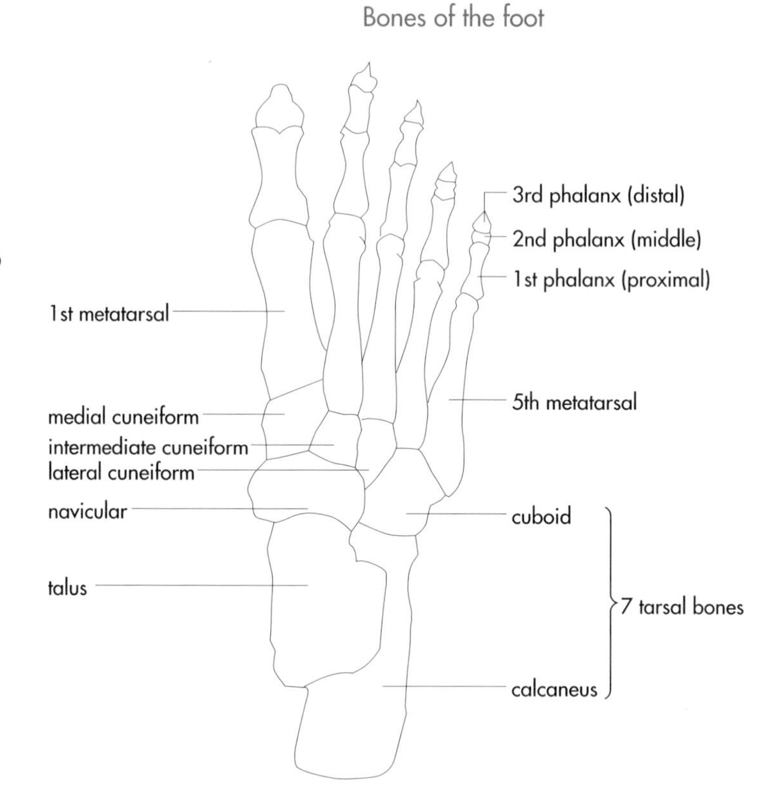

The arches of the foot

The **arches** of the foot are created by the formation of the bones and joints, and supported by ligaments. These arches support the weight of the body and help to preserve balance when we walk on uneven surfaces.

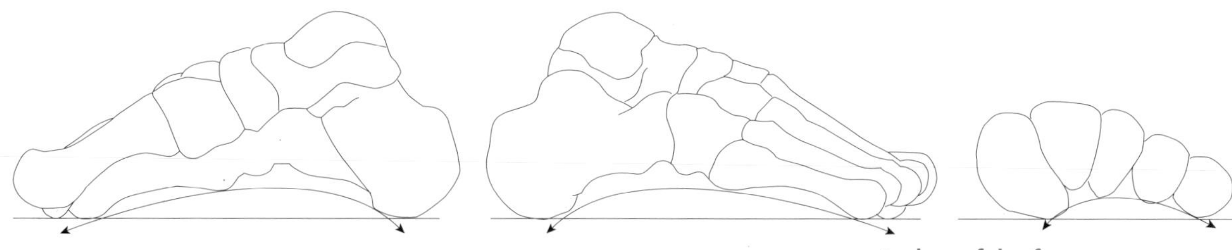

Arches of the foot
left Medial longitudinal arch
centre Lateral longitudinal arch
right Transverse metatarsal arch

The bones of the lower leg

The lower leg is made up of 2 long bones, the **tibia** and the **fibula**. These bones have joints with the upper leg (at the knee) and with the foot (at the ankle). Having two bones in the lower leg – as with the forearm – allows a greater range of movement to be achieved at the ankle.

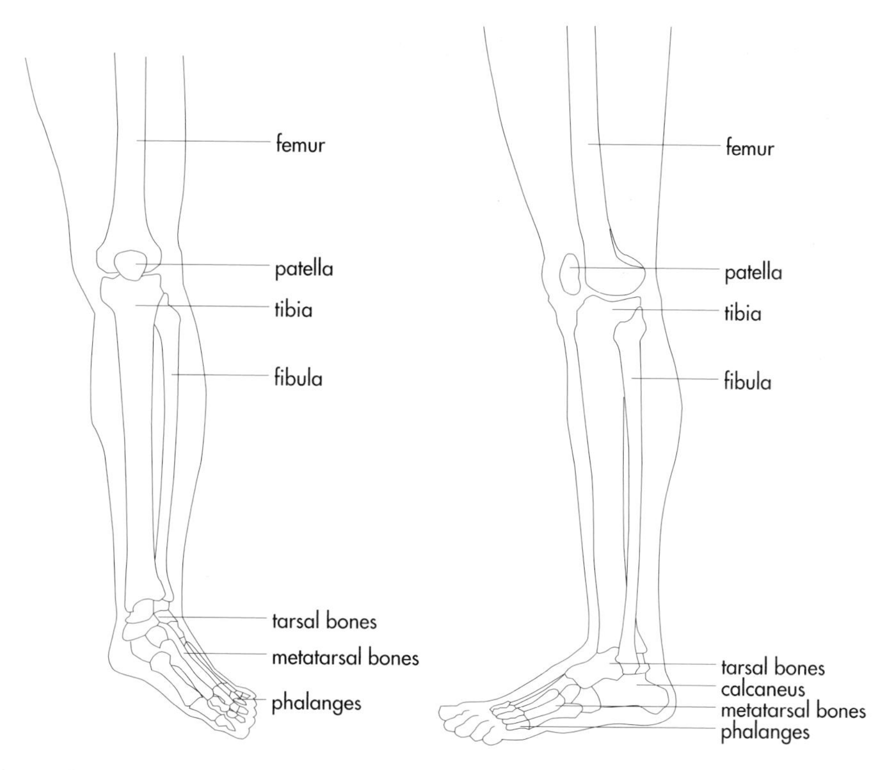

Bones of the lower leg

The muscles of the foot and lower leg

The muscles of the foot work together to help move the body when walking and running. In a similar way to the movement of the hand, the foot is moved primarily by muscles in the lower leg; these pull on tendons, which in turn move the feet and toes.

The arteries of the foot and lower leg

The skin and nails of the feet are nourished by a system of arteries which bring blood to the tissue.

When it is cold, and when the circulation is poor, insufficient blood reaches the feet and they feel cold. Severe circulation problems in the feet may lead to **chilblains**.

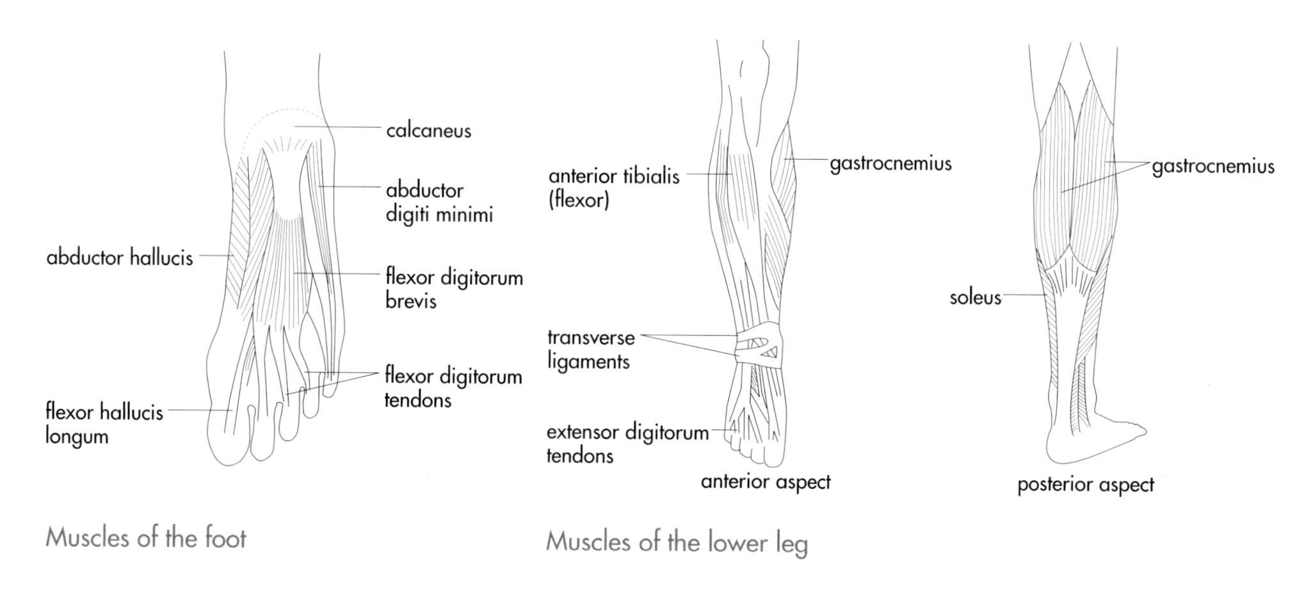

Muscles of the foot

Muscles of the lower leg

posterior tibial artery
medial plantar artery
lateral plantar artery
plantar arch
plantar metatarsal arteries
plantar digital arteries

Arteries of the foot

popliteal artery
posterior tibial artery
anterior tibial artery
peroneal artery
to lateral and medial plantar arteries
calcanean branches of posterior tibial artery
calcanean branches of peroneal artery
popliteal artery
anterior tibial artery
dorsalis pedis artery
digital arteries (dorsal)
metatarsal arteries (dorsal)

Arteries of the lower leg

RECEPTION

When a client makes an appointment for a pedicure treatment, the receptionist should advise the client how long the treatment will take. This will include sufficient time for the nail enamel to dry before replacing footwear.

Ask the client whether she is currently receiving treatment from the chiropodist for conditions such as verrucas or athlete's foot. These would contra-indicate treatment: the receptionist should advise the client to wait until the condition has cleared.

Allow 45 minutes for a pedicure.

CONTRA-INDICATIONS

If a client has any of the following conditions, a pedicure treatment must not be carried out:

- *Athlete's foot.*
- *Verrucas.*
- *Diabetes.*
- *Cuts or abrasions* on the feet or legs.
- *Infectious skin diseases*, such as ringworm.

EQUIPMENT AND MATERIALS

- *Foot bowl (1).*
- *Towels (5).*
- *Small bowls (3).*
- *Sanitising fluid in a glass jar.*
- *Dry cottonwool.*
- *Surgical spirit.*

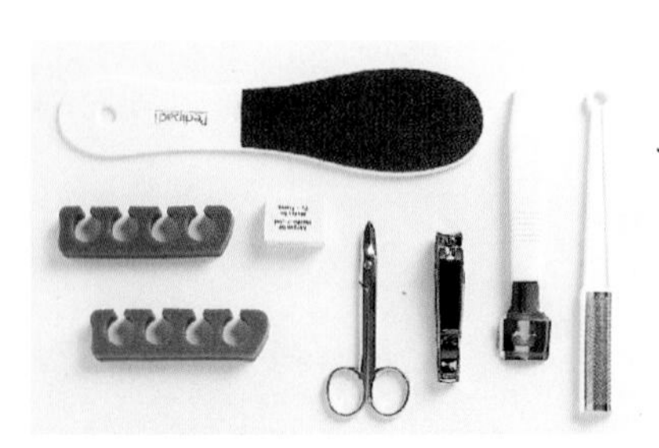

Pedicure equipment

Ellisons

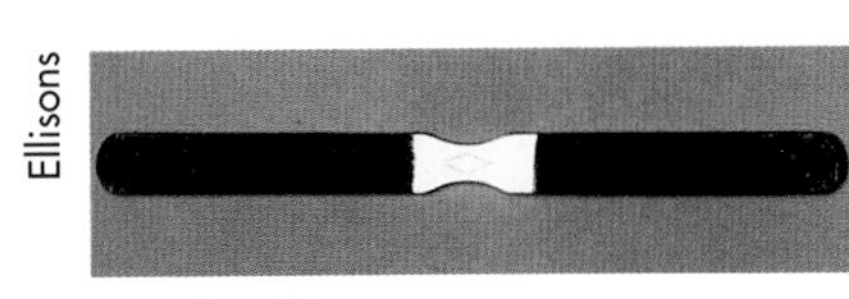

A callus file

Ellisons

A foot spa

- *Nail enamel remover.*
- *Scissors or toenail clippers.*
- *Emery board.*
- *Cuticle massage cream.*
- *Cottonwool-tipped orange sticks (6).*
- *Cuticle remover.*
- *Cuticle knife.*
- *Cuticle nippers.*
- *Cuticle oil.*
- *Callus file or exfoliating scrub.*
- *Massage lotion.*
- *Base coat.*
- *Coloured enamel.*
- *Top coat.*
- *Tissues.*
- *Client's record card.*
- *Foot spa* – optional.

Preparing the working area

All metal instruments should be sterilised in the autoclave prior to use. Non-metal instruments should be sanitised by immersing them in a suitable disinfecting fluid. Prepare the equipment neatly on a trolley so that everything you need is to hand and the client need not be disturbed during treatment.

Place a towel on the floor between you and the client. The foot bowl containing warm soapy water should be placed on this towel.

Towels should be placed on your lap: one is for protection, the other is for drying the client's feet. Keep the other towels close by, for wrapping the client's feet.

When cutting the client's nails and removing hard skin, a tissue should be placed on your lap and then removed before continuing treatment.

PLANNING THE TREATMENT

The procedure for this is discussed with that for manicure – see page 236.

PREPARING THE CLIENT

Ensure that the client has a comfortable chair at the correct height, so that you can work comfortably and the client can enjoy her treatment.

Before treatment begins, ask the client to remove her tights or socks, and any clothing that might restrict her lower leg movement, such as jeans or trousers. Cover her upper legs with a clean towel. This will help her to be more comfortable and allow you to work without restriction.

Discuss the treatment with the client to discover her needs. Record relevant information on the client's record card.

Treatment plan

After analysing the client's nails and adjacent skin, a treatment plan should be considered. In order to correct any problems the client should attend the salon weekly. She should also be advised of the appropriate treatment preparations to use at home, so as to support the salon treatment.

- **Revitalising foot spa agents** These may be in tablet form or as a foaming soak. They are dissolved in warm water, in which the feet are then immersed.
- **Exfoliator** This is used following immersion of the feet in the foot spa. It removes surface dead skin cells, preventing the formation of callus tissue.
- **Massage lotion** This is a massage preparation which includes refreshing essential oils such as peppermint. It is recommended for the relief of tired, aching feet.
- **Foot mask** A mask may be applied to cool and to refresh the feet. Booties may be worn while the mask penetrates the epidermis.
- **Foot gel or spray** This may be applied to create an immediate cooling effect.

Salon Systems

Pedicure products

THE PEDICURE PROCEDURE

1. Wash your hands.
2. Wipe both feet with cottonwool soaked in surgical spirit (separate pieces for each foot).
3. Soak both feet in warm water to which a mild antiseptic liquid soap has been added.
4. Take out the left foot and towel dry it.
5. Remove any existing nail enamel, and check again for contra-indications below the nail plate.
6. Cut the toenails straight across, using toenail clippers or scissors.
7. File the nails smooth with the coarse side of the emery board.

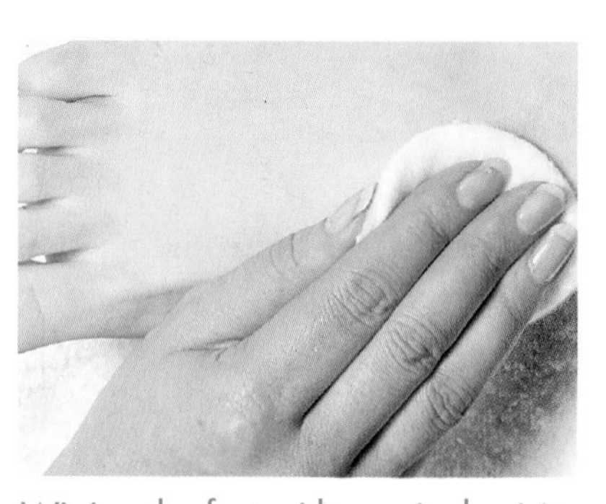

Wiping the feet with surgical spirit

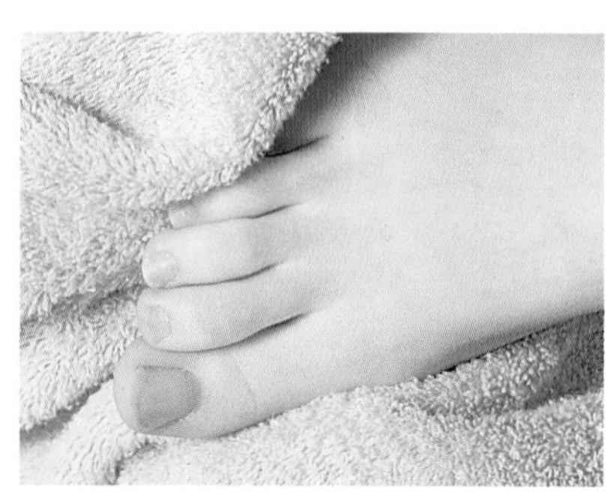

Drying the feet

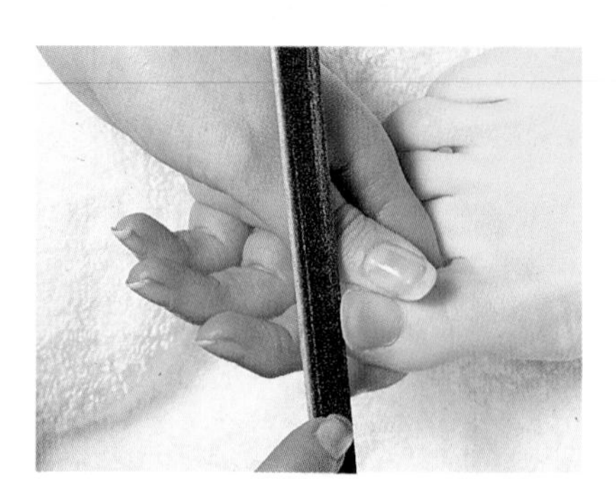

Filing the toenails

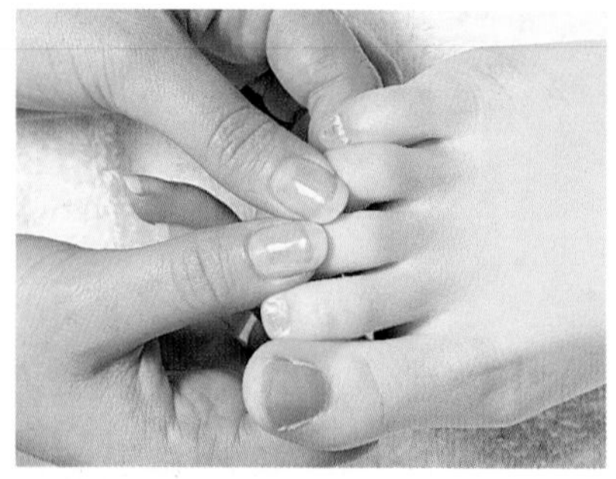

Applying cuticle massage cream

Applying cuticle remover

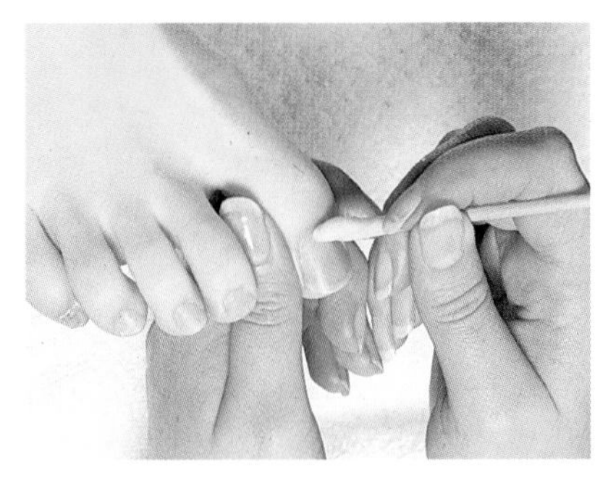
Pushing back the cuticles

Using a cuticle knife

Using cuticle nippers

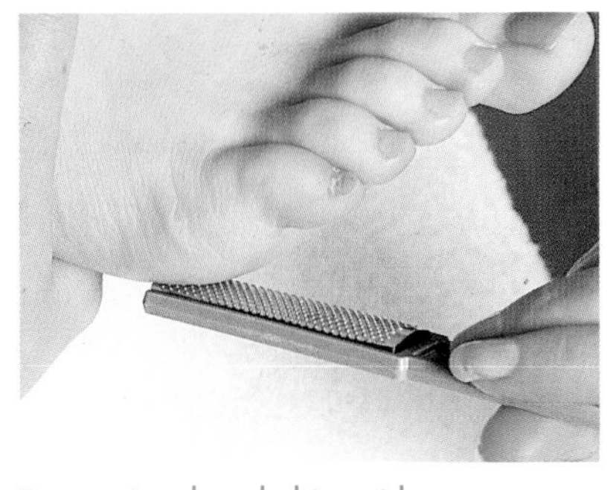
Removing hard skin with a rasp

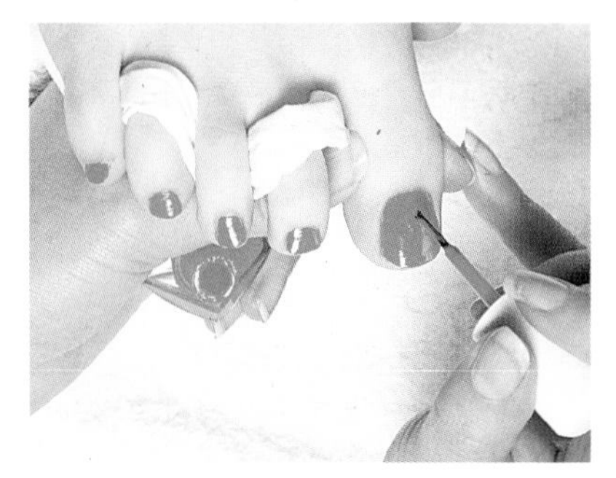
Applying enamel to the toenails

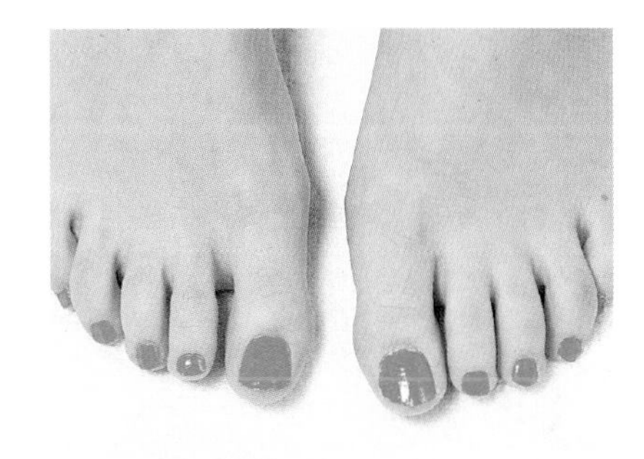
The completed pedicure

8 Apply cuticle massage cream.
9 Place the foot back in the water.
10 Remove the right foot and repeat procedures **4–9**.
11 Dry the right foot and apply cuticle remover.
12 Push back the cuticles with a cottonwool-tipped orange stick or hoof stick.
13 Clean under the free edge with a separate tipped orange stick.
14 Use the cuticle knife and nippers where indicated.
15 Wipe off the remaining cuticle remover with cottonwool, and file the nails again if necessary.
16 Apply cuticle oil.
17 Remove any hard skin. This may be done with exfoliating cream or a rasp, depending on the severity of the condition.
18 Wrap the foot in a dry towel and place it on the floor.
19 Repeat procedures **11–18** for the left foot.
20 Remove the foot bowl from the working area.
21 Perform a foot and lower leg massage.
22 Remove any grease from the nail plates with nail polish remover on cottonwool.
23 Place tissue between the toes.
24 Apply base coat, enamel and top coat where indicated.

TIP

A *foot spa* may be used instead of an ordinary foot bowl, adding a touch of luxury to the pedicure treatment.

Cutting and filing toenails

Toenails should be cut straight across, using nail clippers or strong sharp scissors, then filed smooth using the coarse side of the emery board.

> **HEALTH & SAFETY: CUTTING TOENAILS**
> Never cut toenails down at the sides. This increases the chance of ingrowing toenails, and may lead to infection.

Removing hard skin

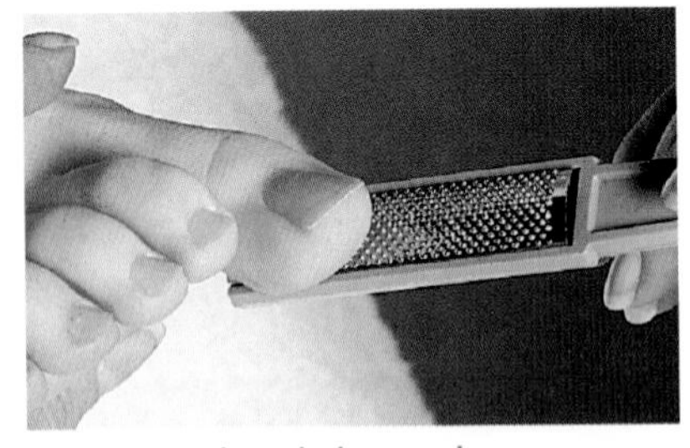

Removing hard skin with a rasp

Hard skin develops on the feet as a form of protection, either from friction from footwear or from standing for long periods of time.

It is therefore not advisable to remove *all* the hard skin from an area, as this would remove the protective pad. Hard skin should be removed only to improve the appearance of the feet. Hard-skin build-up that causes pain or discomfort should be referred to a chiropodist for treatment.

Excess hard skin may be removed from the feet in a number of ways, including abrasive pastes, pumice stones, callus files, chiropody sponges, and corn planes. Abrasive pastes should be used with a deep circular massage movement: they are ideal when only a very small build-up of hard skin is present. Files, pumice stones and the rest should be used with a swift stroking movement in one direction only (similar to buffing). Sawing back and forth would lead to friction, and discomfort for the client.

Always finish off a hard-skin removal procedure with a lot of moisturiser or handcream, to soften the newly exposed skin.

Foot massage

As with a manicure massage, the pedicure massage is carried out near the end of the treatment, prior to nail enamelling. The pedicure massage includes the foot and the lower leg, and offers benefits to the client comparable with those of the manicure massage.

Foot massage sequence

1. **Effleurage from the foot to the knee** Use long sweeping strokes from the toes to the knee, moving on both the back and the front of the leg.

   Repeat step **1** *a further 5 times.*

2. **Thumb frictions to the dorsal aspect of the foot** Use the thumbs, one in front of the other, and move backwards and forwards in a gentle sawing action. Move from the toes to the ankle, then slide back down to the toes.

   Repeat step **2** *a further 2 times.*

3 **Thumb frictions to the plantar aspect of the foot** Use the same movement as in step **2**, but on the sole of the foot, moving from the toes to the heel.

*Repeat step **3** a further 2 times.*

4 **Toe circulating** Supporting the base of the toes with one hand, hold the toes individually with the other and move them through their full range of movement, first clockwise and then anticlockwise.

*Repeat step **4** a further 2 times.*

Effleurage to the lower leg

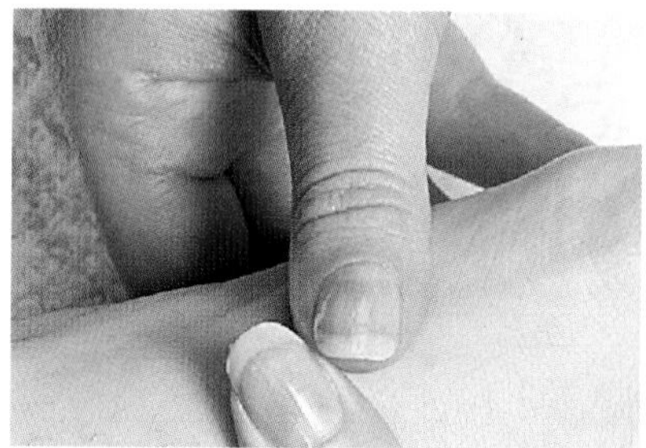
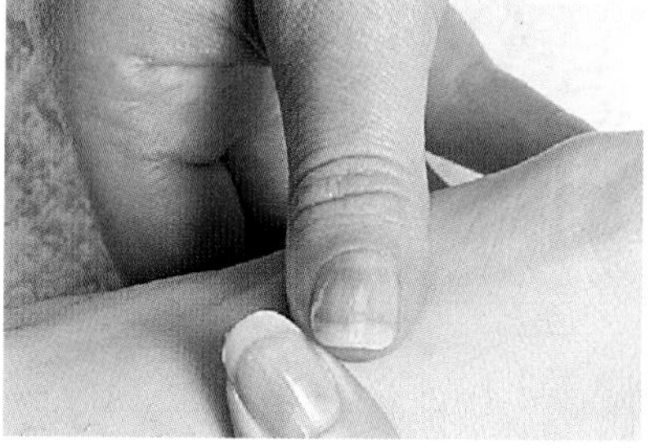

Thumb frictions to the dorsal aspect of the foot

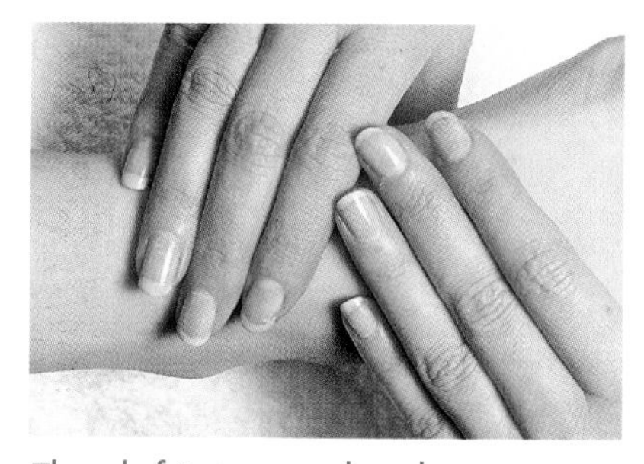

Thumb frictions to the plantar aspect of the foot

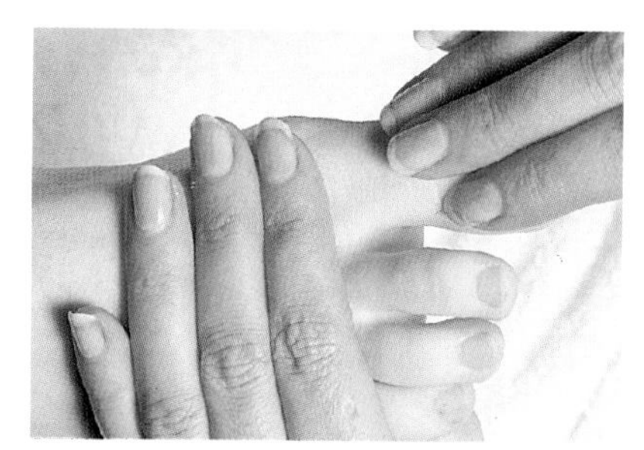

Toe circling

Palm kneading to the plantar surface of the foot

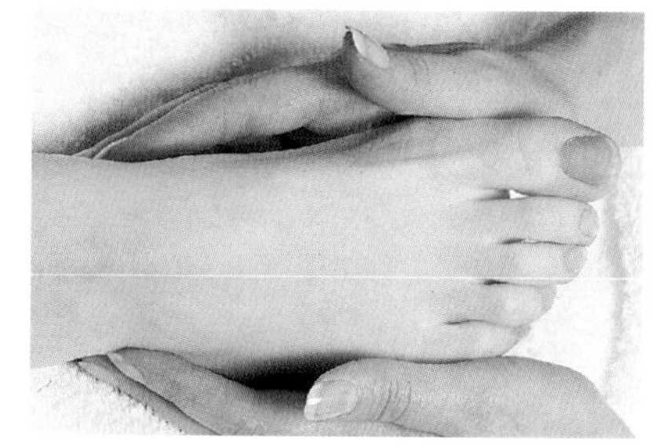

Finger kneading around the malleolus

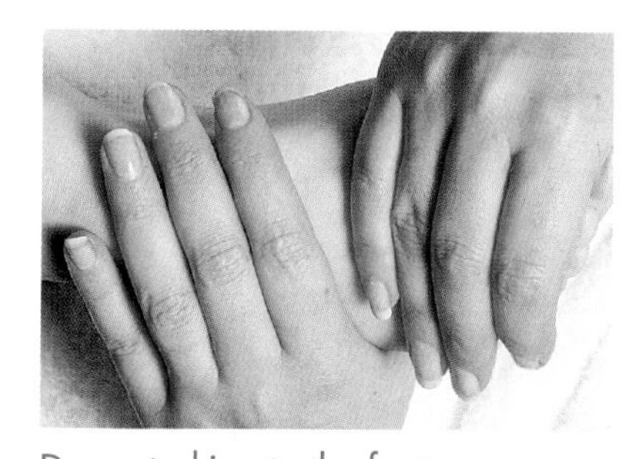

Deep stroking to the foot

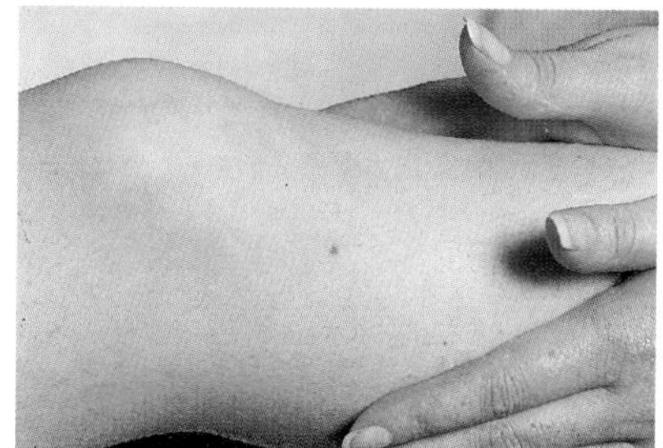

Effleurage from foot to knee

5 **Palm kneading to the bottom surface of the foot** Place the heel of the hand into the arch of the foot and massage with deep circular movements.

*Repeat step **5** a further 5 times.*

6 **Finger kneading around the ankle bone** Using two fingers of each hand, use small circular movements to knead around the ankle bone. Massage both sides of the ankle bone at the same time.

*Repeat step **6** a further 2 times.*

7 **Deep stroking to the top and bottom of the foot simultaneously** Cup the foot with the whole of the hand, so that the thumb is on the sole of the foot and the fingers on top of the foot. Firmly stroke the foot from toe to ankle, using alternate hands.

*Repeat step **7** a further 5 times.*

8 **Effleurage from the foot to the knee** Use the same movement as in step **1**.

*Repeat step **8** a further 5 times.*

CONTRA-ACTIONS

These are the same as for manicure – see page 244.

AFTERCARE AND ADVICE

Offering aftercare advice at the end of a pedicure treatment will help the client to look after her feet between salon visits, and is also an ideal opportunity to recommend retail products.

Aftercare advice will differ slightly for each client, but it will basically be as follows:

- Change socks or tights daily.
- Apply moisturising lotion to the feet after bathing.
- Ensure that the feet are thoroughly dry after washing, especially between the toes.
- Apply talc or a special foot powder between the toes to help absorb moisture.
- Go barefoot wherever it is safe and practical to do so.
- Ensure that footwear fits properly.
- Avoid wearing high heels for long periods of time.
- If any pain is felt in the feet, visit a chiropodist.

Exercises for the feet and ankles

As part of the home-care advice given to a pedicure client, **foot exercises** should be mentioned – these can play a very important role in keeping the client's feet healthy. They help:

- to stimulate circulation;
- to keep joints mobilised, allowing a greater range of movement in the toes and ankles;
- to keep muscles strong, reducing the chance of fallen arches (flat feet).

Here are some examples of exercises:

1. Sitting on a chair with the feet flat on the floor, raise the toes upwards and then relax.
2. Stand on tiptoes, and relax down again.
3. Sitting on a chair with the feet flat on the floor, lift one leg slightly and draw a circle with the toes so that the ankle moves through its full range of movement.
4. **Dorsiflexion**: bending the foot backwards towards the body.
5. **Plantar flexion**: pointing the foot down towards the ground.
6. **Inversion**: moving the foot inwards towards the middle of the body.
7. **Eversion**: moving the foot out towards the side of the body.

KNOWLEDGE REVIEW

Pedicure

Anatomy

1 Why does the nail bed contain nerve endings?
2 What is the function of the nail plate?
3 What is the function of the arches of the feet?
4 How does the foot move?

Preparation

1 Why is it important that the client is correctly prepared for treatment?
2 List *three* contra-indications to pedicure.
3 Why is it important to check the client's feet for contra-indications before starting the treatment?

Application

1 Why should you remove grease from the nail plate prior to enamelling?
2 Why should toenails be cut straight across and not oval, like fingernails?
3 Describe the methods available for removing hard skin from the feet.

Aftercare

1 State the general aftercare advice that you would give to a pedicure client.
2 What advice would you give to a client with calloused heels?
3 List *three* retail products that you would recommend to a pedicure client.
4 How can foot exercises be of benefit to the client?

Health, safety and hygiene

1 How should a stainless steel cuticle knife be sterilised?
2 What precautions should be taken when using a cuticle knife?
3 Why is it important that the client observe hygiene procedures being carried out?

Case study

1 For each of the clients below, suggest a pedicure treatment routine. Include details of the treatment, the products used, relevant home-care advice, the aims of the treatment, and the probable cause of the client's condition.
   (a) A middle-aged hairdresser who has very hard, cracked skin on the soles of her feet around both heels.
   (b) An elderly man who has little movement in his ankle joints and slightly distorted joints in his toes.
   (c) A shopworker who has tired, aching feet and swollen ankles.

CHAPTER 9

Removing and lightening hair

ANATOMY AND PHYSIOLOGY OF HAIR AND SKIN

THE HAIR

The structure and function of hair and the surrounding tissues

A hair is a long, slender structure which grows out of, and is part of, the skin. Each hair is made up of dead skin cells, which contain the protein called keratin.

Hairs cover the whole body, except for the palms of the hands, the soles of the feet, the lips, and parts of the sex organs.

Hair has many functions:

- *scalp hair* insulates the head against cold, protects it from the sun, and cushions it against bumps;
- *eyebrows* cushion the browbone from bumps, and prevent sweat from running into the eyes;
- *eyelashes* help to prevent foreign particles entering the eyes;
- *nostril hair* traps dust particles inhaled with the air;
- *ear hair* helps to protect the ear canal;
- *body hair* helps to provide an insulating cover (though this function is almost obsolete in humans), has a valuable sensory function, and is linked with the secretion of sebum onto the surface of the skin.

Hair also plays a role in social communication.

The structure of hair

Most hairs are made up of three layers of different types of epithelial cells: the *medulla*, the *cortex*, and the *cuticle*.

The **medulla** is the central core of the hair. The cells of the medulla contain soft keratin, and sometimes some pigment granules. There is usually no medulla in thinner hairs.

The **cortex** is the thickest layer of the hair, and is made up of several layers of closely-packed elongated cells. These contain pigment granules and hard keratin.

The **cuticle** is the protective outer layer of the hair, and is composed of thin, unpigmented flat cells. These contain hard keratin, and overlap from the base to the tip of the hair.

TIP

A strand of hair is stronger than an equivalent strand of nylon or copper.

It is the **pigment** in the cortex that gives hair its colour. When this pigment is no longer made, the hair appears white. As the proportion of white hairs rises, the hair seems to go 'grey': in fact, however, each individual hair is either coloured as before, or white.

The parts of the hair

Each hair is recognised by three parts: the *root*, the *bulb*, and the *shaft*:

- the **root** is the part of the hair that is in the follicle;
- the **bulb** is the enlarged base of the root;
- the **shaft** is the part of the hair that can be seen above the skin's surface.

The structure of skin

ACTIVITY: THE FUNCTION OF HAIR

What do you think is the *purpose* of hairs standing on end? Humans are not very hairy: what sign is there when this mechanism is working?

Each hair grows out of a tube-like indentation in the epidermis, the **hair follicle**. The walls of the follicle are a continuation of the epidermal layer of the skin.

The **arrector pili muscle** is attached at an angle to the base of the follicle. Cold, aggression or fright stimulates this muscle to contract, pulling the follicle and the hair upright.

The **sebaceous gland** is attached to the upper part of the follicle; from it, a duct enters directly into the hair follicle. The gland produces an oily substance, **sebum**, which is secreted into the follicle. Sebum waterproofs, lubricates and softens the hair, and the surface of the skin; it also protects the skin against bacterial and fungal infections. The contraction of the arrector pili muscle aids the secretion of sebum.

A cross-section of the hair follicle and hair

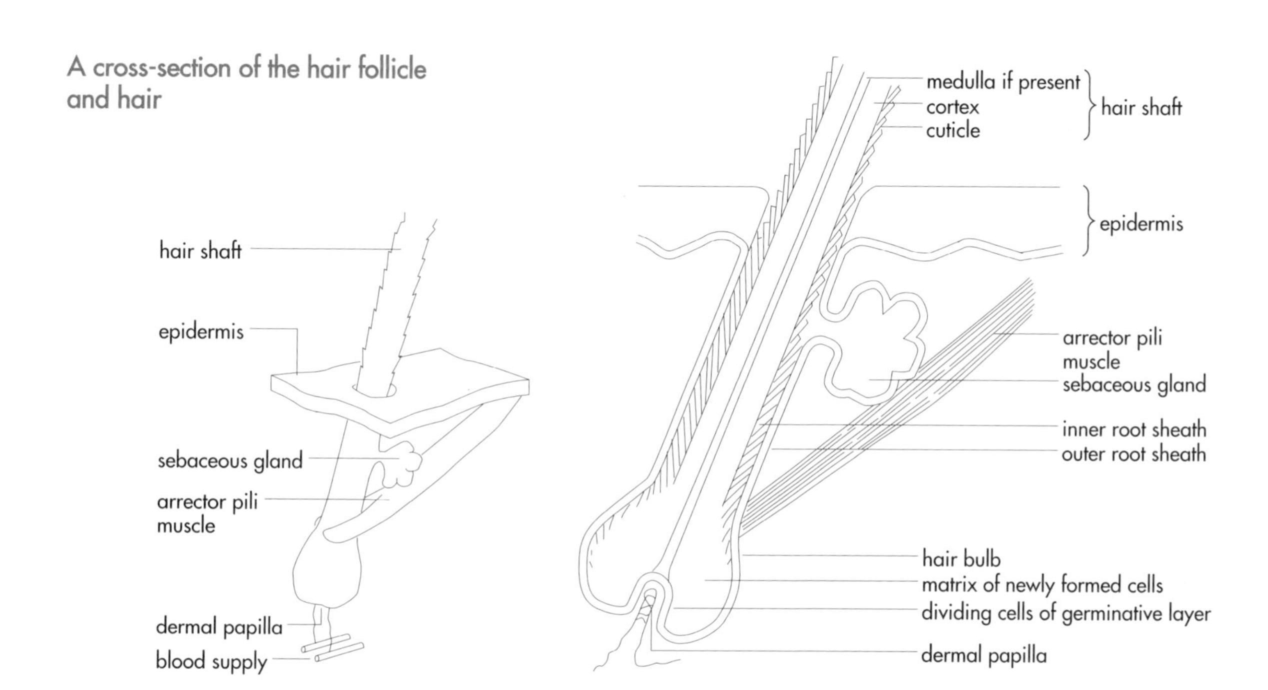

There are two types of **sudoriferous** (sweat) **glands:** the *eccrine* and the *apocrine* glands. The **eccrine glands** are simple sweat-producing glands, distributed over most of the body's surface and responsive to heat. The **apocrine glands** are larger and deeper.

Associated with the hairs in the groin and the underarm, their ducts open directly into the hair follicles near to the surface of the skin. These glands, which are under hormonal control, become active at puberty. They also become more active in response to stress. Apocrine glands produce sweat which decays with a characteristic smell.

The **dermal papilla**, a connective tissue sheath, is surrounded by a hair bulb. It has an excellent blood supply, necessary for the growth of the hair. It is not itself part of the follicle, but a separate tiny organ which serves the follicle.

The **bulb** is the expanded base of the hair root. A gap at the base leads to a cavity inside, which houses the papilla. The bulb contains in its lower part the dividing cells that create the hair. The hair continues to develop as it passes through the regions of the upper bulb and the root.

The **matrix** is the name given to the lower part of the bulb, which comprises actively dividing cells from which the hair is formed.

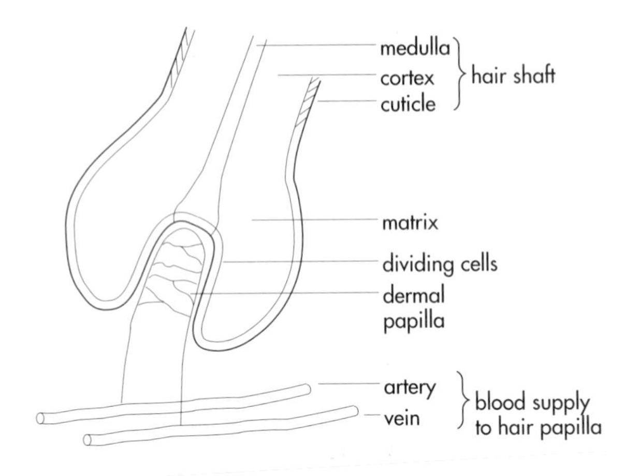

The hair bulb

The hair follicle

The hair follicle extends into the dermis, and is made up of three sheaths: the *inner epithelial root sheath*, the *outer epithelial root sheath*, and the surrounding *connective-tissue sheath*.

The **inner epithelial root sheath** grows from the bottom of the follicle at the papilla; both the hair and the inner root sheath grow upwards together. The inner surface of this sheath is covered with cuticle cells, in the same way as the outer surface of the hair: these cells lock together, anchoring the hair firmly in place. The inner root sheath ceases to grow when level with the sebaceous gland.

The **outer epithelial root sheath** forms the follicle wall. This does not grow up with the hair, but is stationary. It is a

TIP

When hairs break off due to incorrect waxing technique, they will break at the level at which they are locked into the follicle by the cells of the inner root sheath.

continuation of the growing layer of the epidermis of the skin.

The **connective-tissue sheath** surrounds both the follicle and the sebaceous gland, providing both a sensory supply and a blood supply. The connective-tissue sheath includes, and is a continuation of, the papilla.

The *shape* of the hairs is determined by the shape of the hair follicle – an angled or bent follicle will produce an oval or flat hair, whereas a straight follicle will produce a round hair. Flat hairs are curly, oval hairs are wavy, and round hairs are straight. As a general rule, during waxing curly hairs break off more easily than straight hairs.

TIPS

An angled follicle may cause the hair to be broken off at the angle during waxing, instead of being completely pulled out with its root. If this happens, broken hairs will appear at the skin's surface within a few days.

By causing damage to the follicle and changing its shape, waxing can cause the regrowth of hairs to be frizzy or curled where previously the hairs have been straight.

The nerve supply

The number, size and type of nerve endings associated with hair follicles is related to the size and type of follicle. The follicles of vellus hairs (see below) have the fewest nerve endings; those of terminal hairs have the most.

The nerve endings surrounding hair follicles respond mainly to rapid movements when the hair is moved. Nerve endings that respond to touch can also be found around the surface openings of some hair follicles, as well as just below the epidermis.

The three types of hair

There are three main types of hair: *lanugo, vellus* and *terminal.*

Lanugo hairs are found on foetuses. They are fine and soft, do not have a medulla, and are often unpigmented. They grow from around the third to the fifth month of pregnancy, and are shed to be replaced by the secondary vellus hairs around the seventh to the eighth month of pregnancy. Lanugo hairs on the scalp, eyebrows and eyelashes are replaced by terminal hairs.

Vellus hairs are fine, downy and soft, and are found on the face and body. They are often unpigmented, rarely longer than 20mm, and do not have a medulla or a well-formed bulb. The base of these hairs is very close to the skin's surface. If stimulated, the shallow follicle of a vellus hair can grow downwards and become a follicle that produces terminal hairs.

Terminal hairs are longer and coarser than vellus hairs, and most are pigmented. They vary greatly in shape, in diameter and length, and in colour and texture. The follicles from which they grow are set deeply in the dermis and have well-defined bulbs.

TIP

Some areas of the body – for example, the bikini line and underarm areas – often have terminal hairs with very deep follicles. When these hairs are removed, the resulting tissue damage may cause minor bleeding from the entrance of the follicle. Removal of these deep-seated hairs is obviously more uncomfortable than the removal of shallower hairs.

Hair growth

All hair has a cyclical pattern of growth, which can be divided into three phases: *anagen, catagen* and *telogen*.

Anagen

Anagen is the actively growing stage of the hair – the follicle has re-formed; the hair bulb is developing, surrounding the life-giving dermal papilla; and a new hair forms, growing from the matrix in the bulb.

TIP

A hair pulled out at the *anagen* stage will be surrounded by the inner and outer root sheaths and have a properly formed bulb.

Catagen

Catagen is the stage when the hair separates from the papilla. Over a few days it is carried by the movement of the inner root sheath, up the follicle to the base of the sebaceous gland. Here it stays until it either falls out or is pushed out by a new hair growing up behind it.

This stage can be very rapid, with a new hair growing straight away; or slower, with the papilla and the follicle below the sebaceous gland degenerating and entering a resting stage, telogen.

TIP

A hair pulled out at the *catagen* or *telogen* stage can be recognised by the brush-like appearance of the root.

Telogen

Telogen is a resting stage. Many hair follicles do not undergo this stage, but start to produce a new hair immediately. During resting phases, hairs may still be loosely inserted in the shallow follicles.

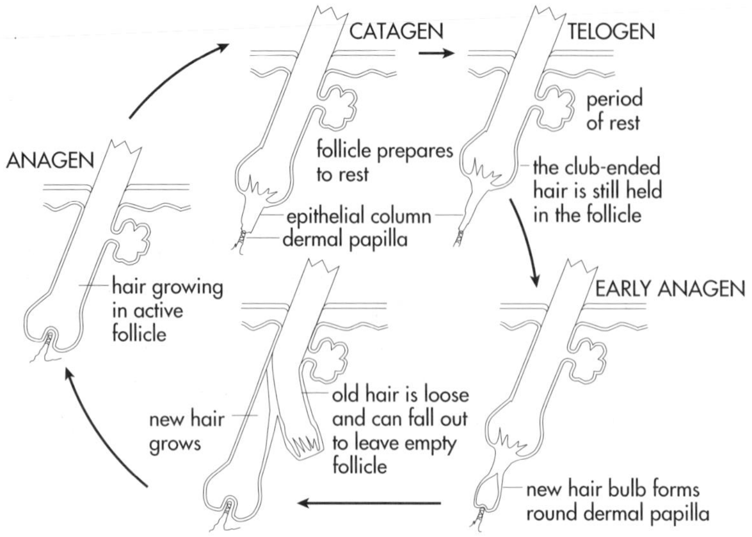

The hair growth cycle

TIP

Because of the cyclical nature of hair growth, the follicles are always at different stages of their growth cycle. When the hair is removed, therefore, the hair will not all grow back at the same time. For this reason, waxing can *appear* to reduce the quantity of hair growth. This is not so; given time, all the hair would regrow. Waxing is classed as a temporary means of hair removal.

Speed of growth

Anagen, catagen and telogen last for different lengths of time in different hair types and in different parts of the body:

- *scalp hair* grows for 2–7 years, and has a resting stage of 3–4 months;
- *eyebrow hair* grows for 1–2 months, and has a resting stage of 3–4 months;
- *eyelashes* grow from 3–6 weeks, and have a resting stage of 3–4 months.

TIP

Vellus hairs grow slowly and take 2–3 months to return after waxing. They can remain dormant in the follicle for 6–8 months before shedding.

After a waxing treatment, body hair will take approximately 6–8 weeks to return.

Because hair growth cycles are not all in synchronisation, we always have hair present at any given time. On the scalp, at any one time for example, 85 per cent of hairs may be in the anagen phase. This is why hair growth after waxing starts within a few days: what is seen is the appearance of hairs that were already developing in the follicle at the time of waxing.

Types of hair growth

Hirsutism

Hirsutism is a term used to describe a pattern of hair growth that is abnormal for that person's sex, such as when a woman's hair growth follows a man's hair-growth pattern.

Hypertrichosis

Hypertrichosis is an abnormal growth of excess hair. It is usually due to abnormal conditions brought about by disease or injury.

Superfluous hair

Superfluous hair (excess hair) is perfectly normal at certain periods in a woman's life, such as during puberty or pregnancy. Terminal hairs formed at these times usually disappear once the normal hormonal balance has returned. Those newly formed during the menopause are often permanent unless treated with a permanent method of hair removal, such as electrical depilation.

Factors affecting the growth, rate and quantity of hair

Hair does not always grow uniformly:

- *Time of day* Hair grows faster at night than during the day.
- *Weather* Hairs grow faster in warm weather than in cold.
- *Pregnancy* In women, hairs grow faster between the ages of 16 and 24, and (frequently) during mid-pregnancy.
- *Age* The rate of hair growth slows down with age. In women, however, facial hair growth continues to increase in old age, while trunk and limb hair increases into middle age and then decreases.
- *Colour* Hair of different colours grows at different speeds – for example, coarse black hair grows more quickly than fine blonde hair.
- *Part of the body* Hair in different areas of the body grows at different rates, as do different types and thicknesses of hair. The weekly growth rate varies from approximately 1.5mm (fine hair) to 2.8mm (coarse hair), when actively growing.
- *Heredity* Members of a family may have heritable growth patterns, such as excess hair that starts to grow at puberty and increases until the age of 20–25.
- *Health and diet* Health and diet are crucial in the rate of hair growth.
- *Stress* Emotional stress can cause a temporary hormonal imbalance within the body, which may lead to a temporary growth of excess hair.
- *Medical conditions* A sudden unexplained increase of body hair growth may indicate a more serious medical problem, such as malfunction of the ovaries; or result from the taking of certain drugs, such as corticosteroids.

The quantity as well as the type of hair present may vary with race:

- *People of Latin extraction* tend to possess heavier body, facial and scalp hair, which is relatively coarse and straight.
- *People of Eastern extraction* tend to possess very little or no body and facial hair growth, and usually their scalp hair growth is relatively coarse and straight.
- *People of Northern European and Caucasian extraction* tend to have light to medium body and facial hair growth, with their scalp hair growth being wavy, loosely curled, or straight.
- *People of Afro-Caribbean extraction* tend to have little body and facial hair growth, but usually their scalp hair growth is relatively coarse and tightly curled.

HEALTH & SAFETY: AFRO-CARIBBEAN CLIENTS
The body hair of Afro-Caribbean clients is prone to breaking during waxing, and to ingrowing hairs after waxing. Skin damage can result in the loss of pigmentation (hypopigmentation).

THE SKIN

The structure and function of skin

Wax depilation, by tearing out many hairs from actively growing follicles, inflicts considerable tissue damage at a microscopic level. This damage stimulates the skin's defences to repair the area and to prevent invasion by bacteria. The visible signs of this process range from a slight redness of the skin (**erythema**), due to increased blood flow to the area, to actual bleeding from the injured follicle, as is often seen when very deep-seated terminal hairs are removed from areas such as bikini lines and the underarms.

TIP

When any cell dies it releases enzymes which liquefy it. The remains are taken away in the bloodstream.

Mast cells are delicate cells scattered throughout the dermis of the body: their function is to detect damage to the skin. If damage does occur – such as hairs being torn out during waxing – the mast cells burst, releasing a chemical called **histamine** into the surrounding tissues.

Histamine

Histamine has two main functions. First, it triggers nerve endings to send impulses to the brain. If a few are triggered, this is registered as an itch or irritation (common after waxing). If many are triggered, actual pain is felt, and the person becomes aware of a problem.

Secondly, histamine also causes dilation of the capillaries in the dermis. This allows increased blood flow through the skin, which is felt as heat and seen as reddening (erythema), either over the whole area or as red pinprick dots around each damaged follicle. The increased blood flow, as well as diluting and dispersing the histamine, removes the remains of the damaged cells and brings materials for repair and infection control – white blood cells, nutrients and oxygen.

The extra fluid delivered to the area can cause minor swelling (**oedema**): this will go away after a few hours, as the damage is healed. Erythema and swelling together are known as **inflammation** in the affected area.

wax depilation
↓
damage to skin cells
↓
breakdown of mast cells
↓
release of histamine
↓
dilation of blood capillaries (erythema)

(breakdown of mast cells → dilation of blood capillaries: HISTAMINE REACTION)

Erythema is accompanied by a rise in temperature of the affected area. More heat added to this area – in the form of more wax, or a heat treatment such as a sunbed – could result in a burn.

HEALTH & SAFETY: IRRITATION AND INFLAMMATION

Irritation is always a signal that something is wrong.
Inflammation indicates that repairs are underway.

Erythema is accompanied by a rise in temperature of the affected area. More heat added to this area – in the form of more wax, or a heat treatment such as a sunbed – could result in a burn.

The damage caused to the skin during waxing can result in a protective reaction by the body: the temporarily increased production of keratinised cells. This is evident as dryness and flaking of the affected skin surface a few days after the treatment. If a build-up of the dead surface skin cells is allowed, these can block the entrance to the follicles. This in turn may cause growing hairs to turn around and become ingrowing.

During a waxing treatment, there is a real risk that these thick, loose, dead layers may be lifted away, sometimes even taking some of the underlying growing layer with them. This may cause damage to the affected area, similar to a shallow graze.

REMOVING HAIR: WAXING

WAX DEPILATION

Wax depilation involves using wax to remove hair temporarily from the face and body. Waxing removes both the visible hair and the root, so the regrowth is of completely new hairs with soft, fine-tapered tips. It will take 4–6 weeks before the client requires the service again.

RECEPTION

When the client is booking her treatment she should be asked whether she has had a wax treatment before in the salon. If she has not, a small area of waxing should be carried out as a test patch, to ensure that she is not sensitive to the technique or allergic to any of the products used. If this test-patch causes an unwanted reaction within 48 hours, then the treatment must not be undertaken.

Advise the client not to apply any lotions or oils to the area on the day of her treatment – these could prevent the adhesion of the wax to the hairs being removed. Ask her also to allow at least one

Treatment	*Warm waxing (minutes)*	*Hot waxing (minutes)*	*Sugaring (minutes)*
Half leg	15	30	20–30
Half leg and bikini	30	45	45–50
Full leg	30–45	60	55
Full leg and bikini	45–60	60–75	60–75
Bikini	10–15	15	15–20
Underarm	10–15	15	15
Half arm	10–15	15–20	15–20
Full arm	20	20–30	20–30
Top lip	5	5–10	5–10
Chin and throat	10	20	15–20
Top lip and chin	15	15–20	15–20
Eyebrows	10	10–15	10–15

week, and preferably two, between any home shaving or other depilatory treatment and a salon waxing treatment. This is to let the hairs grow to a length sufficient to be waxed.

Allow a 4–6-week interval between successive wax depilation appointments. The times to be allowed for wax treatments are as shown in the table.

CONTRA-INDICATIONS

If the client has any of the following, wax depilation must not be carried out:

- *Skin disorders*, such as bruising or recent haemorrhage.
- *Swellings.*
- *Diabetes.*
- *Defective circulation.*
- *Warts or moles.*
- *Scar tissue.*
- *Fractures or sprains.*
- *Varicose veins.*
- *Loss of skin sensation.*

TIP

Waxing a client with a recent well-established suntan may cause the loose sun-damaged epidermis to peel and be lost, along with the hair.

The therapist must not carry out a wax treatment immediately after a heat treatment, such a sauna, or steam or ultra-violet treatments, as the heat-sensitised tissues may be irritated by the wax treatment.

ACTIVITY: CONTRA-INDICATIONS TO WAXING

Briefly describe why each of the contra-indications listed prohibits wax depilation.

EQUIPMENT AND MATERIALS

Types of wax

Warm wax

Warm wax first became available in 1975 and is now the market leader for hair removal. It is clean and easy to use; and because it is disposed of afterwards it is also hygienic. Finally, it is very economical.

Warm wax is used at a low temperature, around 43 °C, so there is little risk of skin burning, and in less-sensitive areas the wax can be re-applied once or even twice if necessary.

Warm wax does not set but remains soft at body temperatures. It adheres efficiently to hairs and is quick to use; treatment is relatively pain-free. It can remove even very short hairs (2.5 mm) from legs, arms, underarms, the bikini line, the torso, the face and the neck.

Warm waxes are frequently made of mixtures of glucose syrup and zinc oxide. Honey (fructose syrup) can be used instead of glucose syrup, and this type is often called **honey wax**.

Hot wax

Hot wax takes longer to heat than warm wax, and is relatively slow to use, taking approximately double the time of a warm waxing.

As hot wax is used at quite a high temperature, 50 °C, extra care must be taken to avoid accidental burns. Because of this risk, hot wax cannot be re-applied to already treated areas.

Hot wax cools on contact with the skin. It contracts around the hair shaft, gripping it firmly. This makes it ideal for use on stronger, short hairs.

Hot waxes for hair removal need to be a blend of waxes and resins so that they stay reasonably flexible when cool. **Beeswax** is a desirable ingredient, and often comprises 25–60 per cent of the finished product. **Cetiol**, **azulen** and **vitamin E** are often added to wax preparations to soothe the skin and minimise possible skin reactions.

Cold wax

Cold wax is used cold straight from its container (although some do in fact need gentle heating). It is spread with a clean spatula, and removed using a Cellophane, muslin or paper strip. This product was designed for home use by the public, and is not generally seen in salons.

Cold waxes are often natural rubber solutions in a volatile solvent. The solvent evaporates from the skin to leave a rubber film with the hairs embedded in it: this is pulled off and thrown away. The adhesion between the rubber and the hair is not sufficient for strong terminal hair or short hair growth.

ACTIVITY: METHODS OF HAIR REMOVAL
Other methods of temporary hair removal include the professional use of threading; and the home use of plucking, cold wax, ready-waxed strips, sugaring, electrical devices, chemical depilatory creams, shaving and pumice powder. Find out how each method works and assess its effectiveness. Are any of them potentially hazardous?

Warm waxing

To carry out warm waxing you will need the following equipment and materials:

- *Wax heater* – with a thermostatic control, a lid and a central bar.

ACTIVITY: WAX HEATERS
Look in current beauty magazines for the different types of wax heater, including roll-ons. Try to see demonstrations of them working. Discuss size, cleaning, safety, hygiene and any other points that you think are important. Which heater would *you* choose, and why?

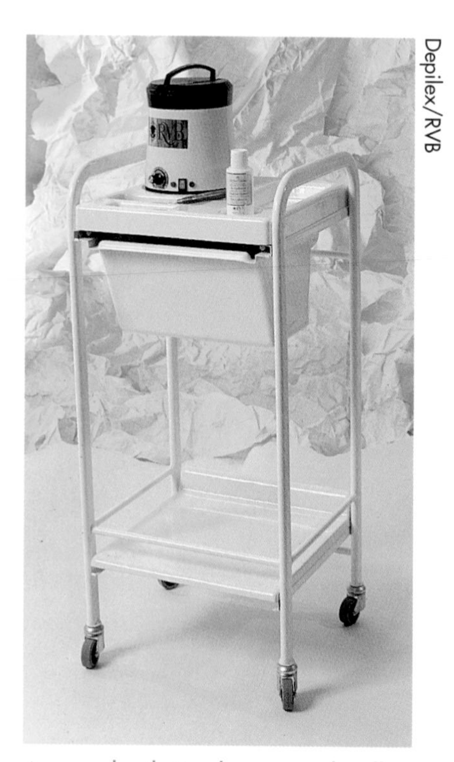

Depilex/RVB

A wax depilation heater and trolley

Equipment and materials

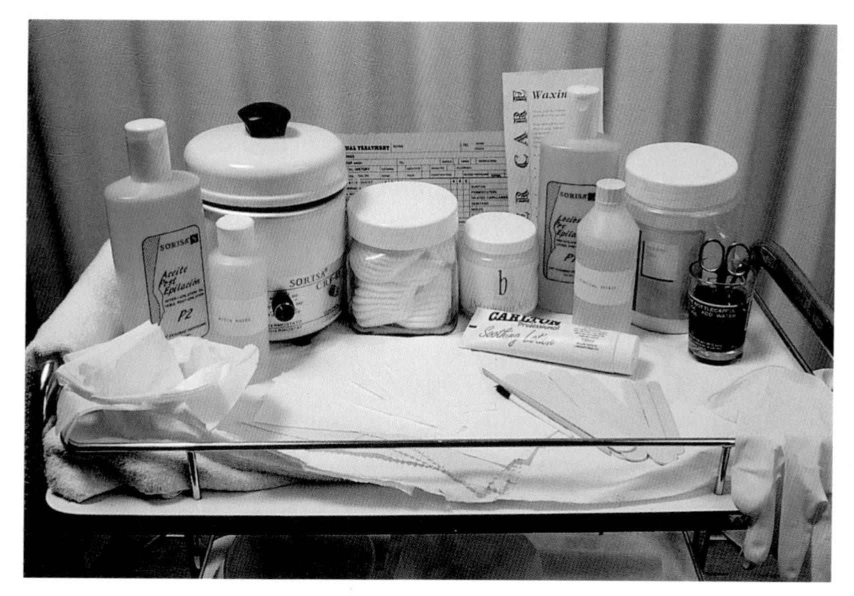

Roll-on wax applicator equipment

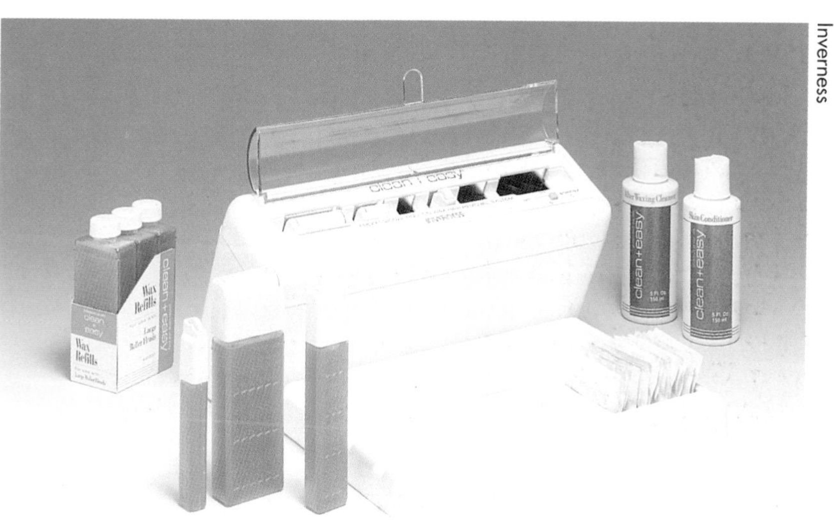

Inverness

- *Couch* – with sit-up and lie-down positions and an easy-to-clean surface.
- *Trolley* – to hold all the necessary equipment and materials.
- *Disposable tissue roll* – such as bed roll.
- *Protective plastic couch cover.*
- *Disposable wooden spatulas* – a selection of differing sizes, for use on different body areas.
- *Wax-removal strips (bonded-fibre)* – thick enough that the wax does not soak through, but flexible enough for ease of work. These strips should be cut to size and placed ready in a container.
- *Talcum powder.*
- *Cottonwool pads* – for cleaning equipment and the client's skin.
- *Facial tissues (white)* – for blotting skin dry.
- *Alcohol-based skin cleanser (professional)* or *antiseptic wipes* – the cleanser is also called a pre-wax lotion.

- *Surgical spirit* – or a commercial cleaner designed for cleaning equipment.
- *Towels (medium-sized)* – for draping around the client.
- *Small scissors (curved)* – for trimming over-long hairs.
- *Disinfecting solution* – in which to immerse small metal implements.
- *Tweezers* – for removing stray hairs.
- *Wax* – see below.
- *After-wax lotion* – with soothing, healing and antiseptic qualities.
- *Mirror (clean)* – for facial waxing.
- *Apron* – to protect workwear from spillages.
- *Disposable surgical gloves.*
- *Bin (swing-top)* – lined with a disposable bin liner.
- *Client's record card.*
- *List of contra-indications.*
- *Aftercare leaflets.*

When choosing a wax, select one with the following qualities:

- It should have a low melting temperature – the wax should be liquid at body temperature (37 °C), and very runny at around 43 °C.
- It should be easy to remove from equipment.
- It should be able to remove short, strong hairs (25 mm).
- It should have a pleasant fragrance or no smell.
- It should not stick to the skin, but only to the hair.

Sterilisation and sanitisation

Disposable waste from waxing may have body fluids on it: potentially it is a health risk. It must be handled, collected and disposed of according to the local environmental health regulations. It is a good idea to wear disposable rubber gloves whilst carrying out the treatment, to protect yourself from body fluids.

HEALTH & SAFETY: AVOIDING CROSS-INFECTION
Never filter hot wax after use: it cannot be used again as it will be contaminated with skin cells, tissue fluid, and perhaps even blood.

ACTIVITY: MAINTAINING HYGIENE
Spreading wax on the client and dipping the used spatula back into the tub creates the possibility of cross-infection between clients. How can cross-infection be avoided, with this method and the use of roll-on wax applicators?

Preparing the cubicle

Before the client is shown through, the cubicle and its contents should be checked to ensure that they are clean and tidy. The bins should have been emptied since the previous client.

Check the trolley to ensure that you have all that you need for carrying out the treatment, and that the wax is both warm and liquid.

The plastic-covered couch should be clean, having previously been washed with hot soapy water or wiped thoroughly with

HEALTH & SAFETY: COUCH COVERS

If the salon chooses to use stretch-towelling couch covers or additional towels to provide extra comfort, these must be provided clean for each client and laundered in hot soapy water after use.

surgical spirit or a professional alcohol-based cleaner. The use of an additional heavy-duty plastic sheet is recommended: this is easier to wipe than the couch, and can be replaced if damaged.

The couch should then be covered and protected with a long strip of paper-tissue bed roll. Place a towel neatly on the couch, ready to protect the client's clothing and to cover her when she has undressed. The tissue should be disposed of after use, and the towel be freshly laundered for each client.

The couch should be in the sit-up position, unless the client is only having her bikini line or underarm areas waxed, in which case it should be flat.

TIP

When a client books for a bikini-line wax, ask her when she comes for her treatment to wear an old pair of high-cut panties.

PREPARING THE CLIENT

After the record card has been completed at reception, the client should be asked to read the list of contra-indications and sign to state that she is not suffering from any of the problems stated. She should then be shown through to the cubicle, and asked to remove specific items of clothing as necessary so that the treatment may be carried out.

ACTIVITY: ENSURING CLIENT COMFORT

Imagine that you are a client who has never had a waxing treatment before. You are shown through to a cubicle and left to 'Get yourself ready, please.' How would you feel? What would you do? What clothing would you think it necessary to remove for each area of waxing?

TIP

Here are some useful explanatory phrases: 'It's a bit like ripping a plaster off, and taking the hairs with it. It isn't so bad, or so many people wouldn't have it done time and time again!'

Ask the client whether it is the first time she has had the treatment: if it is, explain to her that the treatment can be uncomfortable, but it is quick and any discomfort experienced is tolerable.

Be efficient and quick, so that the client does not have to wait. Try to get her talking about something pleasant, such as a holiday, to take her mind off the treatment. Throughout the treatment, reassure her, praising her in order to motivate her to continue with her treatments. Do try to be sympathetic to your client's feelings, and provide support and encouragement when necessary. Waxing, although a necessity for many people, is not a particularly pleasant or relaxing treatment.

1. Use a towel to protect the client's remaining clothing.
2. Wipe the area to be waxed with a professional antiseptic pre-wax cleansing lotion on cottonwool. Blot the area dry with tissues before applying the wax. While wiping the skin, look for contra-indications.

TIP

If there is any bruising on the client's legs gently draw her attention to these bruises, or she might later think that the treatment has caused them.

3 Record any bruising on her record card to avoid potential problems later.

4 If the client's skin is very greasy (she may for example have applied oil before coming to the salon), cleanse it using an astringent lotion such as witchhazel.

5 Immediately before starting the treatment, wash your hands.

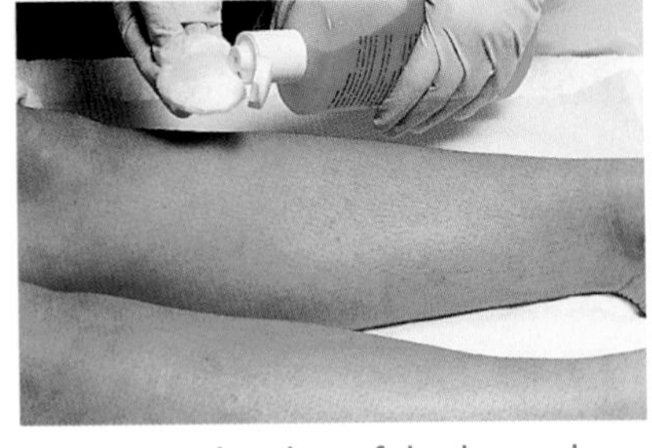

Cleansing the skin of the lower leg

TIP

A woman's pain threshold is at its lowest immediately before and during her period. The hormones which stimulate the regrowth of hair are also at their most active during this period. For these two reasons, avoid wax depilation at this time if possible.

WARM WAXING

General techniques

Warm waxing has a few basic rules which must be followed to ensure a good result.

HEALTH & SAFETY: TEMPERATURE OF WAX

Before applying the wax, check its temperature. Test the wax on yourself first, to ensure that it's not too warm; then try a little on a small visible area of the client (such as the inside of the ankle) before spreading it on other areas.

1 Observe the direction of hair growth. Warm wax must always be *applied with* the direction of hair growth, and *removed against* the direction of growth. This ensures both maximum adhesion between the hair, the strip and the wax, and that the hair will be pulled back on itself in the follicle and thus removed complete with its bulb.

2 Dip the spatula into the wax. Remove the excess on the sides and tip by wiping the spatula against the metal bar or the sides of the tub. Place the strip under the spatula while transferring it to the client: this will control dripping and improve your technique.

TIP

The thinner the wax, the easier the treatment is to carry out and the better the result. Also, less wax and fewer strips are used.

3 Place the spatula onto the skin at a 90° angle, and push the wax along in the direction of the hair growth. Do not allow the spatula to fall forward past 45°. The objective is to coat the area with a very thin film of wax. Quite a large area can be spread with each sweep of the spatula, as warm wax does not set. Do not attempt to smooth out or go over areas on which wax has already been spread, however, as the wax will have become cooler and will not move again, but will drag painfully on the client's skin.

TIP

Wax spilt on carpets can be removed by placing fabric waxing strips or brown paper placed over the spill, and running a warm iron over them.

4 Fold back 20mm at the end of a strip and grip the flap with

the thumb widthways across the strip. The flap should provide a wax-free handle throughout the treatment.

5 Place the strip at the bottom end of the wax-covered area, and make a firm bond between the wax and the strip by pressing firmly along the strip's length and width, following the direction of hair growth. If the strip is placed anywhere but at the *bottom* of a waxed area, the hairs at the bottom of the strip will become tangled together and held in the wax on the area below the strip: the removal of the strip will then be far more painful for the client.

TIP

To take the sting out of the removal, immediately place a hand or finger over the depilated area.

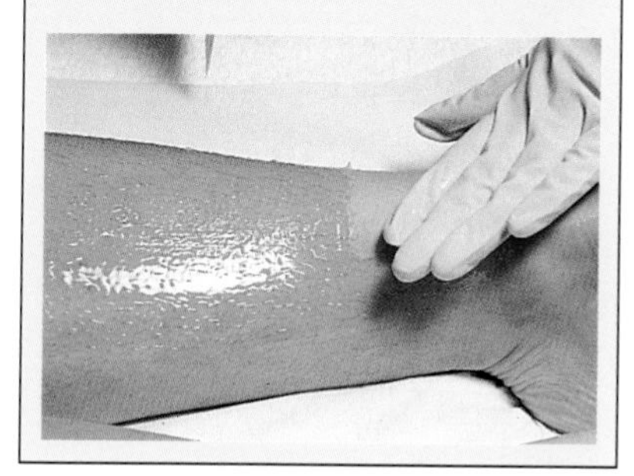

6 Using the non-working hand, stretch the skin to minimise discomfort. Gripping the flap tightly, use the working hand to remove the strip against the direction of hair growth. Use a firm, steady pull. Make sure that the strip is pulled back on itself, close to the skin. (To obtain the correct angle of pull, stand at the side of the client, facing her.) Maintain this same horizontal angle of pull until the last bit of the strip has left the skin: *do not pull the strip upwards at the end of the pull*, as this would break the hairs at the end of the strip and be very painful.

7 The strip may be used many times; in fact it works best when some wax builds up on its surface. When there is too much wax on the surface it will stop picking up more: throw it away and start with a new one.

8 Do not repeatedly spread and remove wax over one area. In particular, wax should not be spread and removed more than twice on sensitive areas such as the bikini line, the face and the underarms. Any remaining stray hairs must instead be removed using sterilised tweezers.

TIP

The faults most commonly seen in warm wax depilation are these:
- spreading the wax too thickly;
- placing the strip in the middle of a wax-spread area, instead of starting at the bottom and working up;
- pulling the strip upwards, away from the leg.

HEALTH & SAFETY: INJURIES

If any injury appears to be more than minor, advise the client to see her doctor straightaway after treatment.

Half leg treatment

The areas of the body where warm-wax hair removal is most frequently used are the lower legs. This is often referred to simply as a **half leg treatment**. A 'pair of half legs' should take 10–15 minutes to treat, and use no more than two or three strips.

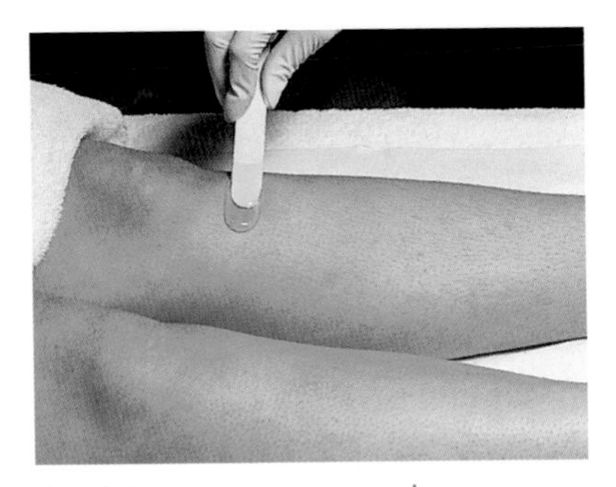
Applying warm wax to the lower leg

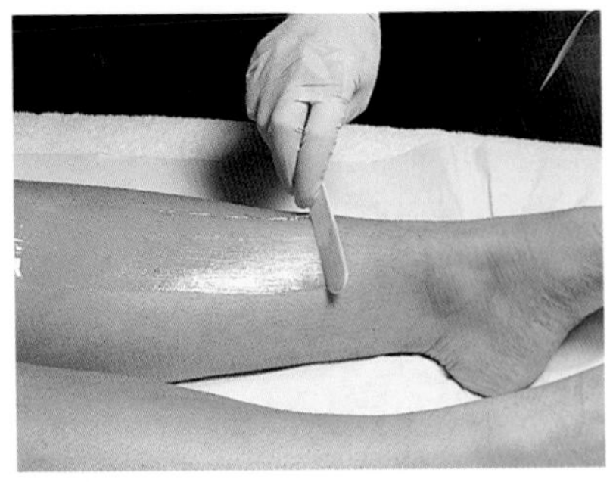
Spreading on wax (to show the angle of the spatula)

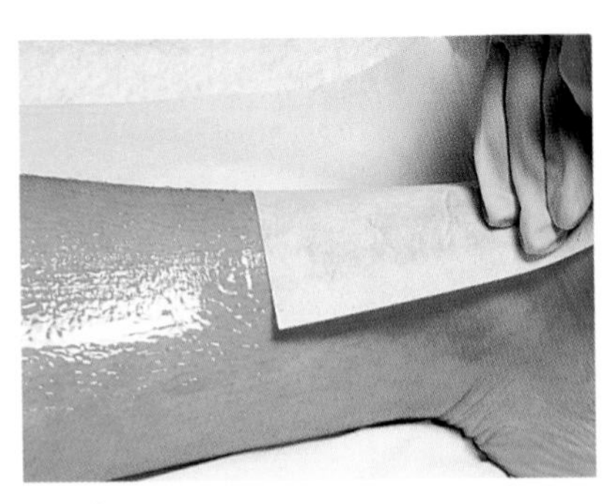
Applying a strip

1 Prior to the treatment, the area to be waxed should be cleansed as previously described.

2 Sit the client on the upraised couch, with both legs straight out in front of her.

3 Spread the wax on the *front* of the leg further away from you. Use three sweeps of the spatula: each sweep should go from just below the knee to the end of hair growth at the ankle.
   Repeat this pattern of spreading on the leg nearer to you. (By spreading the further leg first, you will not have to lean over an already waxed area.)

4 Starting with the nearer leg, remove the wax using the strip. Start at the ankle and work upwards towards the knees.
   Repeat for the other leg.

5 Ask the client to bend her legs to one side. Beginning again with the further leg, spread wax on one *side* of the leg, from the knee to the ankle, using two sweeps of the spatula.
   Repeat for the nearer leg.

6 Starting with the nearer leg, remove the wax from the bottom upwards.
   Repeat for the other leg.

7 Repeat steps **5** and **6** for the other side of the legs.

HEALTH AND SAFETY

It is unhygienic to return a spatula to the wax after it has been in contact with the body part. It is therefore recommended that a new spatula be used for each entry to the wax. This applies to warm, hot and sugar-strip wax.

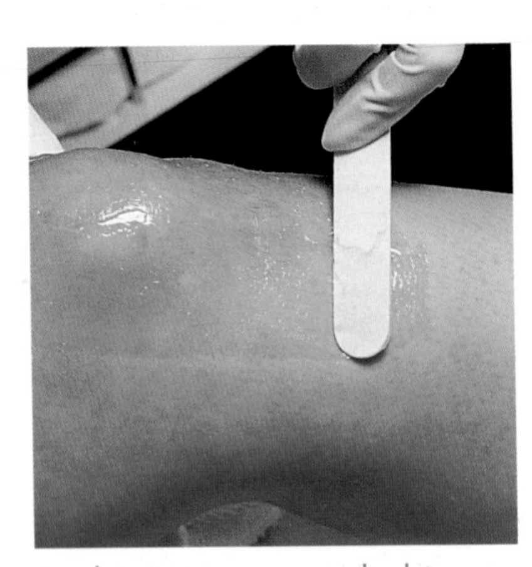
Applying warm wax to the knee

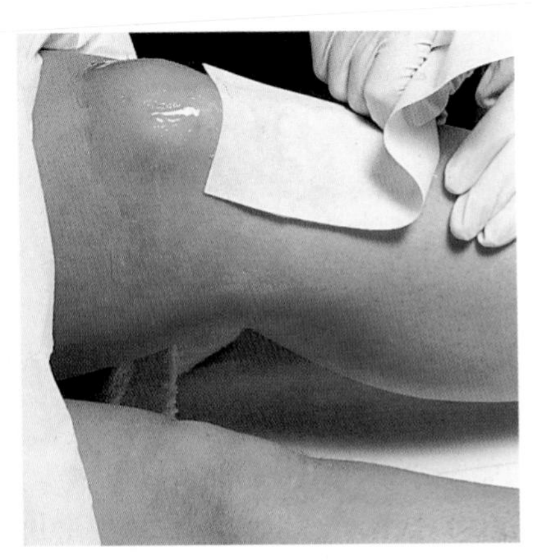
Removing the wax strip from the knee

8 Bend one knee. Spread the wax from just above the knee, downwards.

9 Remove this wax from the bottom upwards, remembering that strips cannot pull around corners effectively. To keep the angle of the pull horizontal, remove wax below the knee first, then that above the knee.

10 Repeat steps **8** and **9** for the other knee.

11 Lower the back rest and ask the client to turn over.

12 On the *back* of the legs, the direction of hair growth is not from the top to the bottom but sweeping at an angle, from the outside to the inside of the calf muscle.
Starting with the further leg, spread wax following the direction of the natural hair growth.
Repeat for the nearer leg.

13 Starting with the nearer leg, remove the wax against the direction of the natural hair growth.
Repeat for the other leg.

14 Finally, apply after-wax lotion to clean cottonwool and apply this to the back of the client's legs. As you apply the lotion, check for hairs left behind: if there are any, remove them using tweezers.
Ask the client to turn over, and repeat application on the front of the legs. Remove any excess using a tissue.

TIP

The most common fault seen in half leg waxing is trying to take too big an area at once over the calf muscle and not supporting the surrounding tissues adequately. This will result in a painful treatment for the client.

Toes

Clients frequently request that their toes be waxed in conjunction with a half leg or full leg treatment. When doing this, follow the normal guidelines for waxing, but be aware that hair may grow in many directions.

Wax spills

If you spill wax on the couch cover, immediately place a quarter-width facial-sized piece of strip on top of the spill. This prevents the wax from damaging the client's clothing when she moves.

Full leg treatment

A **full leg wax** should take 30–45 minutes and from four to six strips should be sufficient.

When doing a full leg wax, follow the same sequence of working as with the half leg. On the thighs, observe the direction of hair growth carefully, as the hair grows in different directions. It is best not to spread wax on too large an area at once, or you may forget the direction of growth. Each direction of hair growth should be treated as a separate area. It is of prime importance that you support the skin on the thighs as you remove the strip – the tissues in this area can bruise very easily. The two essential factors in preventing bruising, pain, and hair breakage are:

- the correct angle of pull;
- adequate support for the tissues.

Different temperatures

Summer heat and winter cold can each give rise to problems with the wax treatment. In summer, the wax stays too warm on the body, becoming sticky and difficult to work with, and tending to leave a sticky residue on the treated areas. In winter, the client's legs may be cold, causing the wax to set too quickly as you spread it, so that it becomes too thick. This prevents the efficient removal of both the wax and the hair growth.

To some extent these problems can be overcome by using thicker waxes with higher melting points in the summer, and thinner waxes with lower melting points in the winter.

TIP

Thick lumps of wax stuck to the legs (the winter problem) can be removed by pulling more slowly than normal, or by reversing the direction of pull on the strip. A sticky residue on the legs (the summer problem) must be removed using after-wax lotion.

Bikini-line treatment

A **bikini-line treatment** takes 5–15 minutes, and requires a new strip for each section to ensure effective hair removal from this delicate area.

1. Treat one side at a time. It is best if the client lies flat, as the skin's tissues are then pulled tighter, but the treatment can be carried out in a semi-reclining position if the client prefers. Bend the client's knee out to the side, and put her foot flat against the knee of the other leg. This is sometimes referred to as the **figure-four position**.
2. Tuck a tissue along and under the lower edge of the client's briefs. Raise this edge and agree with the client where she wants the final line to be. Hold the briefs slightly beyond this line, and ask the client to place her hand on top of the tissue to hold everything in place. This leaves you with both hands free, one to pull the strip and one to support the tissues.
3. Cleanse and dry the areas to be waxed.
4. Using sterilised scissors, trim both the hair to be waxed and the adjacent hairs down to about 5–12 mm in length. This is essential to avoid tangling, pulling and pain, and to prevent wax going onto hair that you do not want to remove.
5. Spread and remove the wax in two or three separate and distinct areas, the number depending on the directions of hair growth.
6. Use half of the strip length to remove the wax. Do not cover

HEALTH & SAFETY: BRUISING

Bruising is neither normal nor acceptable, but a sign of faulty technique.

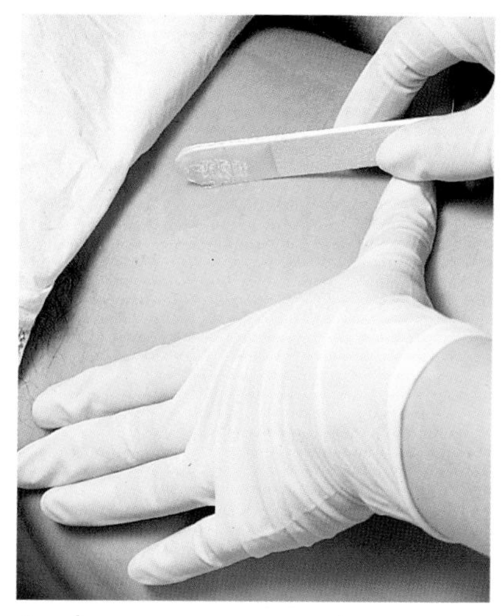
Applying warm wax to the bikini line

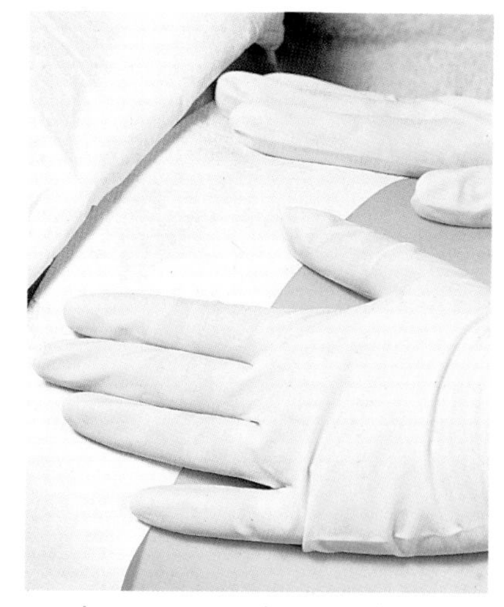
Applying a strip (showing the support of the skin area)

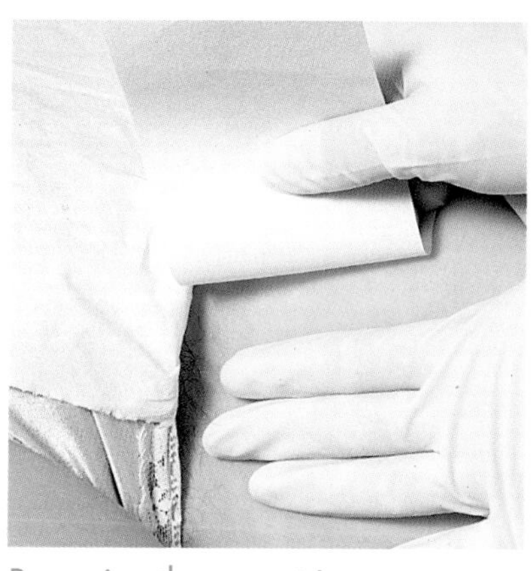
Removing the wax strip

the whole area and tear it off at at once! As soon as an area is completed, apply after-wax lotion. If necessary, use a clean tissue to remove excess lotion.

7 With both legs straight out in front on the couch, place a tissue along the top edge of the client's briefs, against the abdomen. Lower the briefs as necessary to expose just the hair to be removed – check this with your client. Usually the direction of hair growth here sweeps in from the sides and then up to the navel.

8 Trim the hair as before.

9 Apply and remove the wax in small sections, carefully observing hair growth. On completion, apply after-wax lotion.

Underarm treatment

An **underarm treatment** should take 5–15 minutes and two strips, one for each underarm.

1 With her bra still on, ask your client to lie flat on her back with her hands behind her head, elbows flat on the couch.

2 Cleanse both underarms with prewax lotion on clean cottonwool; blot with a tissue.

3 Place a protective tissue under the edge of the bra cup on the side away from you. Ask the client to bring her opposite arm down and over, and to pull the breast away from the underarm being waxed and across towards the middle of her chest. This pulls the tissues tight, making the treatment a lot more comfortable for her; it also leaves you with both hands free, one to pull and one to support.

4 Underarm hair usually grows in two main directions. Observe the directions of hair growth, then apply and remove the wax separately for each small area.

TIP

If certain types of water-soluble waxes are being used, perspiration may cause problems: it may prevent the wax from gripping the hair. Bear this in mind when selecting the wax to use on areas liable to perspiration such as the underarm and the top lip.

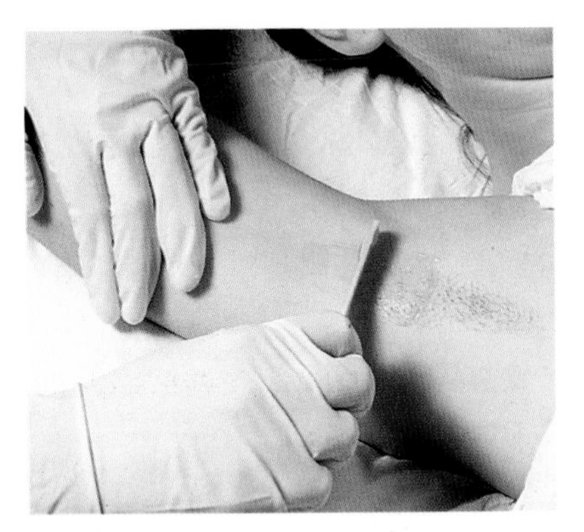

Applying warm wax to the underarm (showing support)

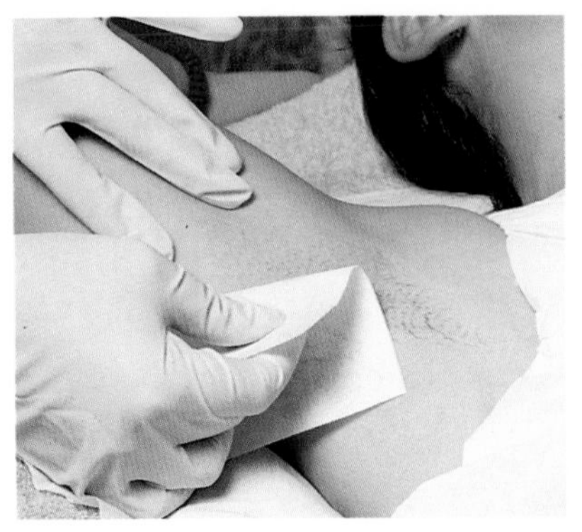

Removing the wax strip

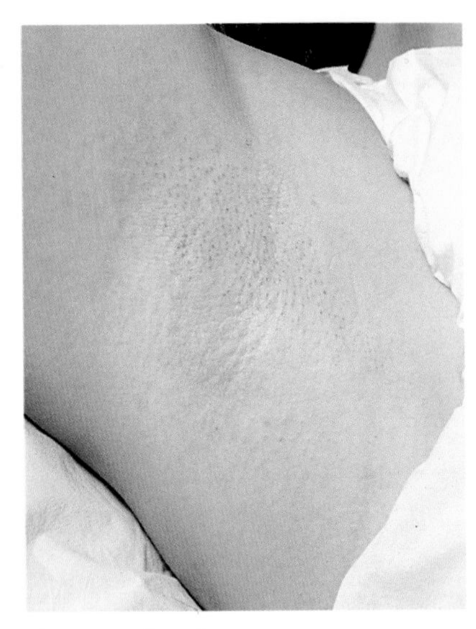

The completed area

5 Apply after-wax lotion to the treated area. Check for stray hairs; remove these with sterile tweezers.

6 Blot any excess cream with a clean tissue.

TIP

Both bikini line and underarm waxing can be uncomfortable, especially if the hair growth is thick – always bear this in mind when carrying out the treatment. Some slight bleeding can be expected as the hairs in this area are very strong and have deep roots. Any waste contaminated by blood must be disposed of hygienically in a sealed bag.

Arm treatment

Depilation in this area should take 10–15 minutes for a **half arm treatment** and 20–30 minutes for a **full arm treatment**. Half the length of the strip should be used.

Arms are usually waxed with the client in the sitting position, her general clothing being protected with a towel. Sleeve edges can be protected with tucked-in tissues; ideally, though, upper outer clothing should be removed.

The roundness of the arm means that in order to effect a horizontal pull the work must be done in short lengths. Other than this, follow the general rules for waxing.

Face treatment

The **face treatment** must always be approached with extra care as facial skin is more sensitive than skin elsewhere on the body. Faulty technique can result in the top layer of skin being removed. (If this happens, a scab will form after about a day and the mark will take days to heal and fade completely.)

A **top-lip wax** should take approximately 5 minutes, a **chin and throat wax** approximately 10 minutes, and a **lip and chin wax** approximately 15 minutes. A removal strip of no more than one-eighth normal size should be used on the face.

TIP

Previously bleached hair tends when waxed to break off at skin level. Clients should be told not to bleach facial hair if it is to be waxed.

HEALTH & SAFETY: SURGICAL SPIRIT

Surgical spirit is too harsh for use on the face.

1. The client should be in a semi-reclining position, with her head supported and a clean towel draped across her shoulders to protect her clothing. A clean headband can be used to keep the hair away from the face.
2. Cleanse and wipe over the area using an antiseptic cleansing lotion. Blot it dry with a tissue.
3. Spread and remove wax for small areas at a time, paying close attention to the direction of natural hair growth. (For example, you might need to spread one-half of the top lip with wax; remove this in three or four narrow strips; repeat the process on the other half; and finally treat the central section immediately under the nose.)

Do not allow wax to build up on the facial wax strips – such a build-up could lead to skin removal. Use a new strip for each area.

HEALTH & SAFETY: THE LIPS

The lips are extremely sensitive. To avoid possible irritation, do not allow the wax to come into contact with them.

Eyebrow treatment

Eyebrows, as a part of the face, are treated accordingly (see above). An **eyebrow wax** should take approximately 5–10 minutes.

1. Study the eyebrows and decide upon their final shape and proportions.
2. Brush the eyebrows and separate the unwanted hair from the line of the other hairs.
3. Using a small spatula, apply a thin film of wax to the unwanted hairs in a small area. Remove the wax using a clean strip.

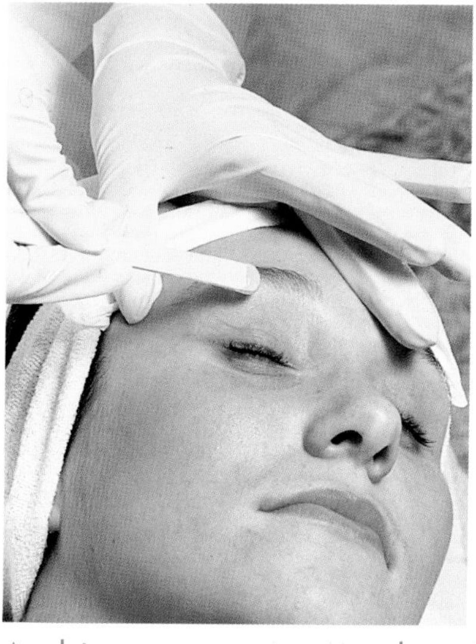

Applying warm wax to an eyebrow

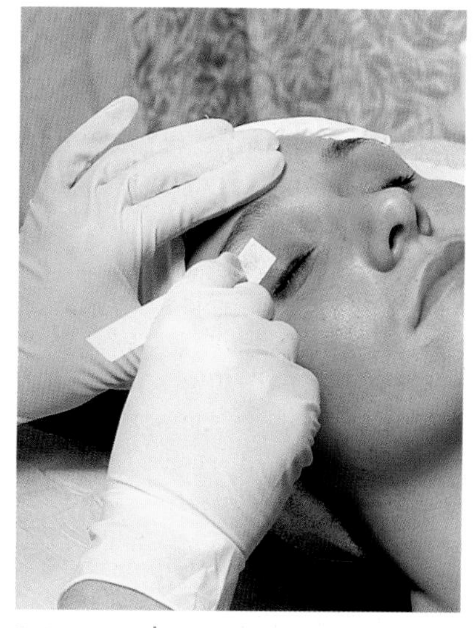

Removing the wax strip

Repeat in different areas, using a clean strip each time, until all the unwanted hairs have been removed.

4 Apply a soothing antiseptic cream and use sterile tweezers to remove any stray hairs.

TIP

Other areas of the neck and face can be treated by wax depilation – for example to tidy up a haircut at the neckline, or to remove sideboards – provided that you follow the general guidelines for facial treatments.

HEALTH & SAFETY: SENSITISATION

Under no circumstances should wax be applied to facial hair, underarm hair or bikini-line hair more than twice during any one treatment. If after this any hairs remain, they must be removed with tweezers. The delicate skin in these areas readily becomes sensitised.

Regrowth

Because of the cyclical nature of hair growth and the fact that follicles will be at different stages of their growth cycle when the hair is removed, the hair will not all grow back at the same time. Waxing can therefore appear to reduce the *quantity* of hair growth. This is not so, however, and the hair will all grow back eventually: waxing is therefore classed as a temporary form of hair removal.

Nevertheless, certain bodily changes (such as ageing), when *combined* with waxing, can result in permanent hair removal. This effect is so erratic and unpredictable, though, that waxing cannot reliably be sold as a permanent method of hair removal.

CONTRA-ACTIONS

Four contra-actions are quite common with waxing:

- ingrowing hairs;
- removal of skin;
- burns – both wax and friction burns;
- erythema – increased blood flow to the skin, giving a slight redness.

Ingrowing hairs

Ingrowing hairs can arise in three ways:

- *Over-reaction to damage* An excessive reaction by the skin and the follicle to the 'damage' produced by depilation may cause extra cornified cells to be made. These may block the surface of the follicle, causing the newly growing hair to turn around and grow inwards.
- *Overtight clothing* If after the treatment the client wears

clothing that is too tight, this too can block the follicle.
- *Dry skin* Likewise dry skin can cause blockage of the follicle.

Ingrowing hairs can usually be recognised to be one or other of three types:

- *A flat thread growing along beneath the surface of the skin* Identify the tip (the pointed end); then pierce the tissues over the root end with a sterile needle. Free the tip and leave it in place so that the follicle can heal around it.
- *A coiled ingrowing hair* This looks like a small black spot or dome on the skin. If this is gently squeezed and rolled between the fingertips, using a tissue, it will release the coiled ingrowing hair and some hardened sebum. If the hair does not fall out, it should be left in place (as above).
- *An infected ingrowth* If the trapped sebum or hair starts to decay, either of the two preceding forms can become infected or begin to display an immune response. The area first becomes red (irritation); then an infected white dome-shaped pustule develops. Release the trapped tissue (as above), and cover the affected area – which may bleed, or leak tissue fluid – with a sterile non-allergenic dressing.

HEALTH & SAFETY: HYGIENE

When carrying out any of these procedures, wear rubber gloves. Both client and therapist must observe hygienic procedure, for example you should wipe over the area with an antiseptic before touching it.

Skin removal

If the upper, dead, protective cornified layer of the skin is accidentally removed during a treatment – leaving the granular layer of the living, germinative layer exposed – the skin should be treated as if it has been burnt. Cool the area immediately by applying cold-water compresses for 10 minutes. Dry it carefully; then apply a dry, non-fluffy dressing to protect the area from infection. The dressing should be worn for 3–4 days, and the area then left open to the air. (Antiseptic cream by itself should be used only when the injury is very minor.) Medical attention should be sought if necessary.

Burns

A burn should be treated as 'skin removal' (above). If blisters form, they should not be broken – they help prevent the entry of infection into the wound. Medical attention should be sought.

Erythema

Erythema is a visible redness, accompanied by an increase in warmth on the surface of the skin. It derives from increased blood flow through the capillaries near the surface of the skin, caused by the histamine reaction after waxing (page 265). Ask the client to follow the recommended aftercare advice.

AFTERCARE AND ADVICE

An **after-wax lotion** should be applied, using clean cottonwool, at the end of the treatment. This breaks down any wax residue, helps to guard against infection and irritation, and takes away any feelings of discomfort. Encourage your client to continue with the use of such a lotion at home for a few days: it will protect against dryness, discomfort and ingrowing hairs.

Advise the client against wearing any tight clothing (such as tights or hosiery) over the waxed areas for the 24 hours following a treatment. Such clothing could lead to irritation and ingrowing hairs.

If the client suffers from ingrowing hairs, she should **exfoliate** her skin every 4–7 days, starting two or three days after the treatment. Exfoliation prevents the build-up of dead skin cells on the surface of the skin; these would otherwise block the exit from the follicle and cause a growing hair to turn back on itself and grow inwards. Ingrowing hairs should be freed and, if possible, left in place so that the follicle exit will re-form around the hair's shaft.

Advise the client that for the 24 hours following her treatment she should not apply any talcum powders, deodorants, anti-perspirants, perfumes, self-tanning products or make-up over the treated areas. Any of these products could block the pores or cause irritation or allergy reactions on the temporarily sensitised area. During this time she must use only plain, unperfumed soap and water on the treated area.

For the same 24-hour period she must not apply heat or ultra-violet treatments – hot baths, for example, or the use of a sunbed or sauna – as these would add to the heat generated in the skin following the treatment and would probably cause discomfort or irritation. She must also refrain from touching or scratching the area, so as to avoid infecting the open follicles.

TIP

Aftercare leaflets should contain this information: these can be given to the client at the end of the treatment to remind her what to do at home.

ACTIVITY: AFTERCARE LEAFLETS

Discuss why aftercare leaflets should be given to clients, as well as giving them the advice verbally.

HOT WAXING

In **hot waxing**, the wax is applied at a higher temperature. The hairs embed in the wax and are gripped tightly as the wax cools and contracts. When the wax is pulled away, it removes the hair from the base of the follicle.

TIP

When selecting a wax, choose one that does not go brittle when cool. A wax sold as small bars is preferable as this melts quickly.

Equipment and materials

The equipment and materials required for hot waxing are the same as for warm waxing (page 269), except that:

- a *wax heater* suitable for hot wax should be selected;
- *wax-removal strips* are not necessary.

Some people prefer to apply the hot wax with disposable brushes rather than spatulas, but either can be used.

Carrying out the treatment: hot waxing

General technique

1. Ensure that the area to be waxed is clean and grease-free.
2. Apply a small amount of talc, against the direction of hair growth. This is to make the hairs stand on end and to stick more firmly into the wax.
3. Keep the same order of work as for warm waxing.
4. Apply wax in strips approximately 5 cm wide and 10 cm long, with about 5 cm distance between the strips.
5. Carefully observe the direction of hair growth and the size of the area to be waxed.
6. Test the heat of the wax on the inside of the wrist.

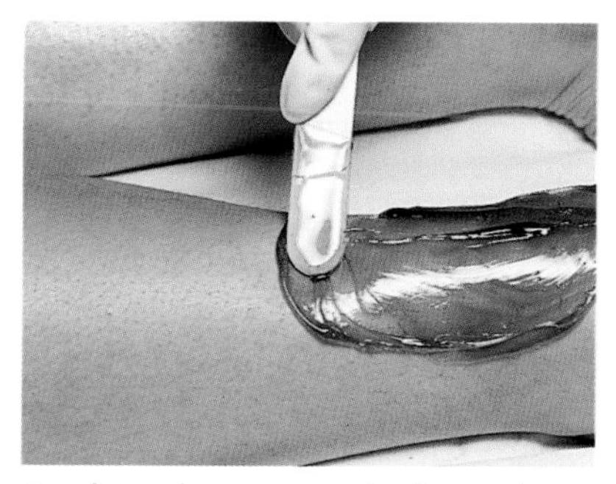

Applying hot wax to the lower leg

TIP

If the wax is *not hot enough* when it is applied, it will not contract effectively around the hair and will therefore not grip it properly. This may result in poor depilation and possible hair breakage.

If on the other hand the wax is allowed to *overheat*, it may cause burns. Also, the quality of the wax will deteriorate and the wax will become brittle as it cools.

7. Using either a disposable spatula or a brush, apply a layer of wax about 5 cm × 10 cm *against* the direction of hair growth. Apply a second layer *with* the direction of growth; and a third layer *against* the direction of growth. (If two or three strips are applied at the same time, you can work faster.)

   Keep the edges of the wax thicker than the middle, to make it easier to remove. Overlap the lower edge by about 2 cm onto a hair-free area: this makes it less painful later, when you lift the edge to make a lip to pull.

   Curl up the lower end of the wax to make a lip, and press and mould the wax firmly onto the skin.
8. Leave the wax for a minute or so, to cool. The wax has to cool sufficiently to grip the hairs, but not so much that the wax becomes brittle and breaks on removal. As it sets, it starts to lose its gloss: it should be removed when this happens and while it is still pliable.

9 Support the area below the wax, grasp the lip, and tear the wax off the skin *against* the direction of hair growth in one movement (as with warm-wax removal). Immediately press or firmly stroke the area with your hand: this takes away some of the discomfort.

10 Check the area for any remaining wax and any stray hairs. Remove using tweezers. (Second applications are not advisable when using hot wax, because of the risk of burning.)

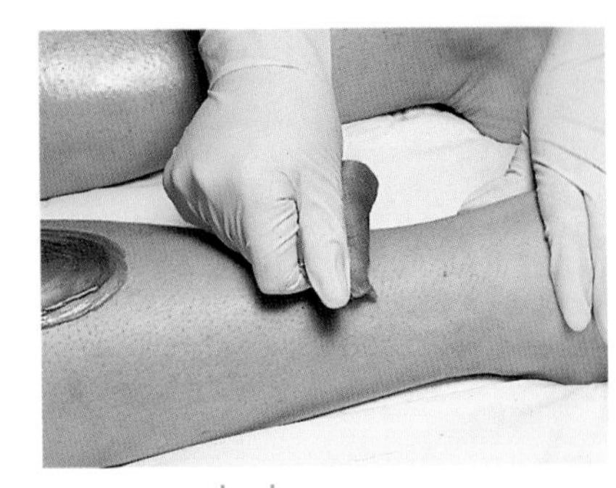

Removing the hot wax

If this general application technique is followed, any area of the body can be depilated – use the same order of work as for warm waxing: observe the direction of hair growth and take into account the body area: use smaller strips in smaller areas; support the skin; and use the correct angle of pull for removal.

TIP

If the wax becomes too cool and brittle for removal, fresh hot wax may be applied with care over the area to soften it.

Aftercare and contra-actions

The contra-actions and aftercare are the same as for warm waxing (pages 280–1 and 282).

REMOVING HAIR: SUGARING

SUGARING AND STRIP SUGAR

Sugaring is an ancient and popular method for hair removal.The superfluous hairs become embedded in a pliable organic paste substance of sugar, lemon and water. The sugar paste is then removed from the skin's surface, against the hair's growth, leaving the skin hair-free.

Sugaring may be applied as a paste or as traditional warm wax being removed with material strips – referred to as **strip sugar**. Sugaring is effective on fine hair growth, and as it contains no chemicals or additives this method is unlikely to cause skin allergy.

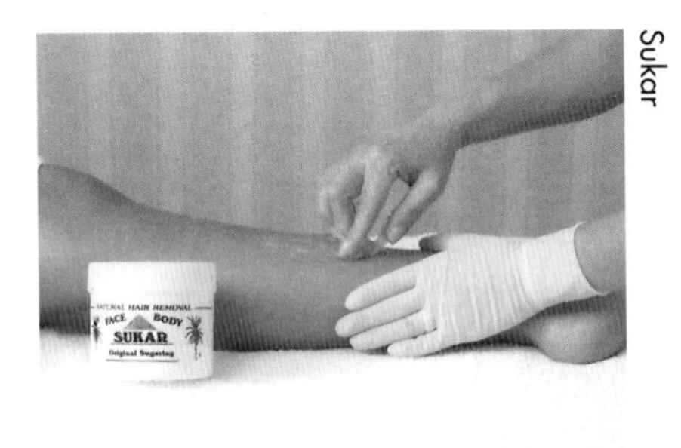

Sukar

Applying sugar paste

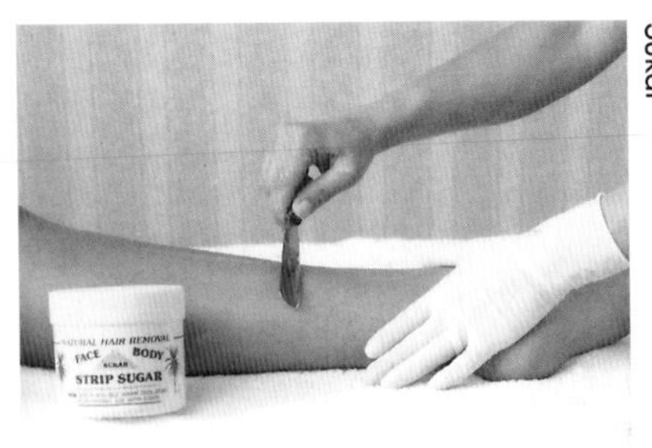

Sukar

Applying strip sugar

RECEPTION

Re-book the client for sugaring hair removal every 4–6 weeks, dependent upon the hair growth rate and area. For effective hair removal, the hairs must be at least 2 millimetres long.

EQUIPMENT AND MATERIALS

The equipment and materials required for sugaring are the same as for warm waxing (pages 268–70), except for the following differences.

Sugar paste

- *Wax heater (with a thermostatic control)* or *microwave* – to heat the paste.

- *Sugaring paste* – either soft or hard, depending on your personal preference and the temperature of the working environment: hard paste is a better choice in warmer temperatures and when working on coarser hair.
- *Talc (purified).*
- *Bowl of clean water* – to remove the sugar paste from the hands, reducing stickiness.

> **TIP**
>
> Talc aids the sugaring treatment technique, by reducing stickiness when working with the paste, and by absorbing perspiration (particularly in areas such as the underarms).

Strip sugar

- *Wax heater (with a thermostatic control)* – to heat the strip sugar.
- *Strip sugar.*
- *Talc (purified).*
- *Disposable wooden spatulas* – a selection of differing sizes, for use on different body areas.
- *Wax-removal strips (bonded-fibre)* – thick enough that the wax does not soak through, but flexible enough for easy working.
- *Disposable surgical gloves.*

Sterilisation and sanitisation

As sugar paste is water-soluble, it is easily cleaned from any surface. However, the wax heater containing the sugar wax must be regularly sanitised.

PREPARING THE CLIENT

1. Cleanse the area to be treated, using an antiseptic cleansing tissue.
2. Blot the skin dry.
3. Apply talc to cover the area: this prevents the sugar from sticking to the skin.

Sukar

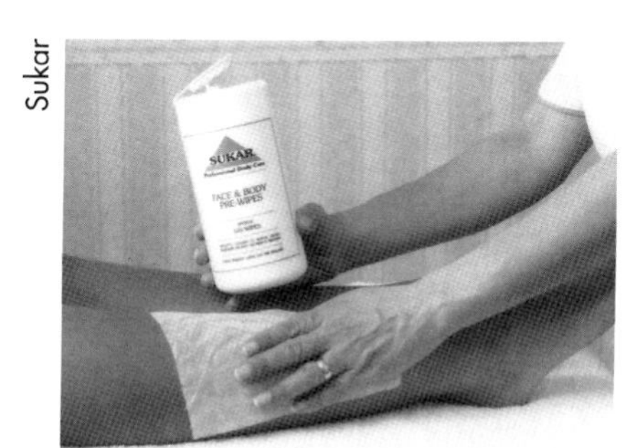

Cleansing with an antiseptic pre-wipe

Sukar

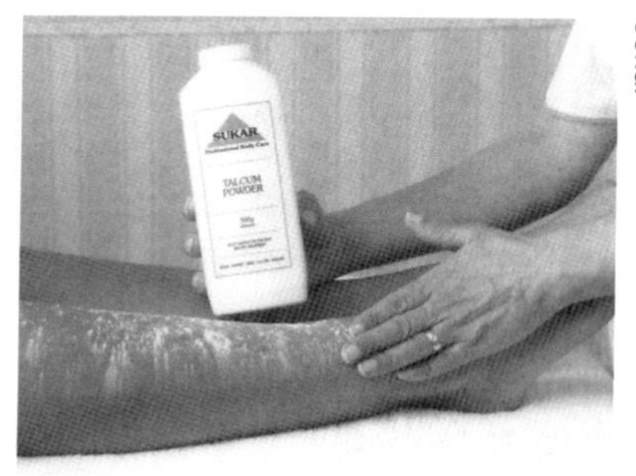

Applying talc

Sukar

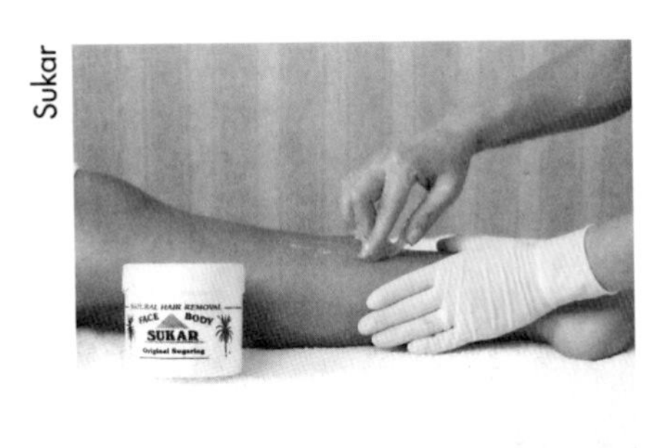

Applying sugar paste

Sukar

Removing sugar paste

Sukar

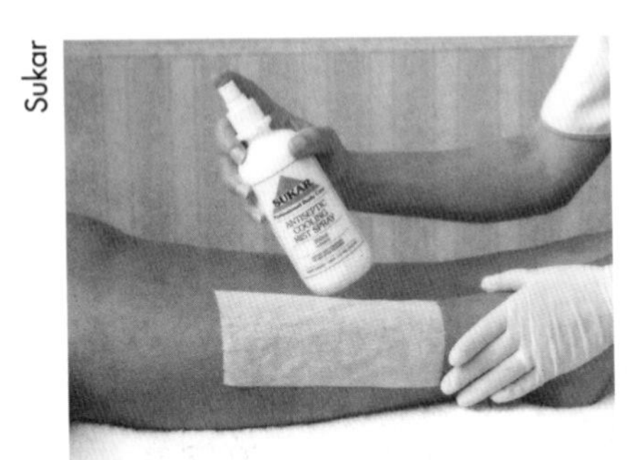

Applying a cooling spray

Sukar

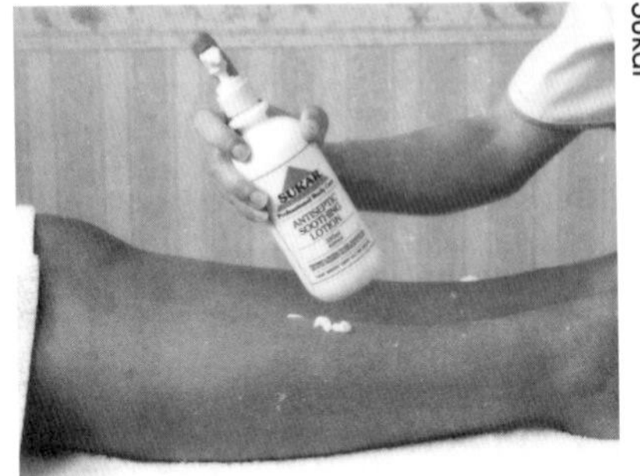

Applying a soothing cream

SUGARING

Sugar paste

The sugar paste adheres to the hair and not to the skin, which allows the sugar to be reapplied to a treatment area. Technique is important and it takes practice and experience to become skilled.

1. Heat the sugar gently, to soften it.
2. Apply the sugar paste to the skin by hand. Select the amount used according to the treatment area. Draw the paste over the treatment area *in the direction of* the hair growth, embedding the hair in the wax.
3. Remove it quickly *against* the hair growth.
4. If necessary, reapply the paste to the area to remove further hairs. Continue this process until no hair remains.
5. After use, discard the paste as it will be contaminated with excess hair and dead skin cells: this affects the ease of paste removal, and presents a risk of cross-infection.
6. To complete the treatment, a cooling treatment spray may be applied to the area, followed by a soothing cream.
7. Record details of the treatment on the client's record card.

TIP

Sugar wax – both strip and paste – may be heated in a microwave to soften it. Check guidelines set by the microwave manufacturer on power and timing.

Strip sugar

This is similar in application and removal to warm wax.

Sukar

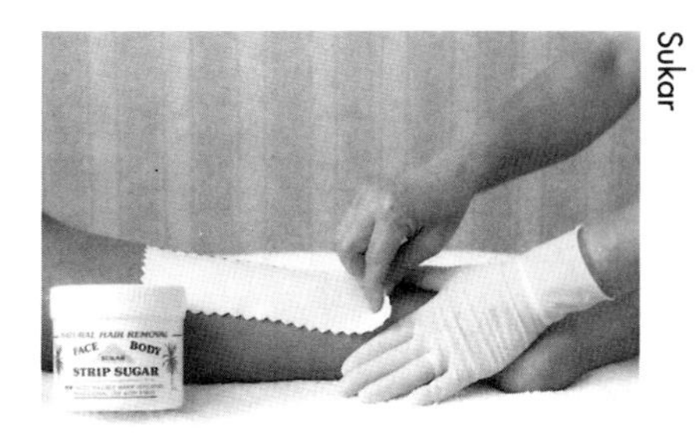

Sukar

Applying strip sugar

Removing strip sugar

HEALTH AND SAFETY: TEMPERATURE
Ensure that the temperature is correct – neither too hot, which may cause burning, nor too cool, which may make working uncomfortable and inefficient.

1. The wax is gently heated until it is fluid.
2. Apply the wax using a spatula *in the direction of* the hair growth to cover the treatment area.
3. Remove the wax *against* the hair growth using a clean strip.
4. To complete the treatment, a cooling treatment spray may be applied to the area, followed by a soothing cream.
5. Record details of the treatment on the client's record card.

CONTRA-ACTIONS

General waxing contra-actions should be considered (see pages 280–1), also:

- *Ingrowing hairs* – caused by incorrect removal technique when using sugaring, or a build-up of dead skin cells over the hair follicle opening.
- *Discomfort* – caused by the wax being too hard, affecting application and efficient removal.
- *Burning*, of the client or the therapist – caused by the sugar paste or the strip sugar being too hot.

AFTERCARE AND ADVICE

General post-waxing advice should be offered (see page 282). Appropriate aftercare retail products may be recommended such as an exfoliating body mitt, or antiseptic moisturising lotion to promote skin healing.

Ellisons

REMOVING HAIR: DISPOSABLE APPLICATORS

DISPOSABLE APPLICATOR WAXING SYSTEMS

This is a hygienic method, using **disposable applicators** which are new for each client. A disposable applicator head screws onto the wax applicator tube in place of a cap. This reduces the possible risk of contamination through cross-infection. The method is considered less messy, as the wax is contained in the tube and is not exposed until application.

Ellisons

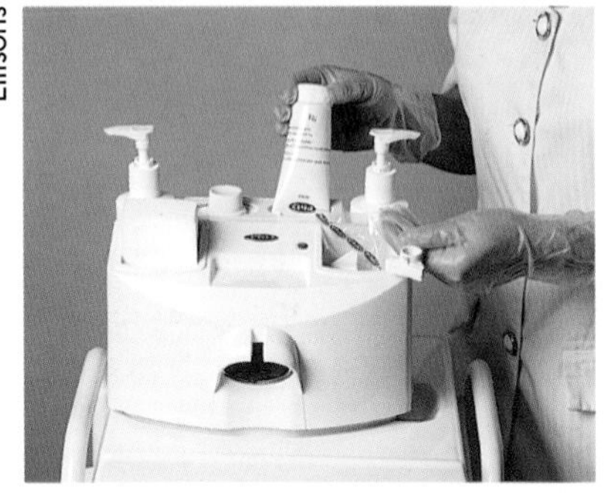

Removing the applicator

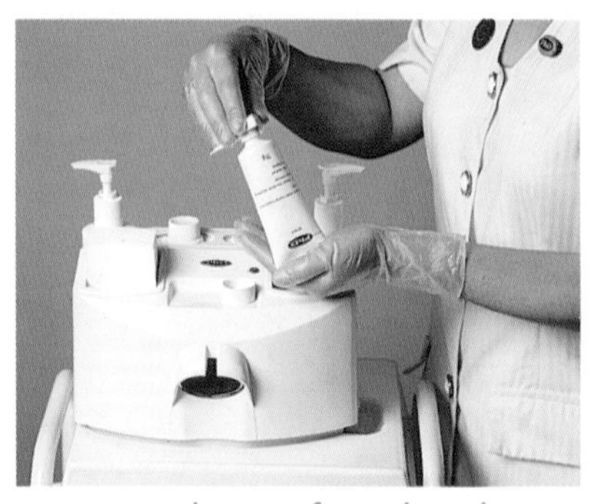

Removing the cap from the tube

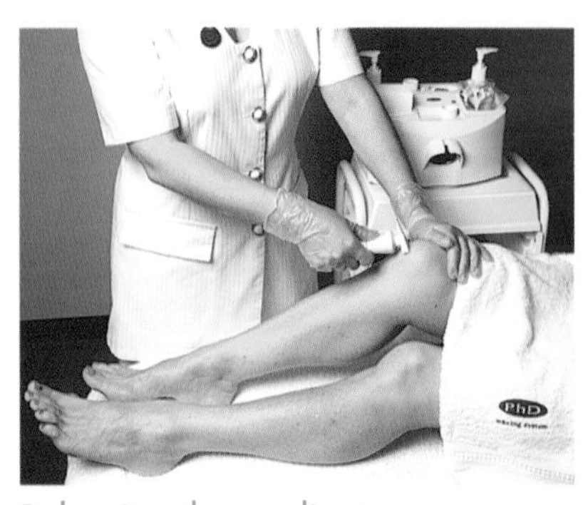

Releasing the applicator

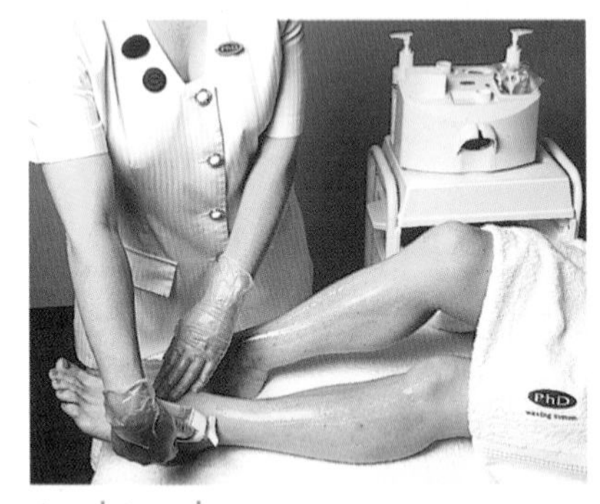

Applying the wax

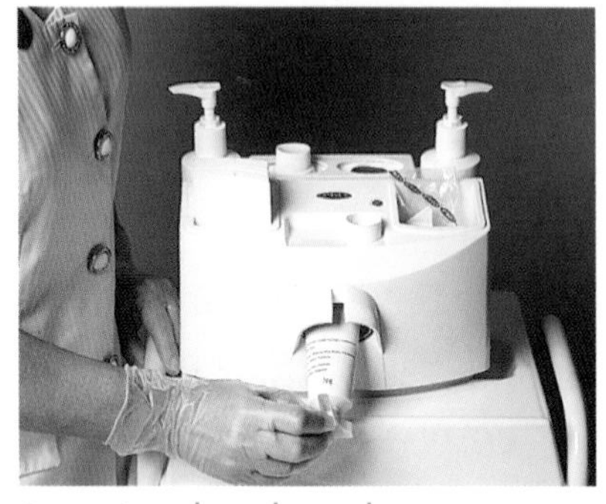

Returning the tube to heat

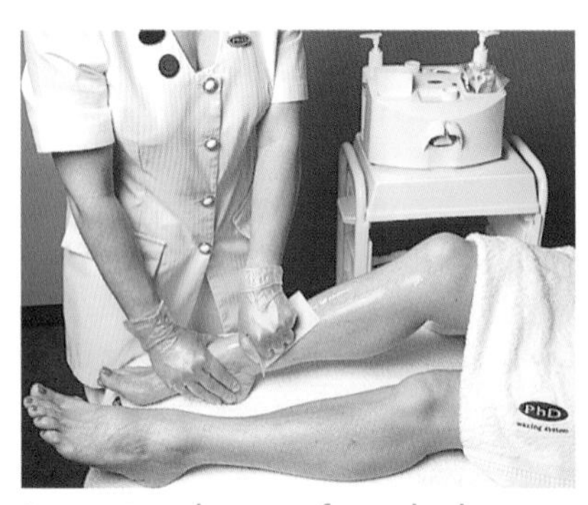

Removing the wax from the leg

Ellisons

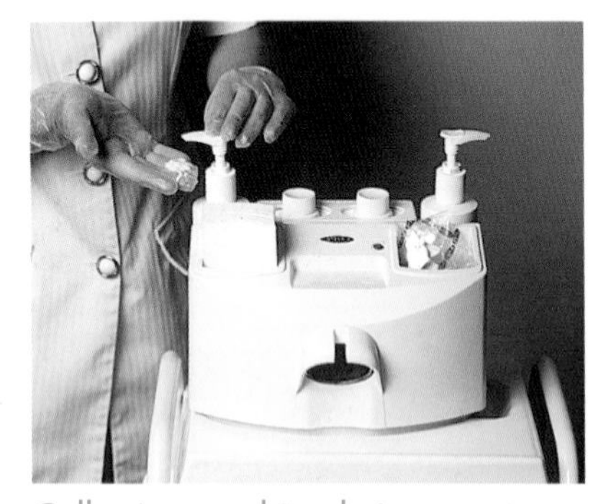

Collecting soothing lotion

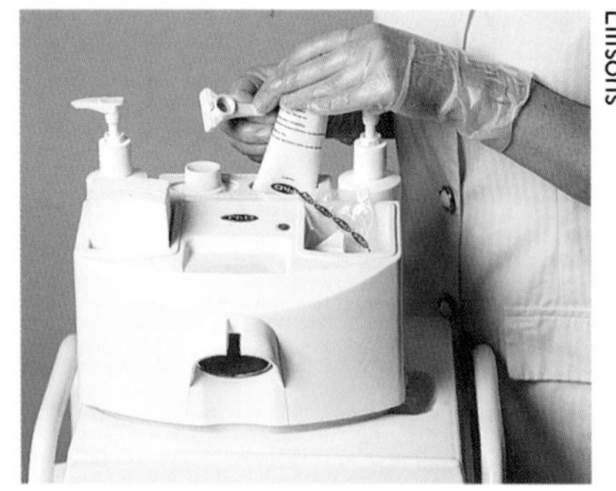

Discarding the applicator

Each tube of wax as needed is heated to working temperature, which minimises the risk of burning.

WAXING THE LEGS

1. Follow general waxing preparation guidelines for the treatment area (see pages 271–2).
2. Remove the disposable applicator from the right-hand heating and storage compartment.
3. Remove the cap from the tube of wax. Attach the applicator head to the tube.
4. Release the applicator by lifting the lever upwards. Squeeze until a small amount of wax appears on the front of the applicator, then apply the wax. Hold the applicator at a 45° angle to the leg, and glide it smoothly down the leg.

5. Apply a thin film of wax to the front of both legs, then press down the closing device on the applicator to stop wax flow.
6. Wipe any wax residue from the front of the applicator and return the tube to heat.
7. Remove wax from the leg, starting at the ankle and working towards the knee. Support the skin with one hand, and firmly stretch it against the removal of the wax. Continue application and removal to the sides and back of the legs.
8. When treatment is complete, apply antiseptic soothing lotion.
9. Record details of the treatment on the client's record card.
10. Remove the applicator from the tube, replacing the cap and returning the tube to the heater. Dispose of the applicator head.

AFTERCARE AND ADVICE

Follow general aftercare and advice guidelines.

LIGHTENING HAIR

Bleaching lightens the hair and is an effective way to disguise pigmented facial and body hair that the client finds unacceptable. The effect lasts up to 4 weeks.

The bleaching products, usually powder and cream, are mixed together and are immediately applied directly to the area to be lightened.

RECEPTION

The client should receive the bleaching treatment every 3–4 weeks. Allow 15–30 minutes when booking this service, according to the client's treatment requirements.

Treatment should not follow any preheating treatment such as facial steaming, as the pores would be open which could cause skin irritation.

Skin tests

HEALTH AND SAFETY: HYDROGEN PEROXIDE

Hydrogen peroxide is a strong skin irritant. Perform a skin test if the client has a possible skin sensitivity. Remember that clients can become allergic or sensitive to ingredients that have previously been satisfactory.

1. Mix a small amount of the product, as directed by the manufacturer.
2. Apply a little of this to the inner arm, approximately 25mm

square. If there is an allergic reaction, remove the bleach product immediately.

3. After 10–15 minutes, remove the product using cool water applied with cottonwool.
4. Check the area over the next 24 hours to assess skin tolerance. Redness, itchiness and swelling indicate a positive (allergic) reaction.
5. Note the test result on the client's record card.

TIP

Ensure the bleach does not come into contact with the client's clothes – these too could be bleached!

CONTRA-INDICATIONS

After completing the record card and inspecting the area, if you have found any of the following do not proceed with the bleaching treatment:

- *Hypersensitive skin.*
- *Dry, sensitive skin.*
- *Cuts and abrasions* in the area.
- *Skin disorder or disease.*
- *Positive reaction (allergy)* to a skin test.
- *Skin erythema.*

EQUIPMENT AND MATERIALS

HEALTH AND SAFETY: BLEACHING MATERIALS

Store the bleach materials in a cool environment away from direct sunlight and other heat sources.

Avoid contact with the eyes. In the case of contact, rinse the eyes thoroughly with purified (distilled) water for 15 minutes.

Wear protective gloves when dealing with any spillages.

- *Couch* – with sit-up and lie-down positions and an easy-to-clean surface.
- *Trolley* – to hold all the necessary equipment and materials.
- *Bleach powder* and *cream hydrogen peroxide* – in facial and body cosmetic formulations.
- *Cleansing cosmetic preparations* – to cleanse the area to be lightened.
- *Barrier cream* – to protect the surrounding areas.
- *Damp cottonwool* – to remove cleansing preparations and excess bleach from the skin and hair.
- *Bowl (clean)* – to hold the clean cottonwool.
- *Facial tissues (white)* – for blotting skin dry.
- *Headband* – to protect the scalp hair when performing facial bleaching.
- *Towels (medium-sized)* – to protect the client's clothing.
- *Non–metallic mixing dish* – metal would speed the decomposition of the product.
- *Disposable spatula* for mixing and application.

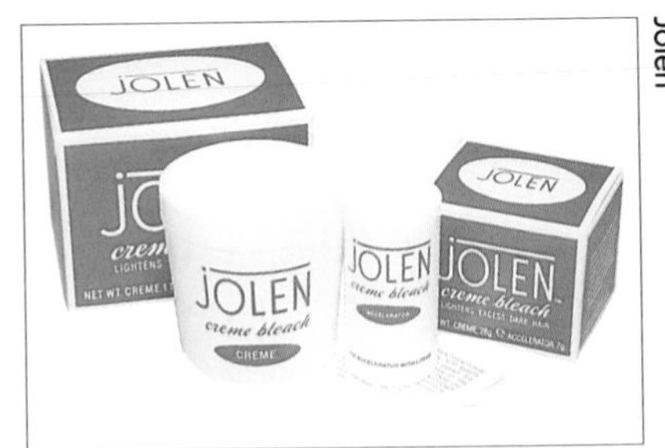

Jolen

Bleaching products

Equipment and materials

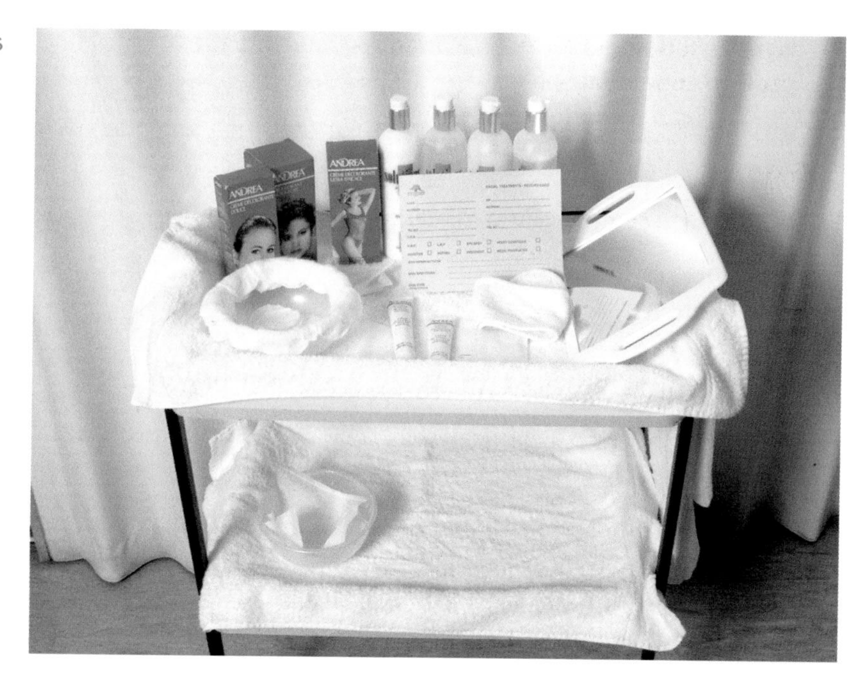

Ellisons

Soothing lotion

- *Mirror (clear)* – for facial bleaching to show the client the effect produced for acceptability.
- *Soothing lotion* – to apply to the skin following bleaching.
- *Disposable surgical gloves* (if desired).
- *Swing-top bin* – lined with a disposable bin liner, for waste.
- *Client's record card.*

Sterilisation and sanitisation

Use disposable equipment where possible to avoid cross-infection. Wear protective gloves during application and removal of the bleach, if you wish, to protect your hands from possible chemical contact.

Preparing the cubicle

Before the client is shown through to the cubicle, it should be checked to ensure that the required equipment and materials are available and the area is clean and tidy.

The couch or chair should be protected with a long strip of disposable tissue-paper bedroll, or a freshly laundered bedsheet and a bath towel. A small towel should be placed neatly at the head of the couch, ready to be draped across the client for protection. (The paper tissue will need replacing and the towels laundered for each client).

PREPARING THE CLIENT

Facial hair lightening

1. Protect the client's clothing with clean towels. A headband should be worn to protect the client's hair. Drape a small towel across her chest.

2. Cleanse and tone the area. Blot the skin dry with a clean facial tisssue. Ensure that the area is free of grease, which would form a barrier to the hair-lightening products.
3. For facial bleaching, the client should be slightly elevated.

TIP
The lightening process will vary according to the coarseness and colour of the client's hair. You should always consider this.

Body hair lightening

1. Protect the client's clothing in the area with clean towels or disposable paper tissue.
2. Cleanse the area thoroughly, with a cleansing agent and clean cottonwool, to remove dead skin cells and any body oils or cosmetic lotions. Blot the skin dry with a clean facial tissue.
3. Position the client comfortably, in a position in which you are able to treat the area easily.

TIP
Advise the client that the part of the *skin* treated will be lightened slightly also. As this affects only the surface dead skin cells of the epidermis, the effect is temporary – usually 12–24 hours.

LIGHTENING THE HAIR

Facial hair

1. Place the required amount of powder and cream in a non-metallic dish, following the manufacturer's instructions. Mix well to a smooth opaque cream.

HEALTH AND SAFETY: BLEACHING PRODUCTS
Always mix the ingredients according to manufacturer's instructions.
A barrier cream may be applied to protect the surrounding area from the bleaching products.

2. Using a clean, sanitised spatula, apply the bleach to the area. Cover the hair evenly.
3. Time the application accurately, taking into account:
   - hair coarseness and pigmentation;
   - skin type and tolerance.
4. If necessary – because the hair is very dark – apply the bleach a second time.
5. When the hair is sufficiently bleached, remove the bleaching products thoroughly, using a spatula and dampened cottonwool.

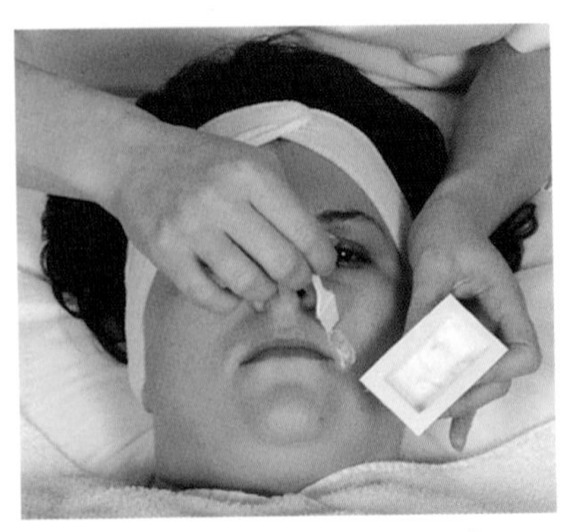
Applying bleach to the upper lip

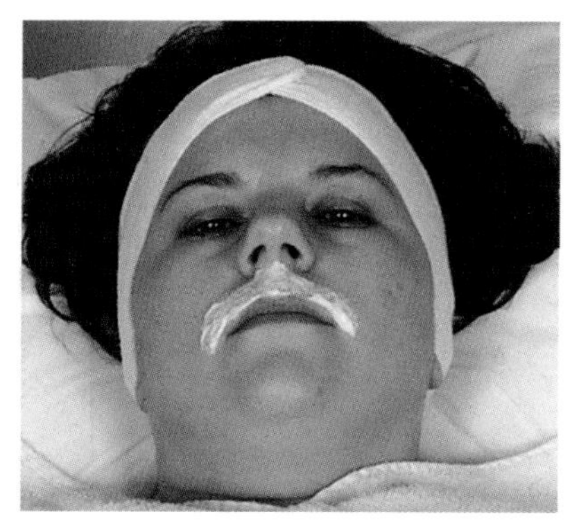
Processing

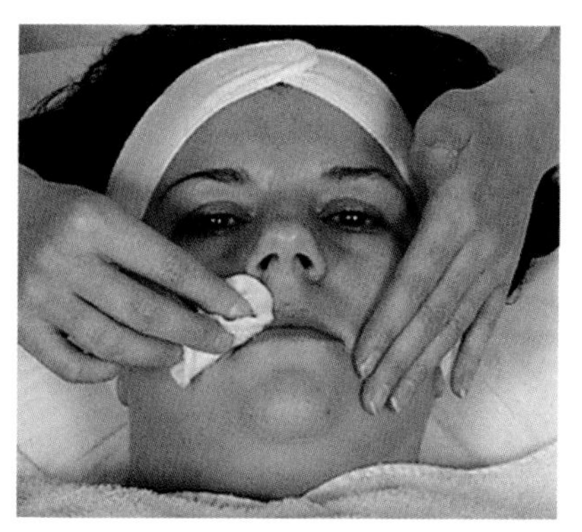
Removing the bleach

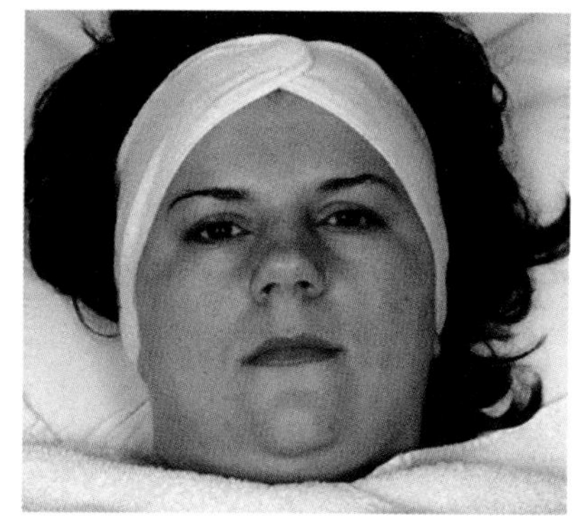
The finished result

6 Dry the area with a facial tissue and apply an unperfumed moisturising agent.
7 Show the client the finished result in the mirror.
8 Discard the bleach and clean the mixing container.
9 Record the treatment details on the client's record card.

HEALTH AND SAFETY: BLEACH
Never spread the bleach onto other surrounding areas unnecessarily.

Eyebrows

1 The brows lighten very quickly. Lighten the brows to complement the colour of the client's scalp hair.
2 Show the client the finished result in the mirror.
3 Discard the bleach and clean the mixing container.
4 Record the treatment details on the client's record card.

HEALTH AND SAFETY: EYE CARE
Take great care to avoid the bleach mixture entering the eyes or sensitising the delicate eye tissue.

Body hair: arms and legs

1 Mix sufficient bleach for one arm or one leg.
2 Allow the bleach to process, for approximately 15 minutes. If the hair is not sufficiently lightened, mix and apply the product for approximately a further 10 minutes.
3 Remove the mixture.
4 Mix new bleach and apply this to the other limb. Process and remove.
5 Discard the bleach and clean the mixing container.
6 Record the treatment details on the client's record card.

CONTRA-ACTIONS

- *Irritation* – caused by skin intolerance. If irritation occurs, wash off the bleaching product immediately with clean water and apply a suitable soothing product. If irritation continues, obtain medical attention.
- *Erythema or skin burn* – caused by over-exposure to the bleaching agents, or using bleaching agents at an incorrect strength for the body part.

HEALTH AND SAFETY: ULTRA-VIOLET LIGHT
On areas that have previously been bleached, ultra-violet light may cause a reaction and hyperpigmentation in the skin.

AFTERCARE AND ADVICE

A moisturising, soothing agent should be applied following treatment. For the next 24 hours the client should avoid possible sensitisers such as perfumed products and ultra-violet light.

KNOWLEDGE REVIEW

Physiology

1. Referring to the cross-section of the hair follicle, briefly describe the function of each numbered area.
2. What is the soft downy hair found on the face called?
3. What is coarse pigmented hair called?
4. What are the different stages of the hair growth cycle called?
5. What relevance has the hair growth cycle for a wax depilation treatment?
6. When wax depilation is carried out, which layer of the skin would you expect to be removed with the wax?
7. Why does spot bleeding occasionally occur with wax depilation?

Wax depilation treatment

Preparation

1. When observing the area for wax depilation, what conditions would contra-indicate treatment?
2. How should the area for wax depilation be prepared to ensure effective hair removal?
3. To reassure the client, how would you explain the expected post-treatment skin reaction?
4. How would you position the client for wax depilation in the following areas?
   (a) Bikini wax.
   (b) Underarm wax.
   (c) Forearm wax.

Application

1. How long should hair growth be before wax depilation can be carried out?
2. How do hot wax and warm wax differ in relation to:
   (a) application?
   (b) removal?
3. How do sugar paste and strip sugar differ in relation to:
   (a) application?
   (b) removal?
4. How can you prevent skin bruising following wax removal?

Aftercare

1. What aftercare instructions should be given to the client following wax depilation treatment?
2. When would you book a client to return for the following repeat wax depilation treatments, and how much time would you allow for each area?
   (a) Eyebrow wax.
   (b) Full leg wax.
   (c) Bikini wax.
   (d) Underarm wax.
   (e) Forearm wax.
3. If a client suffers from ingrowing hairs, what advice would you give her which could possibly prevent them recurring?

Health, safety and hygiene

1. What hygiene legislation should be followed in relation to wax depilation?
2. What is the recommended temperature of wax when using:
   (a) hot wax?
   (b) warm wax?
3. What additional precautions should you take to ensure that the wax used is at a comfortable temperature?
4. What hygiene precautions should be taken when completing wax depilation? Consider:
   (a) the therapist's safety;
   (b) care of the client.
5. What would be undesirable post-treatment skin reactions?

Case studies

1. How would you explain the procedure for a bikini wax treatment to a nervous client who has not received this treatment before?
2. One week after receiving wax depilation, a client complains of the appearance of stubbly hairs. What could be their cause? How can this be prevented in future?
3. Explain the treatment procedure for wax depilation of the upper lip and chin. Consider preparation of the area, the choice of wax, wax application, removal, and aftercare.

Hair lightening

Preparation

1. How should the area be prepared for hair lightening?
2. When should the lightening agents be prepared for application, and why?
3. What details should be recorded on the client's record card?

Application

1. How should the product be applied?
2. How does treatment time vary for different hair types and treatment areas?
3. If the hair had not lightened sufficiently on removal of the product, what action would you take?

Aftercare

1. What should be applied to the treatment area following treatment?
2. How often should a hair lightening treatment be recommended?
3. What aftercare advice shoud be given to the client following treatment?

Health, safety and hygiene

1. State three contra-indications to this treatment, explaining why treatment cannot be received.
2. What protective clothing or equipment may be worn by the therapist, and why?
3. Why should a client receive a skin test before hair lightening? What would be a positive skin reaction to the skin test?

Case study

1. How does development treatment time vary for the following:
   (a) dark-haired client, upper lip?
   (b) red-haired client, upper lip?
2. To which clients could you promote this service?

CHAPTER 10

Ear piercing

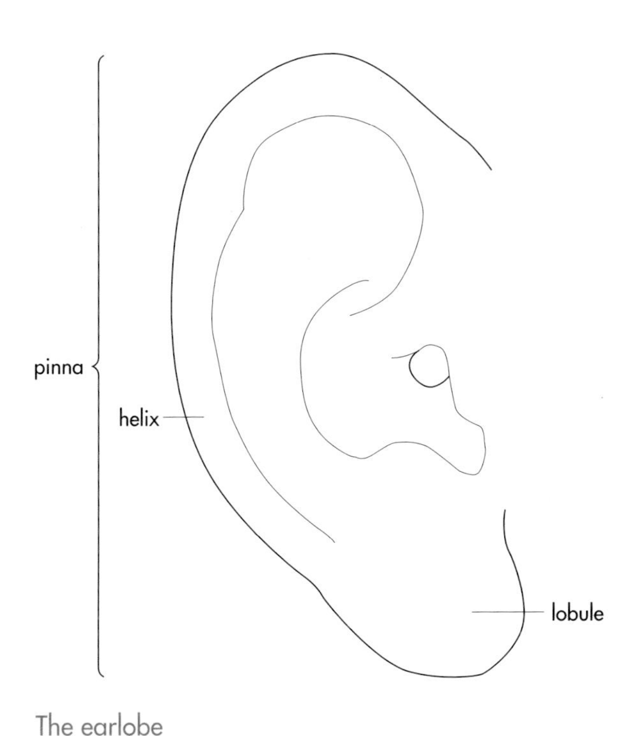

The earlobe

THE EARS

Ear piercing is a quick, profitable and popular salon service, which also has the potential to generate further custom. This may be the client's first visit to a beauty salon, and the service she receives may encourage her to return for further treatments.

The structure and function of the earlobe

The external ear collects sound waves and directs these to the inner ear. The part of the ear that is commonly seen pierced is called the **pinna**, which comprises the **helix** and **lobule**. The helix is composed of cartilage, which does not heal quickly and can form lumpy scar tissue; it is therefore considered unsuitable for piercing. The lobule, in contrast, consists of fibrous and fatty tissue with no cartilage, and is therefore suitable for piercing.

RECEPTION

When making an appointment for this service, allow 15 minutes. Although ear piercing is completed quickly, time must be allowed to complete the client's record card and to give clear concise aftercare instructions.

A record card should be prepared for the client, recording just the information that is relevant to the ear-piercing treatment. Whilst completing the record card you will be able to ascertain whether the client is suited to this treatment – if the client is under 18 years of age, for example, it is necessary for a parent or guardian to sign a **consent form**.

Contra-indications to ear piercing should be checked for. If the area for piercing is unsuitable, politely explain to the client why this is so.

Explain the simple treatment procedure. Most clients will be interested to know what to expect – usually the client's first question is 'Will it hurt?'

Prepare the equipment required for ear piercing, and show the client the range of studs available.

Some salon receptionists are trained to carry out this service. This is practical: many clients will be acting on impulse in deciding to have their ears pierced, and will not have made an appointment.

CONTRA-INDICATIONS

In certain circumstances you should not carry out the ear-piercing treatment. Use your professional judgement to assess the suitability of each client.

- *If a client is particularly nervous* it would be inadvisable to pierce her ears – she might faint or jump whilst the treatment was being carried out, causing incorrect placement of the earring.

If any of the following are present, do not proceed with ear piercing:

- *Diabetes* – the skin is very slow to heal, so infection of the area would be a strong possibility.
- *Epilepsy* – the stress before the treatment and the possible shock incurred during it might induce a fit.
- *Skin disease or disorder.*
- *Hepatitis B or HIV* – though the client may not know that she is a carrier, or may choose not to disclose the fact if she does (hygienic practice is vital, for this reason).
- *Inflammation of the ear.*
- *Open wounds, cuts or abrasions* in the area.
- *Moles or warts* in the area.
- *Circulatory disorders*, such as high or low blood pressure.
- *A predisposition to keloid scarring* – lumpy scar tissue (black skin often forms keloid scar tissue following skin healing).
- *Following an operation* – piercing must not be carried out at the site of a recent operation.

Where there is any contra-indication to ear piercing, the client must seek written permission from her physician before treatment can be carried out.

Children may receive this service only if approval has been given in writing by the parent or guardian, and a disclaimer form signed.

ACTIVITY: SAFETY AND HYGIENE

Research the ear-piercing systems available. Which ones are designed to protect the client's ear, and the gun, from contamination?

Display the literature you have collected in your log book, explaining the various systems' safety features.

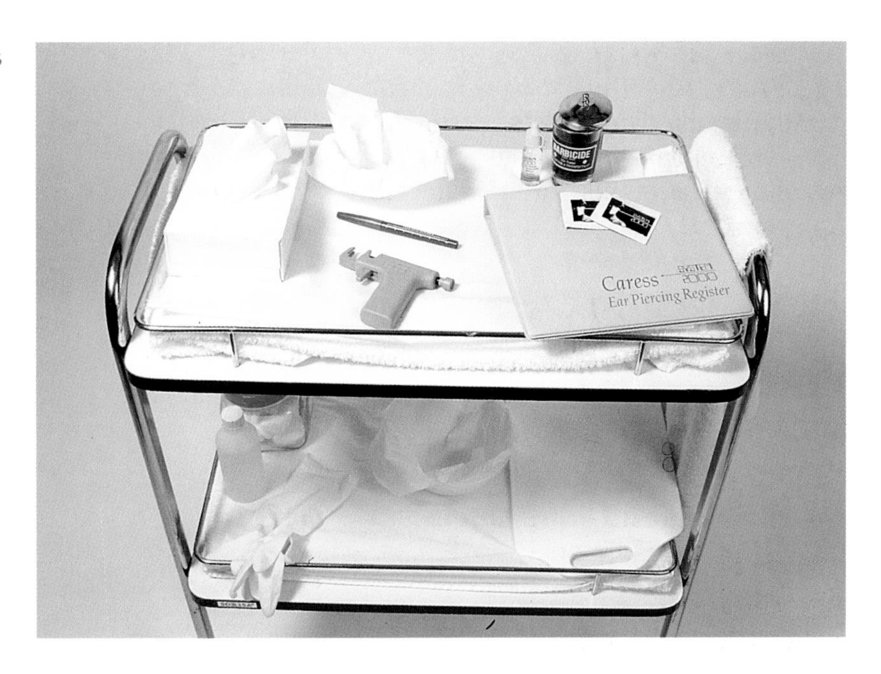

Equipment and materials

EQUIPMENT AND MATERIALS

You will need the following equipment and materials:

- *Ear-piercing gun* – one that complies with health and safety legislation.
- *Studs* – a variety of styles and designs to accommodate differing client preferences.
- *Surgical skin marker pen.*
- *Pre-packed alcohol tissues (2)* or *manufacturer's cleansing solution* – used to cleanse the ear area.
- *Surgical spirit* – for cleaning the gun after use.
- *Clean cottonwool* – to apply the surgical spirit.
- *Disposable rubber gloves* – to ensure a high standard of hygiene and to reduce the possibility of contamination.
- *Waste bowl.*

HEALTH & SAFETY: MARKER PENS

Keep the marker pen in a safe hygienic place – if left out, somebody might mistake it for an ordinary felt-tip pen!

An ear-piercing system

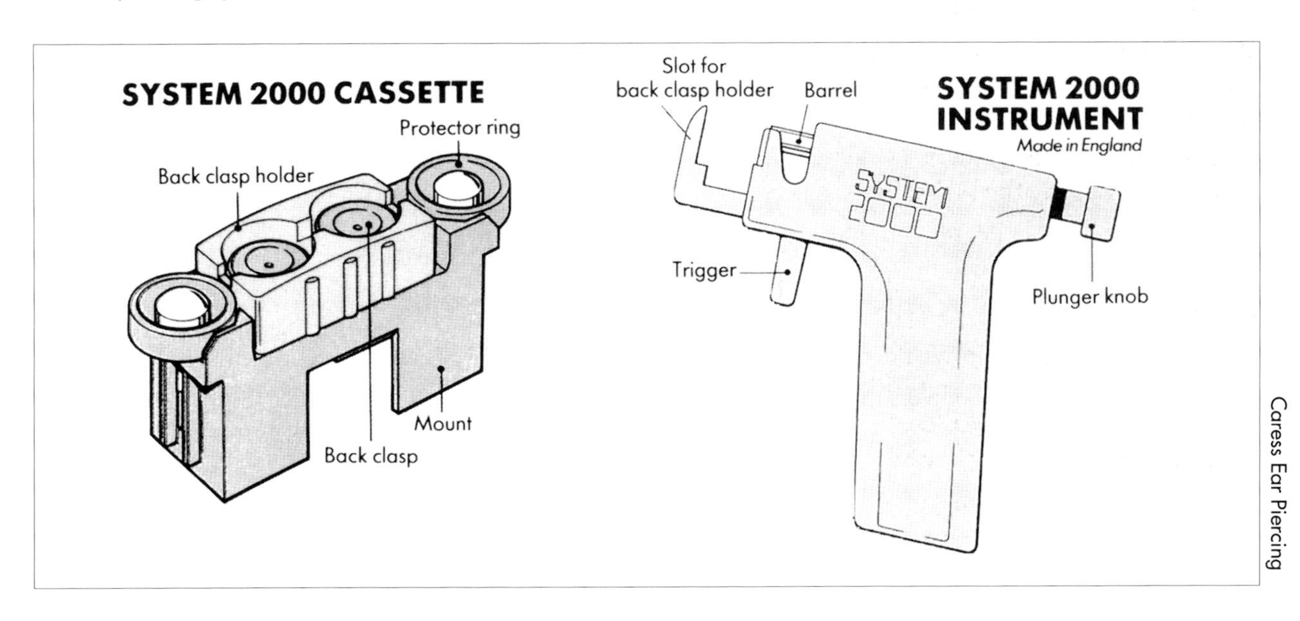

Caress Ear Piercing

- *Pedal bin (with liner).*
- *Headband (clean)* or *clip* – to hold the hair away from the ear during ear piercing.
- *Hand mirror (clean)* – to show the client the proposed placement of the earrings, after marking and after the ear piercing.
- *Client's record card.*
- *Aftercare solution* – either to offer for sale or to include as part of the cost of the service.
- *Aftercare instruction leaflet* – for the client to keep.
- *Ultra-violet light cabinet* – to store the ear-piercing gun between piercing treatments.

Sterilisation and sanitisation

The working area should be kept clean. The floor covering should be such that the surface can be cleaned with a disinfectant and hot soapy water; work surfaces likewise should be washed regularly with hot soapy water and wiped with a disinfectant.

The **Local Government (Miscellaneous Provisions) Act 1982** requires that salons offering any form of skin piercing be registered with the local health authority. This registration includes both the operators who will be carrying out the treatment and the salon premises where the treatment will be carried out.

Premises are inspected by a health inspector, who checks that relevant local bye-laws are being followed. (The bye-laws are to ensure that treatment is carried out in a healthy, safe and hygienic manner.)

If the inspector is satisfied, the salon will be issued with a **certificate of authorisation**; this should be displayed in the reception area. Any breach of the Act or the bye-laws could result in a fine, and permission to carry out the treatment could be withdrawn.

ACTIVITY: PERSONAL CLEANLINESS

Personal cleanliness is a fundamental requirement of the Local Government (Miscellaneous Provisions) Act 1982. Discuss how a high standard of personal cleanliness can be guaranteed.

HEALTH & SAFETY: CROSS-INFECTION

If you have any cuts on your hands or fingers, these should be covered with a clean dressing before you treat the client. It is preferable to wear disposable gloves.

Some salons that offer the ear-piercing service also sell earrings. Those designed for pierced ears should not be tried on by a client, in the interest of health and hygiene. (Although there have been no reported cases of transmission of hepatitis B or HIV in this way, all possible risks should be avoided.) Such earrings

HEALTH & SAFETY: DISPOSAL OF WASTE
All consumable materials used during the ear-piercing treatment should be placed in a covered, lined waste bin, and disposed of in a sealed bag.

TIP
Guidelines for ear piercing are freely available from your local environmental health department.

may instead be attached to a special perspex slide which can be held against the ears to assist in selection.

The gun approved for ear piercing is designed so that it does not come into contact with the client's skin, and is used with pre-sterilised ear studs and ear clasps.

The studs are provided in sterile packs. Many give a date after which their sterility can no longer be assumed; others have a seal that changes colour when the expiry date has been reached.

Preparing the cubicle

Ear piercing can be carried out either in a private cubicle or at reception. Use your discretion to decide which would be more appropriate.

Good ventilation is important – some clients may feel faint following the treatment.

The client should sit on a chair at a comfortable height for you.

PREPARING THE CLIENT

If the position of the stud has not already been marked, do this now.

1. Wash your hands and apply a fresh pair of disposable gloves.
2. Thoroughly cleanse the back and the front of the client's earlobes. Allow the skin to dry. (If moisture is present, the mark will blur.)

HEALTH & SAFETY: SECONDARY INFECTION
As you cleanse the client's earlobes, you may notice that her skin, ear or hair is dirty. You may proceed, but when giving her the aftercare instructions politely and tactfully point out the importance of cleanliness in preventing secondary infection.

3. Holding a mirror in front of the client, discuss with her where the stud should be placed. Mark the position using the sterile pen.

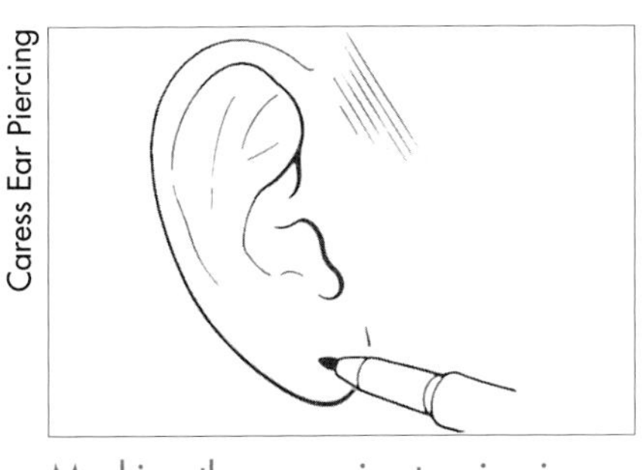
Caress Ear Piercing
Marking the ear prior to piercing

HEALTH & SAFETY: EAR STUDS
Studs are preferable to hooped earrings or sleepers, as dirt is less likely to cling to the stud and infect the ear.

Remember that the ears are not at the same level on each side of the head, and may protrude at different angles. Take time when marking the position of the studs, to ensure that the final appearance is balanced.

> **TIP**
> Try to aim for a central position on the earlobe – this will achieve the best result.

PIERCING THE EARS

The following procedure illustrates the general technique. The details differ according to the ear-piercing gun you are using. Always follow the manufacturer's instructions.

1. Holding the stud pack firmly, remove the backing paper. Take care not to drop the cartridge on the floor! (If you do drop it you will have to throw it away.)
2. Pull back the plunger knob on the back of the gun, until it is fully extended – you will hear it click.
3. Remove the plastic cartridge from the package, holding it by the plastic mount. To avoid contamination, make sure that you do not touch the stud or backing clasp.

   There are two parts, which must be separated. One part holds the back clasps: this is positioned in the slot. Push the cartridge down until it will go no further. The second holds the studs: position this against the stud barrel of the gun, which places a protective plastic ring around the barrel of the gun and stud.

   Gently pull the holder upwards, away from the barrel: this will deposit the stud in the barrel.

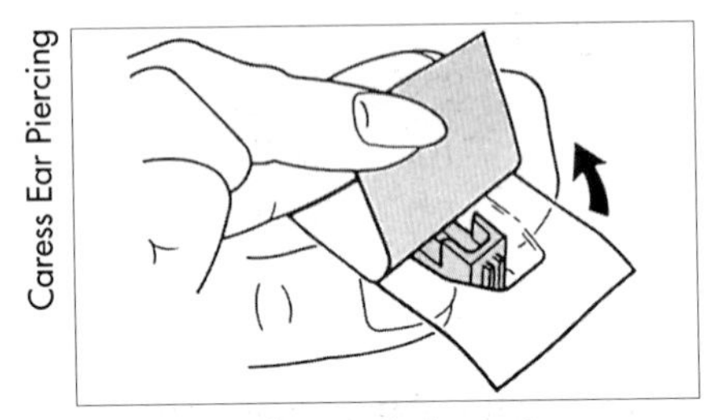

Caress Ear Piercing

Removing the studs from the sealed pack

Caress Ear Piercing

Preparing the gun

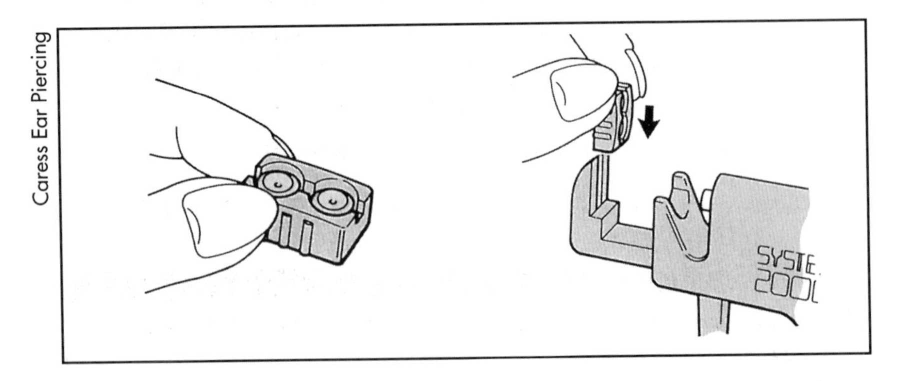

Caress Ear Piercing

left Removing the plastic cartridge holding the back clasps
right Positioning the back clasps in the gun

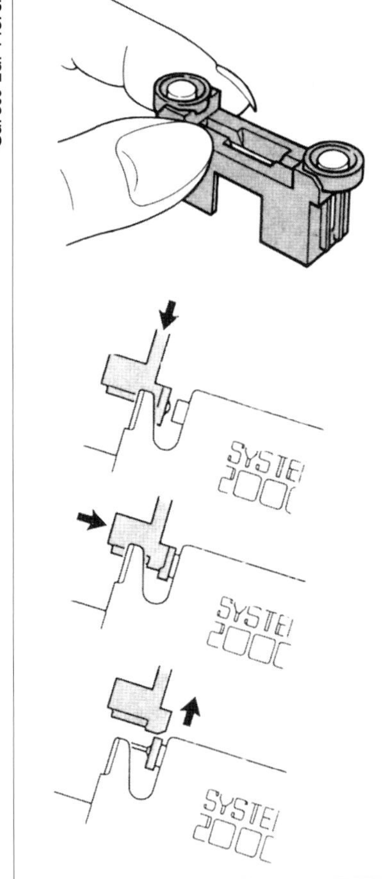

Loading the studs

> **TIP**
>
> Always hold the gun either horizontally or backwards. Never point the gun downwards once it has been loaded – if you do, the studs will fall out.

4 Gently hold the first ear to be pierced in the non-working hand.

Holding the gun horizontally, position the stem of the stud so that it is accurately placed over the mark on the earlobe.

Gently squeeze the trigger until it stops. Check that the point of the stud is still in the correct position. If it is, squeeze the trigger again. The ear will be pierced, with the back clasp placed on to the back of the stud.

Check that the earring and back clasp are securely connected.

Holding the ear gently, move the gun down from the lobe.

> **TIP**
>
> The gun should not point down or up when piercing. If the angle of piercing is incorrect, the earring will hang forward, sideways or backwards.

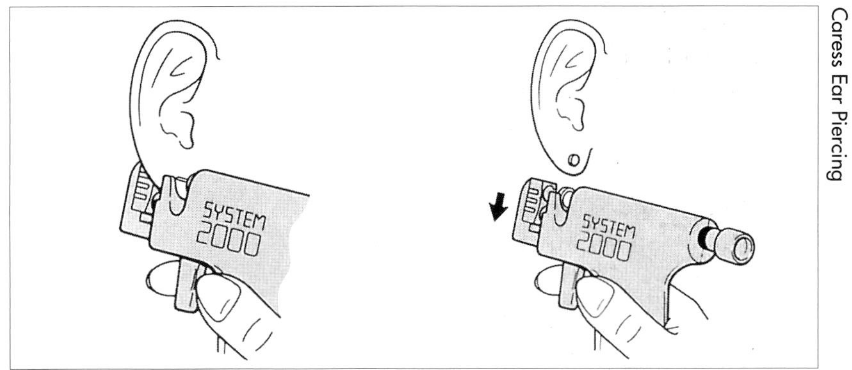

Piercing the ear

Caress Ear Piercing

> **TIP**
>
> If the client has fat lobes explain that the studs may feel a little tight.

5 Hold the gun upside down and discard the plastic ring into the waste bin.

6 Pull back the plunger knob and insert the next stud and protective ring.

Holding the gun in one hand, grip the back-clasp holder (using the mount in the other hand) and remove the holder from the gun. Invert the holder and place it back into the gun, with the remaining clasp in the top position.

7 Now pierce the second ear, repeating stage **4**.

8 Using the hand mirror, show the client her pierced ears.

Caress Ear Piercing

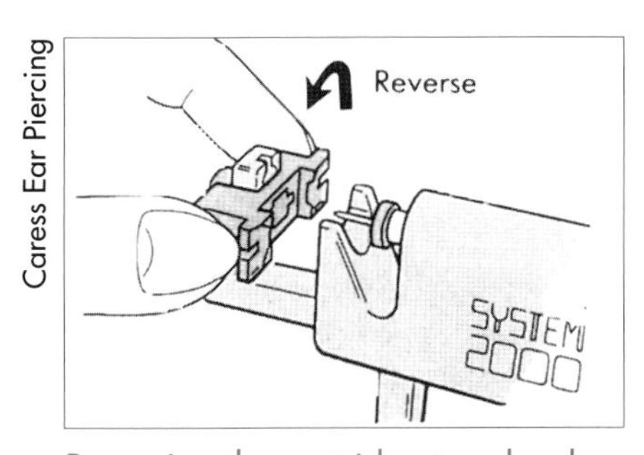

Reversing the cartridge to reload

> **TIP**
>
> To avoid anxiety to the client, pierce the second ear without hesitation, yet safely.

TIP

Allow the client to sit still for a few minutes following the treatment. Offer her tea or coffee. Some clients may be very anxious and need time to relax.

9 Discuss the aftercare instructions with the client (see below).

10 When the service has been completed, invert the gun, eject the protective ring, and remove the empty back-clasp holder. Dispose of the plastic cartridges into the waste bin.

11 Clean the gun using clean cottonwool and surgical spirit, and place it in the ultra-violet sterilising cabinet.

12 Dispose of all used products in the covered waste bin.

ACTIVITY: DIFFERING PROCEDURES

The procedure for loading the different ear-piercing guns varies. The procedure given above is one example. Referring to the literature you have collected on the different guns, note ways in which the various treatment procedures differ.

CONTRA-ACTIONS

Redness, swelling, inflammation and the exudation of serum (weeping) all signify that the ear is infected. If this occurs, the client should contact the salon. Depending on what she describes, it may be possible to give her adequate instructions over the telephone, or it may be necessary to make an appointment for the therapist to look at the ear. If following action by the therapist the infection persists, the client should be advised to contact her physician. Infection usually results from incorrect aftercare, removing the studs too early, or wearing cheap earrings.

Closed holes are caused by removing the studs too early, or by not continuing to wear earrings after the removal of the original studs.

Keloids – overgrowths of scar tissue – sometime occur at the site of ear piercing. If a client suffers from keloids, advise her not to have her ears pierced more than once – further keloid tissue could develop, giving an unsightly appearance.

Fainting may occur if the client was particularly nervous beforehand. The shock of the ear-piercing treatment may cause a short period of unconciousness due to insufficient blood flow to the brain.

HEALTH & SAFETY: FAINTING

Loosen any restrictive clothing. Reassure the client, and position her lying flat with her feet raised higher than her head, or sat with her head bent forwards between her knees. Instruct her to breathe deeply and slowly. Increase ventilation in the area.

HEALTH & SAFETY: SECONDARY INFECTION

If the client's ears become slightly red, indicating a possible infection, she should cleanse the area more frequently.

HEALTH & SAFETY: AFTERCARE LOTION
A professional manufacturer's aftercare lotion should be provided for each client following an ear piercing which is protected by the manufacturer's product liability insurance.

AFTERCARE AND ADVICE

After the ear-piercing treatment the client should be given clear instructions on how to care for her ears to prevent secondary infection. The studs must not be removed for six weeks. Invite the client to return to the salon after the six-week period for you to remove the studs.

Remind the client always to wash her hands before cleaning the ear area. (If someone else is going to clean the ears, she also must wash her hands with soap and warm water before touching the client's ears.) The ears should be cleaned twice daily, using an **aftercare lotion**. The lotion may be applied as directed by the manufacturer. This should be applied to clean cottonwool, and the cottonwool squeezed to allow the lotion to run around the stud, at the front of the ear and then at the back. Whilst the ears are being bathed with the cleansing solution, the stud should be rotated by holding it firmly at the front.

HEALTH & SAFETY: ANTISEPTIC LOTIONS
Remind the client that if an antiseptic lotion is applied to the ear area, it must be diluted as directed or skin burning may occur.

Even if the client cleanses her ears effectively, infection could still occur in other ways. These include the following:

- Long nails harbour germs. Infection of the ear can occur whilst the client is turning the stud.
- The client may touch the stud and ear at times other than when cleaning the ear.
- Shampoo and dirty water may collect around the ear whilst shampooing the hair. Remind the client to rinse the ear thoroughly with clean water after shampooing.
- The client should protect her ears when applying hair lacquer or perfume, to avoid sensitising the ear.
- Following the removal of the studs, the client should only wear gold earrings. Cheap fashion earrings should be worn only for short periods of time, or allergic reactions and ear infections may occur.

After oral instructions have been given, the client should be provided with an aftercare leaflet containing these instructions.

BODY PIERCING

Body piercing is currently very fashionable. Common areas for piercing include the nipples, the eyebrows, the nose and the navel. Note that body piercing is inappropriate for clients under 17 years of age, as the body is still growing.

Beauty therapy salons that offer ear piercing are often asked for body piercing also. Body piercing should be undertaken only if you have had thorough training, however, and only if you have the necessary specialised equipment and appropriate insurance cover.

HEALTH AND SAFETY: BODY PIERCING

Before performing body piercing, get approval from your local environmental health authority.

Remember that body piercing is inappropriate to clients under 17 years of age.

TIP

Always check with your insurance company which areas of the body you are able to pierce.

Equipment

The equipment for piercing is either a **body-piercing gun**, which inserts a hollow reed, or the **needle and clamp** method. With the latter, the clamp reduces circulation in the area, thereby anaesthetising it, and the needle is inserted into the skin. The opening made by the needle is then filled with a ring of high-quality non-reactive surgical steel or gold (above 16 carat).

Aftercare

Special care should be taken of the area for one month following piercing. Ideally the jewellery must not be changed for 4–6 months to avoid tissue damage and possible infection. If infection occurs, causing excessive redness, swelling or a discharge, the client should return to the salon.

KNOWLEDGE REVIEW

Ear piercing

Preparation

1. How would you describe to a client the treatment procedure of ear piercing to ensure that she was confident about both the procedure and your expertise?
2. What conditions would contra-indicate an ear-piercing treatment?
3. How can you ensure that the position of the ear stud will be correct?

Application

1. What part of the earlobe is recommended for piercing?
2. Which parts of the earlobe are considered unsuitable, and why?
3. What should be recorded on the client's record card following the ear-piercing treatment, and why is this of importance?
4. What is a normal skin reaction following ear-piercing treatment?

Aftercare

1. What aftercare instructions should be discussed with the client? Why are these of importance?

Health, safety and hygiene

1. If the client failed to follow the aftercare instructions, what complications could occur?
2. When purchasing an ear-piercing gun, what should be considered?
3. What does the Local Government (Miscellaneous Provisions) Act 1982 require of those who carry out ear piercing?
4. How can you prevent infection when carrying out an ear-piercing treatment?
5. How would infection of the earlobe be recognised?
6. If infection of the earlobe were to occur, what action should you take?

Case studies

1. What factors should be considered when piercing the earlobes of the following clients?
   (a) A black client.
   (b) A nervous client.
   (c) A client with fat ear lobes.
   (d) A client with diabetes.
   (e) A client with senile skin.
   (f) A client under the age of 16 years.
   (g) A client who is allergic to nickel.

Assessment

This section of the book will provide you with the background information and explain the principles surrounding the assessment of **competence**.

Before NVQs became available, trainees who wanted to gain nationally recognised qualifications would have to practise the technical skills, study science and theory, and then, after a period of training, be expected take both practical and theory examinations. Depending on how they performed on the day, they either passed or failed. This system is considered to be unfair for vocational qualifications. Some trainees who were expected to pass with flying colours have actually failed because of the pressures exerted by taking the exam. The examination system neither provides a realistic working situation, nor truly reflects the candidate's previous training achievement.

Continuous assessment of performance provides a suitable alternative. Monitoring and recording the trainee's progress towards achievement (occupational competence) is now providing the currently preferred route to certification.

Trainees are enrolled with an **approved centre** and then undergo a period of initial assessment to identify their training needs. Following this assessment, a personal training and assessment plan is devised. This plan will state where, when and how the training will be delivered, who will be responsible for monitoring, training and assessing, and what will be the expected timescale for training and reviewing of progress.

This detailed plan provides a tailor-made training programme which allows the trainee to:

- practise the required skills; and
- acquire the essential supporting knowledge.

Then, at the point where trainees can consistently perform the required tasks to standard and can demonstrate that they have the required knowledge, they are deemed competent.

WHAT IS THE REQUIRED STANDARD?

The **standard** could be described as 'the level of competence required in order to perform the task'. The currently nationally recognised standards have been devised by the Hairdressing and Beauty Industry Authority (HABIA). This National Training Organisation is made up of representative bodies within the beauty industry. It is responsible for defining the standards of competence at each of the NVQ levels. After the standards have been agreed they are submitted to the Qualifications and Curriculum Authority (QCA) for their approval and subsequent **accreditation** (accreditation is the giving of QCA's formal approval but to the National Training Organisation's national standards).

The standards are written specifications of how certain tasks or functions are to be performed. They are used in two ways to assess trainees:

- by observing their performance of practical ability within the specified conditions of the range;
- by questioning candidates to find out their basic understanding and knowledge of the practical task.

Each standard is make up of the following components:

- element title;
- performance criteria;

- range statement;
- essential knowledge.

Level 2 in Beauty Therapy consists of nine mandatory units and one additional unit. The **mandatory units** are the basic essentials of the NVQ, and they must all be completed.

Each unit can be achieved and certificated separately. This means that candidates may accumulate units of competence:

- for specific work-related tasks;
- for part-certification;
- for full Level 2 certification;
- in any order;
- at their own pace of learning;
- within the workplace or at a training centre.

Element title

Units consist of various numbers of individual standards called **elements.** The element title states the function or task that has to be performed competently.

Performance criteria

A standard specifies not only *what* has to be done, but also *how* it is to be done. It sets out a list of **performance criteria** for each task. These are concise statements of procedural functions, the specific steps that should be taken during the performance of the task if the standard is to be met. They are used as a checklist when the candidate is being observed during assessment.

Range statements

To enable the trainee to gain competence in a variety of situations the standard also states the **range** of circumstances for which competence must be demonstrated. For example, if a trainee were asked to perform a brow tint, the effect created would depend on the client's natural hair colour. Although the same methods and techniques are often required with different clients, the desired results are not always the same. The candidate therefore needs to show competence in a broad range of contexts – skin types, nail conditions, hair removal techniques and so on – which imply a wide variety of conditions and complexity for the given task.

Essential knowledge and understanding

This refers to the candidate's awareness of why the task is done in a certain way. For example, in improving the appearance and condition of the nails and adjacent skin, the trainee will need to show a clear understanding of:

- the importance of working to commercially acceptable time limits;
- the reasons for the recommended nail shape;
- the effect of incorrect use of manicure and pedicure tools.

The trainee will need to know methods of:

- completing clients' records;
- carrying out manicure and pedicure techniques correctly;
- minimising waste.

The trainee will need to know:

- commercially acceptable timings of the manicure and pedicure process;
- the range of and use of products available for home care;
- how to match suitable nail shapes to hand shapes;
- how to protect hands and nails.

THE ASSESSMENT ACTIVITY

Why do we assess?

Assessment isn't just a matter of finding out if a trainee can 'pass' a test. There are four main reasons for an assessment:

- to check whether a trainee can perform the job competently to the required national standard;
- to identify where further improvement can be made;
- to monitor and record progress towards competence;
- to maintain accurate records of the trainee's achievement.

How do we assess?

Assessment is carried out by the collection and evaluation of evidence by a wide variety of techniques. But regardless of the technique used, only one of three possible assessment decisions can be made:

- the trainee is competent;
- the trainee is not yet competent;
- the trainee has provided insufficient evidence for the assessor to infer competence.

Observation of performance

A trainee's performance evidence should be sufficient to convince an assessor that it consistently meets the stated performance criteria and that all items within the range are covered. The minimum numbers of observed competent performances are stated in the assessor's element specific guidance.

In addition to this, virtually all the perfomance evidence must be a result of the trainee's own endeavours and should be a result of real work activities. However, some exceptions for simulated activity will be allowed, for example in observing emergency proecdures.

Oral questioning

Most of the evidence of competence will arise from the observed performance. However, it may be necessary to supplement this by providing back-up information about the task or by answering specific questions that relate to 'Essential knowledge and understanding' requirements.

Other supplementary evidence

This may include a variety of sources, such as:

- your own notes from your work log;
- witness testimonies from responsible people at work;
- video recordings of yourself at work;
- photographs of yourself at work;
- extracts from salon systems (if available);
- records of clients' comments (if applicable);
- copies of relevant remarks about your work.